Earthquake Processes: Physical Modelling, Numerical Simulation and Data Analysis
Part II

Edited by
Mitsuhiro Matsu'ura
Peter Mora
Andrea Donnellan
Xiang-chu Yin

2002

Springer Basel AG

Reprint from Pure and Applied Geophysics
(PAGEOPH), Volume 159 (2002), No. 10

Editors:

Prof. Mitsuhiro Matsu'ura
University of Tokyo
Bunkyo-Ku
113-0033 Tokyo
Japan
e-mail: matsuura@eps.s.u-tokyo.ac.jp

Andrea Donnellan
Jet Propulsion Laboratory, NASA
4800 Oak Grove Drive
Pasadena, CA 91109-8099
USA
e-mail: donnellan@jpl.nasa.gov

Prof. Peter Mora
University of Queensland
QUAKES, Dep. Of Earth Sciences
4072 Brisbane, Qld
Australia
e-mail: mora@quakes.uq.edu.au

Prof. Xiang-chu Yin
China Academy of Sciences
Laboratory of Nonlinear Mechanics
Institute of Mechanics
Beijing 100080
China
e-mail: yinxc@btamail.net.cn

A CIP catalogue record for this book is available from the Library of Congress,
Washington D.C., USA

Deutsche Bibliothek Cataloging-in-Publication Data

Earthquake Processes: Physical Modelling, Numerical Simulation and Data Analysis / ed. by
Mitsuhiro Matsu'ura ; Peter Mora ; Andrea Donnellan ; Xiang-chu Yin. -
Basel ; Boston ; Berlin : Birkhäuser, 2002
 (Pageoph topical volumes)
 ISBN 978-3-7643-6916-3

ISBN 978-3-7643-6916-3   ISBN 978-3-0348-8197-5 (eBook)
DOI 10.1007/978-3-0348-8197-5

9 8 7 6 5 4 3 2 1

Contents

Pure appl. geophys. 159 (2002) 2169–2171
0033–4553/02/102169–03 $ 1.50 + 0.20/0

❙Pure and Applied Geophysics

Earthquake Processes: Physical Modelling, Numerical Simulation and Data Analysis

PART II

Mitsuhiro Matsu'ura,[1] Peter Mora,[2]
Andrea Donnellan,[3] and Xiang-chu Yin[4]

Introduction

In the last decade of the 20th century, there has been great progress in the physics of earthquake generation; that is, the introduction of laboratory-based fault constitutive laws as a basic equation governing earthquake rupture, quantitative description of tectonic loading driven by plate motion, and a microscopic approach to study fault zone processes. The fault constitutive law plays the role of an interface between microscopic processes in fault zones and macroscopic processes of a fault system, and the plate motion connects diverse crustal activities with mantle dynamics. The APEC Cooperation for Earthquake Simulation (ACES) aims to develop realistic computer simulation models for the complete earthquake generation process on the basis of microscopic physics in fault zones and macroscopic dynamics in the crust-mantle system, and to assimilate seismological and geodetical observations into such models. Simulation of the complete earthquake generation process is an ambitious challenge. Recent advances in high performance computer technology and numerical simulation methodology are bringing this vision within reach.

The inaugural workshop of ACES was held on January 31 to February 5, 1999 in Brisbane and Noosa, Queensland, Australia. Following the fruitful results

[1] Department of Earth and Planetary Science, The University of Tokyo, Bunkyo-ku, Tokyo 113-0033, Japan. E-mail: matsuura@eps.s.u-tokyo.ac.jp

[2] QUAKES, Department of Earth Sciences, The University of Queensland, 4072 Brisbane, Qld, Australia. E-mail: mora@quakes.uq.edu.au

[3] Jet Propulsion Laboratory NASA, 4800 Oak Grove Drive, Pasadena, CA 91109-8099, U.S.A. E-mail: donnellan@jpl.nasa.gov

[4] Laboratory of Nonlinear Mechanics, Institute of Mechanics, China Academy of Sciences, Beijing 100080, China. E-mail: yinxc@btamail.net.cn

in the inaugural workshop [1, 2], the 2nd ACES workshop took place on October 15–20, 2000 in Tokyo and Hakone, Japan. In this workshop, more than 100 researchers in earthquake physics and computational science from 10 countries participated to discuss the integrated simulation-based approach for understanding earthquake processes. The major theme of the 2nd workshop was "microscopic and macroscopic simulation of fault zone processes and evolution, earthquake generation and cycles, and fault system dynamics." The theme was addressed in a series of six regular sessions for microscopic simulation, scaling physics, earthquake generation and cycles, dynamic rupture and wave propagation, data assimilation and understanding, and model applications to earthquake hazard quantification, and an additional special session for collaborative software systems and models [3]. We publish an outcome of the workshop as a set of two special issues (Part I and Part II) of Pure and Applied Geophysics for the six regular sessions. Articles for the additional special session, which cover primarily computer science and computational algorithm, are published separately in Concurrency and Computation: Practice and Experience. The articles in these special issues present a cross-section of cutting-edge research in the field of computational earthquake physics.

Part I collects articles covering two categories; A) micro-physics of rupture and fault constitutive laws and B) dynamic rupture, wave propagation and strong ground motion. Part II gathers articles encompassing two other categories; A) earthquake cycles, crustal deformation and plate dynamics and B) seismicity change and its physical interpretation.

Part II-A assembles articles on earthquake cycles, crustal deformation and plate dynamics. These range from the 3-D simulations of earthquake generation cycles driven by plate motion and interseismic crustal deformation associated with plate subduction to the development of new methods for analyzing geophysical and geodetical data and new simulation algorithms for large amplitude folding and mantle convection with viscoelastic/brittle lithosphere. These results provide us with important elements to construct an integrated realistic simulation model of crustal deformation and earthquake cycles.

Part II-B collects articles on seismicity change and its physical interpretation. These span a theoretical study of accelerated seismic release on heterogeneous faults, numerical simulation of long-range automaton models of earthquakes, and various approaches to earthquake prediction based on underlying physical or combined statistical and physical models for seismicity change. These studies provide us with new and important aspects to understand physical mechanisms underlying spatial and temporal change in seismicity.

Finally, we wish to thank all the participants of the 2nd ACES workshop and the contributors to these special issues, and to gratefully acknowledge financial support for the workshop by STA, RIST, NASA, NSF, ACDISR, ARC, JSPS, SSJ, TMKMF, FUJITSU, HITACHI, IBM, and NEC.

REFERENCES

[1] 1-st ACES Workshop Proceedings (1999), ed. Mora, P. (ACES, Brisbane, Australia, ISBN 1-864-99121-6), 554 pp.

[2] Microscopic and Macroscopic Simulation: Towards Predictive Modelling of the Earthquake Process (2000), eds. Mora, P., Matsu'ura, M., Madariaga, R., and Minster, J-B., Pure and Applied Geophysics, Volume 157, Number 11/12, 1817–2383.

[3] 2-nd ACES Workshop Proceedings (2001), eds. Matsu'ura, M., Nakajima, K., and Mora, P. (ACES, Brisbane, Australia, ISBN 1-864-99510-6), 605 pp.

A. Earthquake Cycles, Crustal Deformation and Plate Dynamics

Pure appl. geophys. 159 (2002) 2175–2199
0033–4553/02/102175–25 $ 1.50 + 0.20/0

Pure and Applied Geophysics

3-D Simulation of Earthquake Generation Cycles and Evolution of Fault Constitutitve Properties

CHIHIRO HASHIMOTO[1] and MITSUHIRO MATSU'URA[2]

Abstract—The earthquake generation cycle consists of tectonic loading, quasi-static rupture nucleation, dynamic rupture propagation and stop, and subsequent stress redistribution and fault restrengthening. From a macroscopic point of view, the entire process of earthquake generation cycles should be consistently described by a coupled nonlinear system of a slip-response function, a fault constitutive law and a driving force. On the basis of such a general idea, we constructed a realistic 3-D simulation model for earthquake generation cycles at a transcurrent plate boundary by combining the viscoelastic slip-response function derived for a two-layered elastic-viscoelastic structure model, the slip- and time-dependent fault constitutive law that has an inherent mechanism of fault restrengthening, and the steady relative plate motion as a driving force into a single closed system. With this model we numerically simulated the earthquake generation cycles repeated in a seismogenic region on a plate interface, and examined space-time changes in shear stress, slip deficits and fault constitutive properties during one complete cycle in detail. The occurrence of unstable dynamic slip brings about decrease both in fault strength and shear stress to a constant residual level. After the arrest of dynamic slip, the breakdown strength drop $\Delta\sigma_p$ of fault is restored rapidly and the process of stress accumulation resumes in the seismogenic region. On the other hand, the restoration of the critical weakening displacement D_c proceeds gradually with time through the interseismic period. The restoration of D_c can be regarded as the macroscopic manifestation of the microscopic recovery process of fractal fault surface structure. Through numerical simulation with a multi-segmented fault model, we examined the effects of viscoelastic fault-to-fault interaction. The effect of transient viscoelastic stress transfer through the asthenosphere is significant as well as the direct effect of elastic stress transfer, and it possibly explains the time lag of the sequential occurrence of large events along a plate boundary.

Key words: Earthquake generation cycle, fault strength restoration, critical weakening displacement, viscoelastic stress transfer.

1. Introduction

Generation of large interplate earthquakes can be regarded as the process of tectonic stress accumulation and release, resulting from relative plate motion. The

[1] Department of Earth and Planetary Science, The University of Tokyo, Hongo 7-3-1, Bunkyo-ku, Tokyo 113-0033, Japan. E-mail: hashi@solid.eps.s.u-tokyo.ac.jp (Present address: Institute for Frontier Research on Earth Evolution, Japan Marine Science and Technology Center, Natsushima-cho 2-15, Yokosuka 237-0061, Japan. E-mail: hashi@jamstec.go.jp)

[2] Department of Earth and Planetary Science, The University of Tokyo, Hongo 7-3-1, Bunkyo-ku, Tokyo 113-0033, Japan. E-mail: matsuura@eps.s.u-tokyo.ac.jp

entire process of earthquake generation cycle may be divided into four stages, such as tectonic loading, quasi-static nucleation, dynamic rupture propagation and stop, and subsequent stress redistribution and fault strength restoration. So far most studies on earthquake generation have focused on the process of dynamic rupture propagation and stop. However, the dynamic rupture process is essentially controlled by the preceding stress accumulation process in seismogenic regions, because the stress released at the time of dynamic rupture is nothing but the stress accumulated in the interseismic period. After the arrest of dynamic rupture, the fault should be restrengthened to generate the next event. Therefore, to understand earthquake generation, physically-based modelling of the entire process of earthquake generation cycles is essentially important.

A simple spring-slider model for earthquake generation cycles has been proposed by BURRIDGE and KNOPOFF (1967). This model consists of a number of blocks representing fault elements, each of which is subjected to three different kinds of force; that is, leaf-spring force representing tectonic loading, coil-spring force representing interaction between adjacent fault elements, and frictional resistance force acting between the block and a floor. The spring-slider type of models has been often used to understand statistical properties of seismicity by many authors (e.g., CAO and AKI, 1984; CARLSON and LANGER, 1989). In these models, however, only the nearest neighbour interaction is considered.

Another type of simplified earthquake generation cycle models is the fault patch model. This type of models also has been used by many researchers to simulate seismic activity along a complex fault system. For example, STUART (1984/85, 1986) tried to simulate the quasi-periodic occurrence of large earthquakes along the southern San Andreas fault system, California. In this model, the fault system is divided into a number of small fault patches, and interaction between them is represented by quasi-static elastic slip-response functions. Recently, WARD (2000) simulated earthquake generation cycles in the San Francisco Bay area, California, by using a fault patch model on the assumption of plane stress. These models can treat remote interaction between fault patches, but not continuous slip and stress distribution. Therefore, we cannot discuss the details of earthquake generation processes with this type of models.

TSE and RICE (1986) have constructed a realistic 2-D simulation model for earthquake generation cycles at transcurrent plate boundaries by incorporating the rate- and state-dependent friction law into a slip-response function for an elastic plate. Since then, this type of models has been used by many researchers to simulate the process of earthquake generation cycles. For example, STUART (1988) and KATO and HIRASAWA (1997) have applied this type of models to the case of underthrust earthquakes at subduction zones. RICE (1993) and RICE and BEN-ZION (1996) have tried to extend the 2-D simulation model into 3-D simulation models. Recently, BEN-ZION and RICE (1997) and LAPUSTA et al. (2000) have developed a new computational method to treat quasi-static tectonic loading and dynamic rupture propagation in a single framework. In these simulation studies, however, the effect of viscoelastic stress transfer in the asthenosphere is completely ignored.

In the present study, we construct a 3-D model of earthquake generation cycles at transcurrent plate boundaries by considering a realistic fault constitutive law and the effect of viscoelastic stress transfer in the asthenosphere. In Section 2 we show a basic idea for modelling and clarify which elements are essential for constructing the earthquake generation cycle model. In Section 3 we quantitatively describe the basic equations governing the entire process of earthquake generation cycles. In Section 4 we show results of numerical simulation; the process of stress accumulation and release during an earthquake cycle in 4.1, the evolution of fault constitutive properties through the interseismic period in 4.2, and the effect of viscoelastic fault-to-fault interaction in 4.3. In Section 5 we discuss the restoration mechanism of constitutive parameters, the breakdown strength drop $\Delta\sigma_p$ and the critical weakening displacement D_c, in detail.

2. Basic Idea for Modelling

In this section we present a basic idea for modelling the entire process of earthquake generation cycles. In general, the earthquake generation cycle consists of tectonic loading, quasi-static rupture nucleation, dynamic rupture propagation and stop, and subsequent stress redistribution and fault strength restoration. Given the rheological structure of the lithosphere-asthenosphere system and geometry of a plate interface, we can quantitatively describe the entire earthquake generation cycle by combining three basic elements; that is, a slip-response function which relates fault slip with stress changes, a fault constitutive law which governs the progress of slip, and a relative plate motion as driving force. In order to construct a realistic model of earthquake generation cycles, these three basic elements must satisfy several necessary conditions from seismological observations, geodetic observations and laboratory experiments of rock friction.

So far, in most earthquake simulation studies, it has been simply assumed that the tectonic loading due to relative plate motion proceeds uniformly both in space and time. The assumption of uniform loading is clearly contradictory to geodetic observations along the San Andreas fault system in California. For example, LISOWSKI et al. (1991) have reported that interseismic strain buildup concentrates in a 80-km-wide zone parallel to the San Andreas fault in the San Francisco Bay area and southern California, where the 1906 San Francisco earthquake and the 1857 Fort Tejon earthquake occurred, respectively. The bell-shaped profile of strain buildup rates across the San Andreas fault can be reasonably explained by the superposition of a steady horizontal block motion and a backslip (slip deficits) in a shallow portion of the plate interface (SAVAGE and BURFORD, 1973; SAVAGE, 1990). This indicates that the essential cause of tectonic stress accumulation at plate boundaries is in the slip deficits relative to the steady plate motion. On the other hand, THATCHER (1983) has reported gradual decrease in the strain buildup rates with time since the last great earthquakes, indicating significant effects of viscoelastic stress relaxation in the asthenosphere. In order to explain the temporal decay of strain buildup rates, THATCHER (1983) has proposed a lithosphere-asthenosphere coupling model. MATSU'URA and SATO (1997) performed a simple numerical simulation for tectonic loading at transcurrent plate boundaries with a lithosphere-asthenosphere coupling model subjected to steady relative plate motion, and have concluded that stress accumulation on a seismic fault is partly due to the base loading (viscous drag at the base of the lithosphere) and partly due to the edge loading (dislocation pile-ups at horizontal edges of the fault).

These geodetic observations can be consistently explained by a 3-D lithosphere-asthenosphere coupling model as shown in Figure 1. We consider a 100-km-long and 15-km-wide seismogenic region on the infinitely-long vertical plate interface in steady slip at 50 mm/yr. The seismogenic region is locked at $t = 0$. Subsequently, the stress accumulation due to slip deficits in the seismogenic region proceeds with time. Figure 2 shows the pattern of surface velocities around the locked portion at $t = 20$ yr. It should be noted that the far-field velocity pattern is merely horizontal block motion at the relative plate velocity. Figure 3 shows the pattern of shear stress accumulation rates around the locked portion at $t = 20$ yr. The stress accumulation concentrates in a narrow zone around the locked portion. As pointed out by MATSU'URA and SATO (1997), the rate of stress accumulation strongly depends on the length of locked portion and time after the event. Figure 4 displays the profiles of stress accumulation rates perpendicular to the fault strike at the center of the locked portion. From Figure 4(a) we can see that the stress accumulation rate is nearly proportional to the inverse of the length of the locked portion. The temporal decay of stress accumulation rates in Figure 4(b), which is due to viscoelastic stress relaxation in

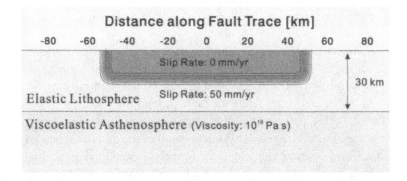

Figure 1

Structure and geometry of a 3-D lithosphere-asthenosphere coupling model. A 100-km-long and 15-km-wide seismogenic region on the vertical plate interface subjected to the relative plate motion at 50 mm/yr is locked at $t = 0$. After $t = 0$ the steady slip proceeds outside the seismogenic region.

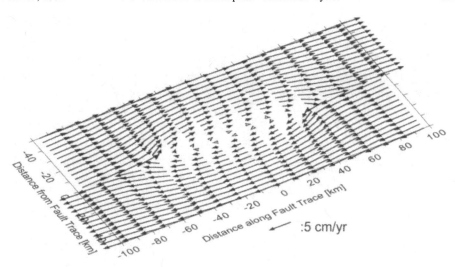

Figure 2
The pattern of surface velocities around the locked portion at $t = 20$ yr. The locked portion extends from −50 km to 50 km along the plate boundary.

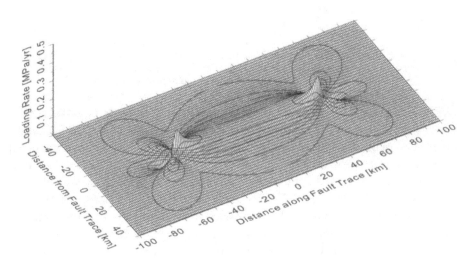

Figure 3
The pattern of shear stress accumulation rates around the locked portion at $t = 20$ yr. The stress accumulation concentrates in a narrow zone around the locked portion, which extends from −50 km to 50 km along the plate boundary.

the asthenosphere, well explains the geodetic observations reported by THATCHER (1983). These results of numerical simulation indicate that the effect of viscoelastic stress transfer through the asthenosphere cannot be neglected in the modeling of earthquake generation cycles.

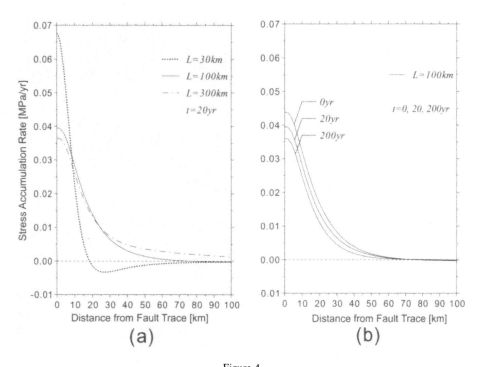

Figure 4

Profiles of stress accumulation rates perpendicular to the fault strike at the center of the locked portion. (a) Dependence of stress accumulation rates on the fault length L. (b) Change in stress accumulation rates with time. The time t is measured from just after the fault was locked.

The second basic element needed for the model construction is a fault constitutive law, which governs the entire process of earthquake generation cycles. Thus far, two different types of laboratory-based constitutive laws have been proposed; one is the slip-weakening type of constitutive law (OHNAKA et al., 1987; MATSU'URA et al., 1992), and another is the rate- and state-dependent type of constitutive law (DIETERICH, 1979; RUINA, 1983). Recently, AOCHI and MATSU'URA (1999, 2002) theoretically derived a slip- and time-dependent constitutive law by integrating the microscopic effects of abrasion and adhesion between statistically self-similar fault surfaces. This constitutive law consistently explains three basic experimental results for rock friction; that is, the slip-weakening in high-speed slip (OHNAKA et al., 1987), the log t strengthening in stationary contact (DIETERICH, 1972) and the slip-velocity weakening in steady slip (DIETERICH, 1978). In this sense the slip- and time-dependent constitutive law can be regarded as a general law that unifies the slip-weakening law and the rate- and state-dependent law. In the slip- and time-dependent constitutive law, as shown in the next section, the essential parameters are the rates of abrasion and adhesion of fault surfaces. The other constitutive parameters, such as the breakdown strength drop and the critical weakening displacement, are process-dependent parameters, changing with fault slip and fault contact time.

In the next section we construct a 3-D simulation model for earthquake generation cycles at a transcurrent plate boundary by combining the viscoelastic slip-response functions obtained by HASHIMOTO and MATSU'URA (2000), the slip- and time-dependent fault constitutive law proposed by AOCHI and MATSU'URA (1999, 2002), and a relative plate motion as a driving force into a single closed system.

3. Modelling of Earthquake Generation Cycles

3.1 Structural Model

To describe the basic elements in a concrete form, first, we need to set a structure model of the lithosphere-asthenosphere system with a plate interface. In the present study the lithosphere-asthenosphere system is modelled by an elastic surface layer overlying a Maxwellian viscoelastic half-space, as shown in Figure 5. The constitutive equation of the elastic surface layer is given by

$$\sigma_{ij} = \lambda^{(1)} \varepsilon_{kk} \delta_{ij} + 2\mu^{(1)} \varepsilon_{ij} \tag{1}$$

and that of the underlying viscoelastic half-space by

$$\dot{\sigma}_{ij} + \frac{\mu^{(2)}}{\eta}\left(\sigma_{ij} - \frac{1}{3}\sigma_{kk}\delta_{ij}\right) = \lambda^{(2)}\dot{\varepsilon}_{kk}\delta_{ij} + 2\mu^{(2)}\dot{\varepsilon}_{ij}, \tag{2}$$

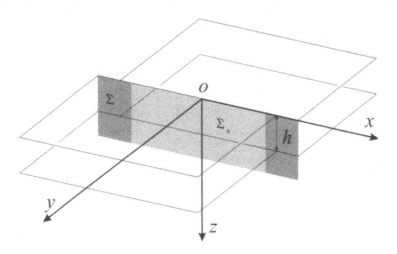

Figure 5

A structural model of the vertical transcurrent plate boundary and a coordinate system. The lithosphere-asthenosphere system is modelled by a 30-km-thick elastic surface layer overlying a Maxwellian viscoelastic half-space. An infinitely-long vertical interface Σ divides the elastic surface layer into two plates being in relative horizontal motion at a constant rate v_{pl}. The boundary condition on the model region Σ_s is given by Eq. (7) in the text. On the plate interface outside Σ_s, we impose uniform fault slip at v_{pl}.

where σ_{ij}, ε_{ij}, and δ_{ij} are the stress tensor, the strain tensor, and the unit diagonal tensor, respectively. The dot indicates differentiation with respect to time, $\lambda^{(i)}$ and $\mu^{(i)}$ ($i = 1, 2$) are the Lamé elastic constants of each medium, and η is the viscosity of the underlying half-space.

The structural parameters used in the present study are given in Table 1. From this table we can calculate P- and S-wave velocities as $V_p = 6.3$ km/s and $V_s = 3.7$ km/s for the lithosphere and $V_p = 7.9$ km/s and $V_s = 4.2$ km/s for the asthenosphere. On the basis of the analysis of uplift data for Lake Bonneville in the Basin and Range province, IWASAKI and MATSU'URA (1982) have estimated the thickness of the elastic lithosphere and the viscosity of the asthenosphere in the western United States to be 30–40 km and 10^{19}–10^{20} Pa s, respectively. In the present study we took these values.

The plate interface is modelled by an infinitely-long vertical plane which divides the elastic surface layer into two plates. Then, following MATSU'URA and SATO (1989), we represent interaction between these two plates by the increase of tangential displacement discontinuity (fault slip) across the interface Σ. The displacement discontinuity (dislocation) is mathematically equivalent to the force system of a double-couple without moment (MARUYAMA, 1963; BURRIDGE and KNOPOFF, 1964), which has no net force and no net torque. Such a property must be satisfied for any force system acting on plate interfaces, since it is the internal force produced by a dynamic process within the earth. Given the structure of the lithosphere-astheno-sphere system and geometry of the plate interface, we can calculate internal viscoelastic stress changes due to a unit step slip on the plate interface by using the mathematical expressions derived by HASHIMOTO and MATSU'URA (2000) on the basis of elastic dislocation theory.

3.2 Basic Equations

Denoting the ij-component of viscoelastic stress change at a point $\mathbf{x}$ and time t due to a unit step slip at a point $\boldsymbol{\xi}$ and time τ by $H_{ij}(\mathbf{x}, t; \boldsymbol{\xi}, \tau)$, we can calculate the internal stress field $\sigma_{ij}(\mathbf{x}, t)$ due to arbitrary fault slip distribution $w(\mathbf{x}, t)$ by using the technique of hereditary integral as

$$\sigma_{ij}(\mathbf{x}, t) = \int_{-\infty}^{t} \int_{\Sigma} \frac{\partial w(\boldsymbol{\xi}, \tau)}{\partial \tau} H_{ij}(\mathbf{x}, t - \tau; \boldsymbol{\xi}, 0) d\boldsymbol{\xi} d\tau. \tag{3}$$

Table 1

Structural parameters used in numerical simulation

	h [km]	ρ [kg/m^3]	λ [GPa]	μ [GPa]	η [Pa s]
Lithosphere	30	3000	40	40	∞
Asthenosphere	∞	3400	90	60	10^{19}

h: thickness, ρ: density, λ and μ: Lamé elastic constants, η: viscosity

In the present case, where pure strike-slip on a vertical fault plane is treated, only the yx-component of the stress tensor is needed. Thus, for simplicity, we use σ and H instead of σ_{yx} and H_{yx} hereafter. The concrete mathematical expression of H_{yx} is given in HASHIMOTO and MATSU'URA (2000).

Now we decompose the fault slip w into the steady plate motion at a constant rate v_{pl} and its perturbation u_s as

$$w(\mathbf{x}, t) = v_{pl}t + u_s(\mathbf{x}, t). \tag{4}$$

Substituting this expression into Eq. (3), we obtain

$$\sigma(\mathbf{x}, t) = v_{pl} \int_{-\infty}^{t} \int_{\Sigma} H(\mathbf{x}, t - \tau; \xi, 0) d\xi \, d\tau + \int_{-\infty}^{t} \int_{\Sigma} \frac{\partial u_s(\xi, \tau)}{\partial \tau} H(\mathbf{x}, t - \tau; \xi, 0) d\xi \, d\tau, \tag{5}$$

where the first and the second terms on the right-hand side indicate the contributions from the steady plate motion and the slip perturbation, respectively. In the case of transcurrent plate boundaries, the steady-state plate motion is merely horizontal block motion parallel to the strike of the plate boundary, which does not bring about any change in tectonic stress. Therefore, we may regard the first term as a constant (σ_0) in time, and take it as a reference level when we measure stress changes due to the slip perturbation. Then, supposing tectonic stress accumulation due to slip deficits begins at $t = 0$, we can rewrite Eq. (5) as

$$\sigma(\mathbf{x}, t) = \sigma_0(\mathbf{x}) + \int_{0}^{t} \int_{\Sigma_s} \frac{\partial u_s(\xi, \tau)}{\partial \tau} H(\mathbf{x}, t - \tau; \xi, 0) \, d\xi \, d\tau. \tag{6}$$

Here, Σ_s indicates the model region, outside of which steady slip at v_{pl} is assumed.

In our problem both the shear stress σ and the fault slip w are unknown. What we know is the relation between σ and w prescribed by a fault constitutive law. The slip- and time-dependent constitutive law defines shear strength σ_{strg} as a function of fault slip w and contact time t, and therefore we can write boundary conditions on Σ_s of the plate interface in the form of

$$\begin{cases} \sigma(\mathbf{x}, t) = \sigma_{strg}[w(\mathbf{x}, t), t; \mathbf{x}] & (dw/dt > 0) \\ \sigma(\mathbf{x}, t) \leq \sigma_{strg}[w(\mathbf{x}, t), t; \mathbf{x}] & (dw/dt = 0) \end{cases}. \tag{7}$$

The boundary conditions state that if the shear stress σ acting on the fault surface exceeds a peak strength, the fault must slip following the constitutive relation $\sigma = \sigma_{strg}(w, t)$. When the fault is in stationary contact $(dw/dt = 0)$, the shear stress can take arbitrary value determined by Eq. (6) unless it does not exceed the peak strength.

Equations (4), (6) and (7) are coupled with each other. If the concrete form of the fault constitutive law is given, the macroscopic process of earthquake generation cycles is completely governed by these coupled nonlinear equations.

3.3 Fault Constitutive Law

MATSU'URA et al. (1992) have theoretically derived a slip-weakening type of fault constitutive law by integrating the macroscopic effects of abrasion between statistically self-similar rock surfaces. This constitutive law well explains the observed slip-weakening behaviour of rock friction, but it has no inherent mechanism of strength restoration. AOCHI and MATSU'URA (1999, 2002) have developed this slip-dependent law into the slip- and time-dependent law by considering the adhesion of fractured rock interfaces, which brings about the restoration of fault strength.

According to MATSU'URA et al. (1992), the shear strength σ_{strg} of fault can be expressed as a function of fault slip w and contact time t in the form of wavenumber (k) integral:

$$\sigma_{strg}(w, t) = \sigma_0 + c \left[\int_0^\infty k^2 |Y(k; w, t)|^2 \, dk \right]^{1/2}. \tag{8}$$

Here, $|Y(k)|$ denotes the Fourier amplitude of fault surface topography, which changes with fault slip w and contact time t following the differential equation

$$d|Y(k; w, t)| = -\alpha k |Y(k; w, t)| dw + \beta k^2 \left[|\bar{Y}(k)| - |Y(k; w, t)| \right] dt \tag{9}$$

proposed by AOCHI and MATSU'URA (1999, 2002). The first term on the right-hand side of this equation represents the effect of abrasion by fault slip, and the second term represents the effect of adhesion by stationary contact. The parameters α and β are the rates of abrasion and adhesion of fault surfaces, respectively. In this constitutive law the physical properties of faults are prescribed by the abrasion rate α, the adhesion rate β, and the inherent Fourier amplitude $|\bar{Y}(k)|$ of fault surface topography, which will be ultimately realized after stationary fault surface contact of very long duration. In the present study, on the basis of the spectral analysis of rock surface profile data (BROWN and SCHOLZ, 1985; POWER et al., 1987), we assume that $|\bar{Y}(k)|^2$ has a k^{-3}-dependence ($|\bar{Y}(k)|^2 \propto k^{-3}$) over a sufficiently broad wavenumber range.

To see the properties of the slip- and time-dependent constitutive law defined by Eqs. (8) and (9), AOCHI and MATSU'URA (1999, 2002) have examined three extreme cases. The first is the case of high-speed slip ($v \equiv dw/dt \gg 0$). In this case the effect of adhesion is negligible, and so the solution of Eq. (9) is given by

$$|Y(k; w, t_0)| = |Y_0(k)| e^{-\alpha k(w - w_0)}, \tag{10}$$

where $|Y_0(k)| \equiv |Y(k; w_0, t_0)|$ indicates the Fourier amplitude of surface topography at a reference state ($w = w_0$, $t = t_0$), which should be determined from Eqs. (8) and (9), depending on the past history of fault slip. As can be seen from Eq. (10), the process is essentially slip-weakening. The characteristic wavenumber k_c representing the lower limit of self-similarity in fault surface topography (or the upper corner wavelength, $\lambda_c = 2\pi/k_c$, in the power spectrum of fault surface topography) is a key parameter, because the critical weakening displacement D_c has linear dependence on the characteristic weakening displacement $d_c \equiv 1/\alpha k_c$ (MATSU'URA et al., 1992; AOCHI and MATSU'URA, 2002).

The second is the case of stationary contact ($v = 0$). In this case the effect of abrasion is negligible, and so the solution of Eq. (9) is given by

$$|Y(k; w_0, t)| = |\bar{Y}(k)| - [|\bar{Y}(k)| - |Y_0(k)|] e^{-\beta k^2 (t - t_0)}. \tag{11}$$

In stationary contact the Fourier amplitude is gradually restored with time to the inherent state $|\bar{Y}(k)|$ in the order of decreasing wavenumber (increasing wavelength). This process brings about fault restrengthening. The third is the case of steady-state slip ($d|Y(k)|/dt = 0$, $v = $ const). In this case the solution of Eq. (9) is given by

$$|Y_{ss}(k; v)| = \frac{\beta k}{\alpha v + \beta k} |\bar{Y}(k)| \tag{12}$$

which indicates slip-velocity weakening.

The essential parameters controlling the slip- and time-dependent fault constitutive law are the abrasion rate α and the adhesion rate β. Given the values of α and β and the past history of the fault slip, we can calculate the values of constitutive parameters, such as the breakdown strength drop $\Delta\sigma_p$ and the critical weakening displacement D_c at each moment from Eqs. (8) and (9). Here, $\Delta\sigma_p$ is defined as the difference between the peak strength σ_p and a residual strength level at the time of dynamic slip, and D_c is a characteristic displacement required for the shear strength to decrease to the residual level.

4. Results of Numerical Simulations

The coupled nonlinear system of Eqs. (4), (6) and (7) can be numerically solved by rewriting these equations in discrete form both in space and time. Following HASHIMOTO and MATSU'URA (2000), we represent the slip perturbation u_s by the superposition of a finite number ($K \times L$) of the bicubic B-spline functions M_{kl} as

$$u_s(\mathbf{x}, t) = \sum_{k=1}^{K} \sum_{l=1}^{L} a_{kl}(t) M_{kl}(\mathbf{x}). \tag{13}$$

Substituting this expression into Eq. (6), we obtain

$$\sigma(\mathbf{x}, t) = \sigma_0(\mathbf{x}) + \int_0^t \sum_{k=1}^{K} \sum_{l=1}^{L} \frac{\partial a_{kl}(\tau)}{\partial \tau} S_{kl}(\mathbf{x}, t - \tau) \, d\tau \tag{14}$$

with

$$S_{kl}(\mathbf{x}, t - \tau) = \int_{\Sigma_s} M_{kl}(\boldsymbol{\xi})H(\mathbf{x}, t - \tau; \boldsymbol{\xi}, 0)d\boldsymbol{\xi}, \tag{15}$$

where Σ_s denotes a model region on the plate interface. Then, our problem can be reduced to the problem of determining the expansion coefficients $a_{kl}(t)$ stepwise in time so that the shear stress σ and the fault slip $w = v_{pl}t + u_s$ satisfy the given boundary conditions in Eq. (7). To solve this discretized problem we use the Levenberg-Marquardt algorithm for nonlinear least-squares (LEVENBERG, 1944; MARQUARDT, 1963).

4.1 Earthquake Generation Cycles

With the computational algorithm mentioned above we numerically simulate the entire process of earthquake generation cycles. For the numerical simulation, first, we must set the appropriate values of α and β to prescribe the constitutive properties of a seismogenic region on the plate interface. Figure 6 shows the

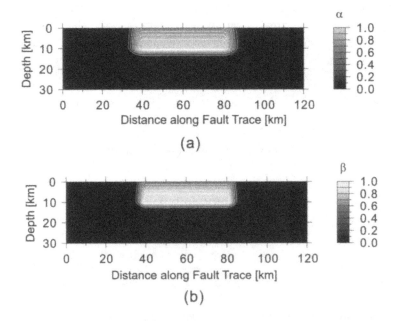

Figure 6
The constitutive properties of the plate interface used in numerical simulation (a single fault system). (a) Spatial variation in the abrasion rate α. The value of α is normalized by its maximum value. (b) Spatial variation in the adhesion rate β. The value of β is normalized by its maximum value. The seismogenic region is characterized by a high abration rate and a high adhesion rate.

spatial variation in α and β on the plate interface, where the seismogenic region is characterized by a high abrasion rate ($\alpha = 0.3$) and a high adhesion rate ($\beta = 10^{-4}\,\mathrm{m}^2/\mathrm{yr}$). The values of α and β gradually increase with depth in the seismogenic region, and rapidly decrease in the brittle-ductile transition zone. In the direction of fault strike the values of α and β are tapered off at the both ends of the seismogenic region. As mentioned in 3.3, α and β are the essential parameters controlling the constitutive properties of faults. For example, if we take the value of α to be very small, then the critical weakening displacement D_c becomes very large. If we take the value of β to be very small, the peak strength σ_p becomes very small. Through the numerical simulation of tectonic stress accumulation with a simple slip-weakening type of fault constitutive law, HASHIMOTO and MATSU'URA (2000) have demonstrated the result that if D_c is very large, accumulated stress is stably released without unstable dynamic rupture. In the present simulation we take the values of α and β to be very small in the surrounding area of the seismogenic region to represent creeping motion there. Figure 7 shows a series of snapshots for the process of stress accumulation and release repeated in the seismogenic region at the recurrence interval of about 26 yrs. Here, the relative plate velocity v_{pl} is taken to be 5 cm/yr. After the unstable slip at $t = 182.7$ yr, the stress accumulation due to slip deficits gradually proceeds in the seismogenic region until $t = 208.4-$ yr, while stable slip with stress

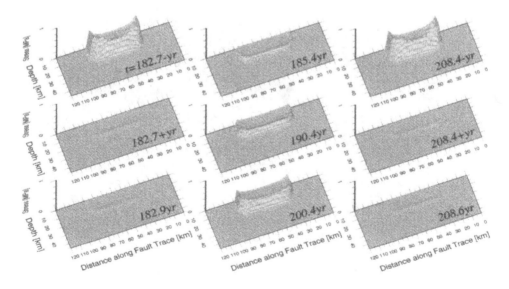

Figure 7

A series of snapshots showing the stress accumulation and release during an earthquake cycle ($t = 182.7+$ yr $\sim 208.4-$ yr). The process of stress accumulation and release is repeated with the recurrence interval of about 26 yrs. The stress concentration at the margin of the seismogenic region is caused by dislocation pile-ups around its edges.

release proceeds in the surrounding area. The stress concentration at the margin of the seismogenic region is caused by dislocation pile-ups around the edges of the seismogenic region. At $t = 208.4$ yr the system becomes unstable, indicating the occurrence of earthquake rupture. Since the algorithm dealing with dynamic rupture propagation (FUKUYAMA et al., 2002) has not yet incorporated into the present simulation model, we permit a jump in fault slip and shear stress, and search a next time-step quasi-static solution ($t = 208.4+$ yr) so that the fault constitutive relation $\sigma = \sigma_{strg}$, calculated from Eqs. (8) and (9), is satisfied. In the present case the dynamic rupture process terminates with the almost complete release of the shear stress stored in the seismogenic region. The average stress drop of the seismic event, evaluated from the stress distribution just before the unstable rupture, is about 2 MPa, which accords with the average stress drop of large interplate strike-slip earthquakes in California and North Anatolia (FUJII and MATSU'URA, 2000).

Figure 8 is a series of snapshots indicating change in slip deficits with time during the same earthquake cycle as in Figure 7. The slip deficits gradually increase with time in the seismogenic region through the interseismic period ($t = 182.7+$ yr $\sim 208.4-$ yr), while stable slip at the relative plate velocity proceeds in the surrounding area. The approximate 1.3 m slip deficits in the seismogenic region are recovered by abrupt fault slip at the time of unstable rupture ($t = 208.4$ yr).

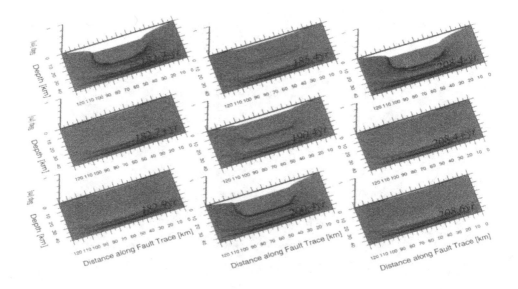

Figure 8
A series of snapshots showing change in slip deficits with time during the earthquake cycle ($t = 182.7+$ yr $\sim 208.4-$ yr). About 1.3 m slip deficits in the seismogenic region are recovered by abrupt fault slip at $t = 208.4$ yr.

4.2 Evolution of Fault Constitutive Properties

As shown in 3.3, the slip- and time-dependent constitutive law has an inherent mechanism of fault strength restoration. In this law the constitutive parameters, the breakdown strength drop $\Delta\sigma_p$ and the critical weakening displacement D_c, change with fault slip w and contact time t. Given the values of α and β and the past history of fault slip, we can calculate the values of $\Delta\sigma_p$ and D_c at each moment from Eqs. (8) and (9). In this subsection we examine the evolution of fault constitutive properties during the complete earthquake generation cycle in Figures 7 and 8.

Figure 9 shows the depth variation in fault constitutive relation at the center of the seismogenic region (distance along fault trace = 60 km) just before the occurrence of unstable dynamic slip ($t = 208.4$ yr shown in Figures 7 and 8). The breakdown strength drop $\Delta\sigma_p$, which is defined as the difference between the peak strength σ_p and a depth-dependent residual strength level at the time of dynamic slip, gradually increases with depth d in the brittle seismogenic region ($d < 12$ km), and rapidly decreases in the brittle-ductile transition zone ($12\,\text{km} < d < 18$ km). The critical weakening displacement D_c is nearly constant (about 0.4 m) in the seismogenic region, but rapidly increases with depth in the brittle-ductile transition zone. On the basis of laboratory experiments, OHNAKA (1992) concluded that D_c is insensitive to temperature T in the range of $T < 300\,°\text{C}$, but has significant temperature dependence in the range of $T > 300\,°\text{C}$. In the case of vertical

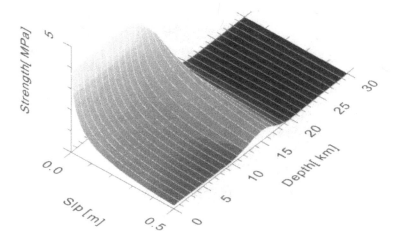

Figure 9

Depth variation in constitutive relation (white lines) at the center of the seismogenic region (distance along fault trace = 60 km) just before the occurrence of unstable dynamic rupture ($t = 208.4-$ yr). The breakdown strength drop $\Delta\sigma_p$ gradually increases with depth d in the brittle seismogenic region ($d < 12$ km), and rapidly decreases in the brittle-ductile transition zone ($12\,\text{km} < d < 18$ km). The critical weakening displacement D_c is nearly constant (about 0.4 m) in the seismogenic region, but rapidly increases with depth in the brittle-ductile transition zone.

transcurrent plate boundaries, the critical temperature of 300 °C corresponds to about 12 km in depth, over which brittle-ductile transition occurs (SIBSON, 1984; MARONE and SCHOLZ, 1988). The result in Figure 9 demonstrates that the simulated constitutive relation realizes the depth dependence of fault constitutive properties estimated from laboratory experiments. The values of $\Delta\sigma_p$ and D_c just before the unstable slip are crucial, because the subsequent dynamic process must be controlled by these constitutive parameters as well as the stress distribution just before the unstable slip. Recently, FUKUYAMA et al. (2002) succeeded in simulating the subsequent dynamic rupture process by using the output of our quasi-static simulation (fault constitutive relation and stress distribution just before unstable slip) as the input of dynamic simulation with the boundary integral equation method (BIEM) developed by FUKUYAMA and MADARIAGA (1995, 1998).

Once unstable dynamic slip occurred, the fault strength decreases rapidly to its residual level. Thus, just after the arrest of dynamic slip, the constitutive relation curve becomes nearly flat. Thereafter, the restoration of fault strength begins. In Figure 10 we show the evolution of fault constitutive relation at a central point of the seismogenic region (distance along fault trace = 60 km and depth = 6 km) during the interseismic period. After the arrest of dynamic slip ($t = 182.7+$ yr), the

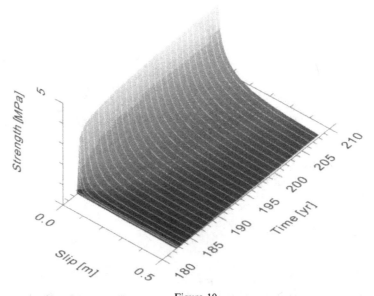

Figure 10
The evolution of fault constitutive relation (white lines) at a central point of the seismogenic region (distance along fault trace = 60 km, depth = 6 km) during the interseismic period from $t = 182.7+$ yr to $t = 208.4-$ yr. After the arrest of dynamic rupture, the breakdown strength drop $\Delta\sigma_p$ is rapidly restored, while the restoration of the critical weakening displacement D_c proceeds gradually with time through the interseismic period.

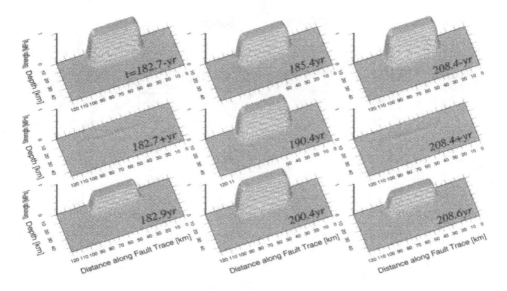

Figure 11

A series of snapshots showing the restoration of peak strength in the seismogenic region during the earthquake cycle ($t = 182.7+$ yr $\sim 208.4-$ yr).

breakdown strength drop $\Delta\sigma_p$ is rapidly restored, while the restoration of D_c proceeds gradually with time through the interseismic period. Just before the occurrence of the next unstable dynamic slip ($t = 208.4-$ yr), the fault has the same constitutive properties as in Figure 9. The series of snapshots in Figure 11 shows the restoration of the peak strength during the same earthquake cycle as in Figures 7 and 8. The peak strength decreases to its residual level by the occurrence of dynamic slip at $t = 182.7$ (208.4) yr. After the arrest of dynamic slip, the restoration of $\Delta\sigma_p$ proceeds rapidly in the seismogenic region. However, the fracture surface energy G_c ($\sim \Delta\sigma_p D_c/2$) is not restored so rapidly, because the restoration of D_c proceeds gradually with time. This indicates that the seismogenic region is still weak in the early stage of the interseismic period, although the peak strength has been restored.

The evolution of fault constitutive properties can be regarded as the macroscopic manifestation of the microscopic break and recovery processes of the fractal fault surface structure. The Fourier amplitude $|Y(k)|$, which represents the roughness of fault surfaces, decreases with fault slip and increases with contact time. The essential point is that this process is scale-dependent. Figure 12 shows the evolution of fault surface roughness during the interseismic period. After the arrest of dynamic slip ($t = 182.7+$ yr), the Fourier amplitude $|Y(k)|$ is gradually restored with time to its inherent fractal amplitude $|\bar{Y}(k)| \propto k^{-3/2}$. The restoration rate of $|Y(k)|$ is faster in larger wavenumber k (smaller wavelength λ), and slower in smaller wavenumber (larger wavelength). This indicates that the smaller-scale fractal structure is restored

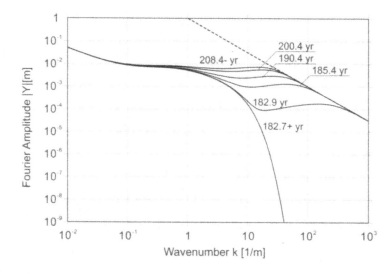

Figure 12

The evolution of fault surface topography at a central point of the seismogenic region during the interseismic period ($t = 182.7+$ yr $\sim 208.4-$ yr). This process corresponds to the evolution of the fault constitutive relation shown in Figure 10. After the arrest of dynamic slip ($t = 182.7+$ yr), the Fourier amplitude $|Y(k)|$ is gradually restored with time to its inherent fractal amplitude, $|\bar{Y}(k)| \propto k^{-3/2}$, which is indicated by the broken line.

more rapidly. The key to understanding the evolution of fault constitutive properties is the decrease of k_c (increase of λ_c) with time during the interseismic period. It should be noted that $\lambda_c (= 2\pi/k_c)$ gives the upper fractal limit of fault surface topography, and the critical weakening displacement D_c is in proportion to λ_c (MATSU'URA et al., 1992; OHNAKA, 1996). Therefore, the decrease of k_c (increase of λ_c) with time means the increase of D_c with time.

4.3 Viscoelastic Fault-to-Fault Interaction

After the arrest of dynamic slip, tectonic stress accumulation proceeds again with time in the seismogenic region, as demonstrated in 4.1. In this subsection we examine the effects of transient viscoelastic stress transfer after the sudden fault slip through the simulation of earthquake generation cycles for a multi-segmented fault system.

We represent the multi-segmented fault system by introducing a narrow stable creeping zone which divides the single seismogenic region into two parts. The seismogenic regions are characterized by large α and β ($\alpha = 0.3$, $\beta = 10^{-4}$ m²/yr), as shown in Figure 13. On the other hand, the creep zone is characterized by very small α and β. The fault system consists of a 30-km-long segment and 50-km-long segment, separated by a 10-km-long creep zone.

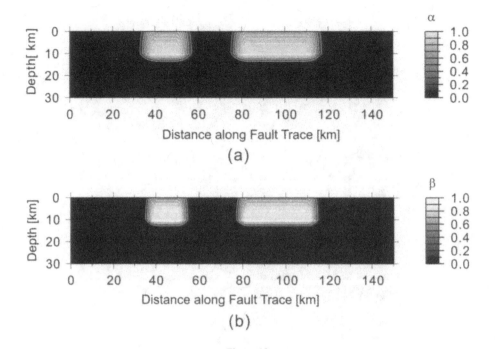

Figure 13
The constitutive properties of the plate interface used in numerical simulation (a two faults system). (a) Spatial variation in the abrasion rate α. The value of α is normalized by its maximum value. (b) Spatial variation in the adhesion rate β. The value of β is normalized by its maximum value. The fault system consists of a 30-km-long segment and 50-km-long segment, separated by a 10-km-long creep zone. The seismogenic regions are characterized by large α and large β.

Figure 14 is a series of snapshots showing the stress accumulation and release process of the fault system with the small and large segments. After simultaneous dynamic rupture of the small and the large segments at $t = 0$, stress accumulation proceeds again with time in both segments at different rates. At $t = 37.5$ yr the small segment first becomes unstable. The elastic stress transfer associated with the dynamic rupture of the small segment brings about stress increase in the large segment, however the dynamic rupture of the large segment is not triggered at once. The subsequent viscoelastic stress transfer accelerates the stress accumulation in the large segment. Then, after about 2 yrs, the large segment becomes unstable. The subsequent unstable dynamic rupture occurs not only in the large segment but also in the small segment, because the fracture surface energy G_c ($\sim \Delta\sigma_p D_c/2$) of the small segment has not yet sufficiently restored.

In Figure 15, to see the effects of elastic and viscoelastic fault-to-fault interaction in detail, we plot the change in shear stress at the center of each segment for 10 yrs, including the two unstable dynamic events in Figure 14. Stress accumulation proceeds in both segments but at different rates, since the stress accumulation rate has fault-length dependence (MATSU'URA and SATO, 1997). Then, at $t = 37.5$ yr, the

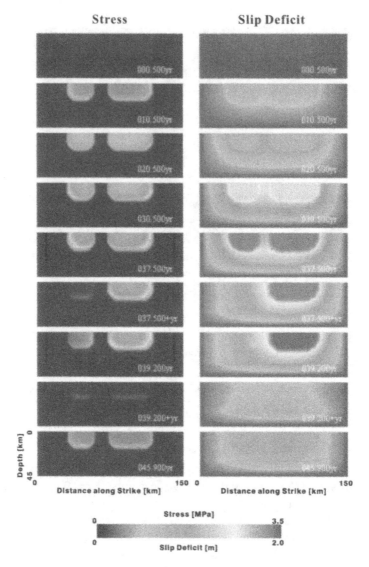

Figure 14

A series of snapshots showing the stress accumulation and release for the 90-km-long fault system with a small (30 km) and a large (50 km) seismogenic region. The small segment first becomes unstable. The dynamic rupture of the small segment accelerates stress accumulation in the large segment through instantaneous elastic stress transfer and subsequent viscoelastic stress transfer. Then, after about 2 yrs, the large segment becomes unstable. The dynamic rupture of the large segment directly triggers the dynamic rupture of the small segment.

small segment first becomes unstable, and its shear stress suddenly decreases to the residual level. The dynamic rupture of the small segment brings about instantaneous stress increase in the large segment, but does not directly trigger the dynamic rupture

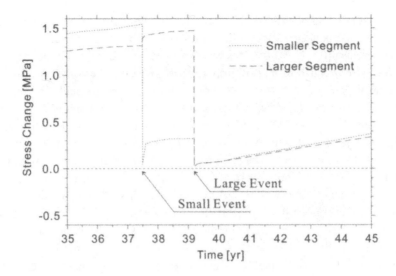

Figure 15
A diagram showing the effects of elastic and viscoelastic fault-to-fault interaction in the fault system with the small (30 km) and large (50 km) seismogenic segments. The dotted line indicates the stress change of the small segment at its center, and the broken line indicates that of the large segment at its center. Stress accumulation proceeds in both segments at different rates. At $t = 37.5$ yr, the small segment first becomes unstable. The dynamic rupture of the small segment brings about instantaneous stress increase in the large segment, but does not directly trigger the dynamic rupture of the large segment. The subsequent stress accumulation proceeds rapidly in both segments, but its rate decreases soon. Then, after about 2 yrs, the large segment becomes unstable.

of the large segment. The subsequent stress accumulation proceeds rapidly in both segments, although its rate soon decreases. This indicates the importance of transient viscoelastic stress transfer through the asthenosphere. Next, after about 2 yrs, the large segment becomes unstable, and the elastic stress transfer associated with the dynamic rupture of the large segment directly triggers the dynamic rupture of the small segment.

5. Discussion and Conclusions

We constructed a realistic 3-D simulation model for earthquake generation cycles at transcurrent plate boundaries by combining three basic elements, the viscoelastic slip-response function (HASHIMOTO and MATSU'URA, 2000), the slip- and time-dependent fault constitutive law (AOCHI and MATSU'URA, 2002), and the steady relative plate motion as a driving force into a single closed system. With this model we numerically simulated the entire process of earthquake generation cycles.

In 4.2, we elucidated the evolution of fault constitutive properties during a complete earthquake generation cycle. At the time of dynamic rupture, fault strength decreases with slip to a constant level. After the arrest of dynamic rupture, the breakdown strength drop $\Delta\sigma_p$ is rapidly restored, while the restoration of the critical weakening displacement D_c proceeds gradually with time through the interseismic period. MATSU'URA et al. (1992) and OHNAKA (1996) have demonstrated the linear dependence of D_c on the inverse of the critical wavenumber k_c, which corresponds to the upper corner wavelength $(\lambda_c = 2\pi/k_c)$ in the power spectrum of fault surface topography. The power spectrum of fault surface topography changes with fault slip and stationary contact time. Therefore, the value of k_c is process-dependent. As shown in Figure 12, the power spectrum amplitude of surface topography suddenly drops in a large wavenumber range at the time of dynamic rupture. Thereafter, the restoration of the power spectrum amplitude proceeds rapidly in larger wavenumber but slowly in smaller wavenumber. Thus, the inverse of k_c gradually increases with time during the interseismic period. This process produces the gradual increase of D_c in Figure 10.

The gradual increase of D_c through the interseismic period gives a possible explanation for the event-scale dependence of D_c, pointed out by AKI (1992), OHNAKA (1996), and SHIBAZAKI and MATSU'URA (1998). From the comparative study of laboratory experiments and seismological observations with a theoretical model of rupture nucleation, SHIBAZAKI and MATSU'URA (1998) have demonstrated that the fracture surface energy G_c, the critical nucleation-zone size L_c, and also the dimension of seismic faults L_f scale with D_c. They have proposed the scaling relation of $D_c \simeq 10^{-5}L_f$. In our simulation, L_f is about 60 km and D_c just before dynamic rupture is about 0.4 m. Consequently the scaling relation proposed by SHIBAZAKI and MATSU'URA (1998) is roughly satisfied.

In 4.3, we examined the effects of viscoelastic fault-to-fault interaction through numerical simulation with a multi-segmented fault model, consisting of the 30-km-long small segment and the 50-km-long large segment. Because of faster stress accumulation in the smaller segment, the smaller segment first becomes unstable. The dynamic rupture of the smaller segment brings about instantaneous stress increase in the larger segment, however it does not directly trigger the dynamic rupture of the larger segment. The subsequent viscoelastic stress transfer through the asthenosphere accelerates the stress accumulation in the larger segment. Then, after about 2 yrs, the larger segment becomes unstable.

There have been many reports on sequential occurrence of large earthquakes along a plate boundary. The occurrence of the 1944 Tonankai and the 1946 Nankaido earthquakes along the Nankai trough, southwest Japan, is one of the good examples. Another example is the migrating earthquake sequence during 1939–44 along the North Anatolian fault. These examples of the sequential occurrence of large interplate earthquakes raise the question of how one big event triggers another big event with a time lag. STEIN et al. (1996) have sought to explain the migrating earthquake sequences along the North Anatolian fault by the change in Coulomb failure stress ΔCFS. In their study, however, the time lag of the sequential events has not been reasonably explained. The essential cause of the time lag is in the viscoelastic transient stress transfer through the asthenosphere, while their ΔCFS analysis was performed in the elastic framework. In the present study, we demonstrated the importance of the viscoelastic fault-to-fault interaction, which supplies a possible explanation for the time lag of migrating sequential events along a plate boundary.

REFERENCES

AKI, K. (1992), *Higher Order Interrelations between Seismic Structures and Earthquake Processes*, Tectonophysics *211*, 1–12.

AOCHI, H. and MATSU'URA, M. (1999), *Evolution of contacting rock surfaces and a slip- and time-dependent fault constitutive law*, Proceedings of the 1st ACES Workshop, ed. P. Mora, APEC Cooperation for Earthquake Simulation, 135–140.

AOCHI, H. and MATSU'URA, M. (2002), *Slip- and Time-dependent Fault Constitutive Law and its Significance in Earthquake Generation Cycles*, Pure Appl. Geophys. *159*, 2029–2046.

BEN-ZION, Y. and RICE, J. R. (1997), *Dynamic Simulations of Slip on a Smooth Fault in an Elastic Solid*, J. Geophys. Res. *102*, 17,771–17,784.

BROWN, S. R. and SCHOLZ, C. H. (1985), *Broad Bandwidth Study of the Topography of Natural Rock Surfaces*, J. Geophys. Res. *90*, 12,575–12,582.

BURRIDGE, R. and KNOPOFF, L. (1964), *Body Force Equivalents for Seismic Dislocations*, Bull. Seismol. Soc. Am. *54*, 1875–1888.

BURRIDGE, R. and KNOPOFF L. (1967), *Model and Theoretical Seismicity*, Bull. Seismol. Soc. Am. *57*, 341–371.

CAO, T. and AKI, K. (1984), *Seismicity Simulation with a Mass-spring Model and a Displacement Hardening-softening Friction Law*, Pure Appl. Geophys. *122*, 10–24.

CARLSON, J. M. and LANGER, J. S. (1989), *Mechanical Model of an Earthquake Fault*, Phys. Rev. *A40*, 6470–6484.

DIETERICH, J. H. (1972), *Time-dependent Friction in Rocks*, J. Geophys. Res. *77*, 3690–3697.

DIETERICH, J. H. (1978), *Time-dependent Friction and the Mechanics of Strike-slip*, Pure Appl. Geophys. *116*, 790–806.

DIETERICH, J. H. (1979), *Modeling of Rock Friction 1. Experimental Results and Constitutive Equations*, J. Geophys. Res. *84*, 2161–2168.

FUJII, Y. and MATSU'URA, M. (2000), *Regional Difference in Scaling Laws for Large Earthquakes and its Tectonic Implication*, Pure Appl. Geophys. *157*, 2283–2302.

FUKUYAMA, E. and MADARIAGA, R. (1995), *Integral Equation Method for Plane Crack with Arbitrary Shape in 3-D Elastic Medium*, Bull. Seismol. Soc. Am. *85*, 614–628.

FUKUYAMA, E. and MADARIAGA, R. (1998), *Rupture Dynamics of a Planar Fault in a 3-D Elastic Medium: Rate- and Slip-weakening Friction*, Bull. Seismol. Soc. Am. *88*, 1–17.

FUKUYAMA, E., HASHIMOTO, C., and MATSU'URA, M. (2002), *Simulation of the Transition of Earthquake Rupture from Quasi-static Growth to Dynamic Propagation*, Pure Appl. Geophys. *159*, 2057–2066.

HASHIMOTO, C. and MATSU'URA, M. (2000), *3-D Physical Modelling of Stress Accumulation and Release Processes at Transcurrent Plate Boundaries*, Pure Appl. Geophys. *157*, 2125–2147.

IWASAKI, T. and MATSU'URA, M. (1982), *Quasi-static Crustal Deformation due to a Surface Load: Rheological Structure of the Earth's Crust and Upper Mantle*, J. Phys. Earth *30*, 469–508.

KATO, N. and HIRASAWA, T. (1997), *A Numerical Study on Seismic Coupling along Subduction Zones Using a Laboratory-derived Friction Law*, Phys. Earth Planet. Inter. *29*, 499–518.

LAPUSTA, N., RICE, J. R., BEN-ZION, Y., and ZHENG, G. (2000), *Elastodynamic Analysis for Slow Tectonic Loading with Spontaneous Rupture Episodes on Faults with Rate- and State-dependent Friction*, J. Geophys. Res. *105*, 23,765–23,789.

LEVENBERG, K. (1944), *A Method for the Solution of Certain Nonlinear Problems in Least-Squares*, Q. Appl. Math. *2*, 164–168.

LISOWSKI, M., SAVAGE, J. C., and PRESCOTT, W. H. (1991), *The Velocity Field along the San Andreas Fault in Central and Southern California*, J. Geophys. Res. *96*, 8369–8389.

MARONE, C. and SCHOLZ, C. H. (1988), *The Depth of Seismic Faulting and the Upper Transition from Stable to Unstable Slip Regimes*, Geophys. Res. Lett. *15*, 621–624.

MARQUARDT, D. W. (1963), *An Algorithm for Least-squares Estimation of Nonlinear Parameters*, Indust. Appl. Math. *11*, 431–441.

MARUYAMA, T. (1963), *On the Force Equivalents of Dynamical Elastic Dislocations with Reference to the Earthquake Mechanism*, Bull. Earthq. Res. Inst., Tokyo Univ. *41*, 467–486.

MATSU'URA, M. and SATO, T. (1989), *A Dislocation Model for the Earthquake Cycle at Convergent Plate Boundaries*, Geophys. J. Int. *96*, 23–32.

MATSU'URA, M. and SATO, T. (1997), *Loading Mechanism and Scaling Relation of Large Interplate Earthquakes*, Tectonophysics *277*, 189–198.

MATSU'URA, M., KATAOKA, H., and SHIBAZAKI, B. (1992), *Slip-dependent Friction Law and Nucleation Processes in Earthquake Rupture*, Tectonophysics *211*, 135–148.

OHNAKA, M. (1992), *Earthquake Source Nucleation: A Physical Model for Short-term Precursors*, Tectonophysics *211*, 149–178.

OHNAKA, M. (1996), *Nonuniformity of the Constitutive Law Parameters for Shear Fracture and Quasistatic Nucleation to Dynamic Rupture: A Physical Model of Earthquake Generation Process*, Proc. Natl. Acad. Sci. USA, *93*, 3795–3802.

OHNAKA, M., KUWAHARA, Y., and YAMAMOTO, K. (1987), *Constitutive Relations between Dynamic Physical Parameters near a Tip of the Propagating Slip Zone during Stick-slip Shear Failure*, Tectonophysics *144*, 109–125.

POWER, W. L., TULLIS, T. E., BROWN, S. R., BOITNOTT, G. N., and SCHOLZ, C. H. (1987), *Roughness of Natural Fault Surfaces*, Geophys. Res. Lett. *14*, 29–32.

RICE, J. R. (1993), *Spatio-temporal Complexity of Slip on a Fault*, J. Geophys. Res. *98*, 9885–9907.

RICE, J. R. and BEN-ZION, Y. (1996), *Slip Complexity in Earthquake Fault Models*, Proc. Natl. Acad. Sci. USA *93*, 3811–3818.

RUINA, A. (1983), *Slip Instability and State Variable Friction Law*, J. Geophys. Res. *88*, 10,359–10,370.

SAVAGE, J. C. (1990), *Equivalent Strike-slip Earthquake Cycles in Half-space and Lithosphere-asthenosphere Earth Models*, J. Geophys. Res. *95*, 4873–4879.

SAVAGE, J. C. and BURFORD, R. O. (1973), *Geodetic Determination of Relative Plate Motion in Central California*, J. Geophys. Res. *78*, 832–845.

SHIBAZAKI, B. and MATSU'URA, M. (1998), *Transition Process from Nucleation to High-speed Rupture Propagation: Scaling from Stick-slip Experiments to Natural Earthquakes*, Geophys. J. Int. *132*, 14–30.

SIBSON, R. H. (1984), *Roughness at the Base of the Seismogenic Zone: Contributing Factors*, J. Geophys. Res. *89*, 5791–5799.

STEIN, R. S., BARKA, A. A., and DIETERICH, J. H. (1996), *Progressive Failure on the North Anatolian Fault since 1939 by Earthquake Stress Triggering*, Geophys. J. Int. *128*, 594–604.

STUART, W. D. (1984/85), *Instability Model for Recurring Large and Great Earthquakes in Southern California*, Pure Appl. Geophys. *122*, 793–811.

STUART, W. D. (1986), *Forecast Model for Large and Great Earthquakes in Southern California*, J. Geophys. Res. *91*, 13,771–13,786.

STUART, W. D. (1988), *Forecast Model for Great Earthquakes at the Nankai Trough Subduction Zone*, Pure Appl. Geophys. *126*, 619–641.

THATCHER, W. (1983), *Nonlinear Strain buildup and Earthquake Cycle on the San Andreas Fault*, J. Geophys. Res. *88*, 5893–5902.

TSE, S. T. and RICE, J. R. (1986), *Crustal Earthquake Instability in Relation to the Depth Variation of Frictional Slip Properties*, J. Geophys. Res. *91*, 9452–9472.

WARD, S. N. (2000), *San Francisco Bay Area Earthquake Simulations: A Step Toward a Standard Physical Model*, Bull. Seismol. Soc. Am. *90*, 370–386.

(Received February 20, 2001, revised June 11, 2001, accepted June 15, 2001)

 To access this journal online:
http://www.birkhauser.ch

Pure appl. geophys. 159 (2002) 2201–2220
0033–4553/02/102201–20 $ 1.50 + 0.20/0

▌Pure and Applied Geophysics

Interplate Earthquake Fault Slip During Periodic Earthquake Cycles in a Viscoelastic Medium at a Subduction Zone

Kazuro Hirahara[1]

Abstract—A 2-D finite-element-method (FEM) numerical experiment of earthquake cycles at a subduction zone is performed to investigate the effect of viscoelasticity of the earth on great interplate earthquake fault slip. We construct a 2-D viscoelastic FEM model of northeast Japan, which consists of an elastic upper crust and a viscoelastic mantle wedge under gravitation overlying the subducting elastic Pacific plate. Instead of the dislocation model prescribing an amount of slip on a plate interface, we define an earthquake cycle, in which the plate interface down to a depth is locked during an interseismic period and unlocked during coseismic and postseismic periods by changing the friction on the boundary with the master-slave method. This earthquake cycle with steady plate subduction is periodically repeated to calculate the resultant earthquake fault slip.

As simulated in a previous study (WANG, 1995), the amount of fault slip at the first earthquake cycle is smaller than the total relative plate motion. This small amount of fault slip in the viscoelastic medium was considered to be one factor explaining the small seismic coupling observed at several subduction zones. Our simulation, however, shows that the fault slip grows with an increasing number of repeated earthquake cycles and reaches an amount comparable to the total relative plate motion after more than ten earthquake cycles. This new finding indicates that the viscoelasticity of the earth is not the main factor in explaining the observed small seismic coupling. In comparison with a simple one-degree-of-freedom experiment, we demonstrate that the increase of the fault slip occurs in the transient state from the relaxed initial state to the stressed equilibrium state due to the intermittent plate loading in a viscoelastic medium.

Key words: Viscoelasticity, interplate earthquake, fault slip, earthquake cycle, FEM, seismic coupling.

1. Introduction

At subduction zones, great interplate earthquakes repeatedly occur and release the accumulated tectonic stress due to plate convergence. It is well known that the average rate of coseismic fault slip over repeated earthquake cycles is smaller than the plate convergence rate estimated from global plate motion models at several subduction zones. The ratio of average fault slip rate to plate convergence, called seismic coupling, varies from place to place, and its relation to other geophysical

[1] Graduate School of Science, Nagoya University, Chikusa, Nagoya 464-8602, Japan. Now at Graduate School of Environmental Studies, E-mail: hirahara@eps.nagoya-u.ac.jp

parameters such as plate convergence rate and age of the subducting plate has been investigated (e.g., PACHECO *et al.*, 1993).

The observation of seismic coupling of less than 100 percent indicates that a portion of relative plate motion is consumed in some form of aseismic slip on the plate interface during earthquake cycles at subduction zones. In fact, geodetic observations using conventional strain meters and recent GPS technology have as yet caught seismically invisible crustal deformations at the earth's surface with wide-range characteristic timescales (e.g., HEKI *et al.*, 1997; KAWASAKI *et al.*, 1998; HIROSE *et al.*, 1999). Owing to the recent development of experimental and theoretical studies of rock friction, understanding of seismic coupling and seismic and aseismic slip on a fault in terms of frictional properties of a plate interface or fault (e.g., SCHOLZ, 1990) has been greatly improved. Since the pioneering work of TSE and RICE (1986) and RICE (1993) at a transcurrent plate boundary, recent numerical experiments based on laboratory inspired friction laws have been extended to model the earthquake cycle at subduction zones (e.g., STUART, 1988; KATO and HIRASAWA, 1997). They have successfully explained the seismic coupling and the recurrence time of interplate earthquakes by assigning a distribution of frictional parameters on the plate interface. Their experiments, however, neglected inelastic properties of the earth, which are thought to be important at subduction zones where thermally anomalous structures exist.

The inelastic behavior of the earth is usually approximated by a viscoelastic rheology. In response to a sudden slip on a fault, the viscoelastic property produces a longer term response in addition to an instantaneous elastic one. This can explain observations of postseismic crustal deformation at the ground surface enduring over decades in addition to the coseismic deformation (e.g., THATCHER *et al.*, 1980). Moreover, viscoelasticity affects deformation in the crust and the mantle wedge caused by steady motion of the subducting plate during earthquake cycles. Most previous studies that considered viscoelasticity have investigated postseismic deformation or the earthquake deformation cycle by prescribing the amount of slip on an earthquake fault or the slip rate on a plate interface using the dislocation model of SAVAGE (1983) (e.g., THATCHER and RUNDLE, 1984; MATSU'URA and SATO, 1989).

The viscoelastic property also affects the amount of earthquake fault slip at subduction zones, which has been noted by WANG (1995). Using a fault lock-and-unlock technique that needs not prescribe the fault slip, he showed that the earthquake fault slip in a viscoelastic medium is smaller than the total relative plate motion, though he examined a single earthquake.

We are employing the general-purpose commercial FEM code "ABAQUS" to perform FEM modeling of repeated earthquake cycles, implementing a particular laboratory derived friction law (e.g., RUINA, 1983) in a laterally heterogeneous viscoelastic medium under gravitation. Through preliminary numerical experiments, we have found an interesting phenomenon, that is, the amount of resultant fault slip increases to a limit with each successive earthquake cycle.

In this study, we investigate only the effect of viscoelasticity on earthquake fault slip. We define a simple artificial earthquake cycle in which the interseismic locked portion of the plate interface undergoes a prescribed friction change from the interseismic period to the coseismic and postseismic periods. Namely, a portion of plate interface is mostly locked in the interseismic period and unlocked in the coseismic and postseismic periods. With this simple earthquake cycle, we investigate the evolution of repeated earthquake fault slip events induced by steady plate subduction.

In addition to this FEM modeling, to understand a physical mechanism for the increase of the earthquake slip, we show a one-degree-of-freedom locked-and-unlocked experiment using a simple spring-dashpot model. This simple numerical experiment suggests that the increase of earthquake fault slip and postseismic slip is a transient phenomenon from the initial relaxed state to the stationary deformed and stressed state due to the subduction of the plate. In comparison of this simple experiment, we examine the result of our FEM modeling.

2. Model of Earthquake Cycle

2.1. 2-D Viscoelastic FEM Model

We construct a 2-D FEM model of northeast Japan, where the Pacific plate is subducting with a dip of about 20 degrees and great interplate earthquakes are repeatedly occurring. The reason for selecting northeast Japan rather than other regions is that the laterally heterogeneous structure in this region is well constrained and is reasonably approximated by a 2-D model. Also, there are several other 2-D viscoelastic FEM models that have been used to investigate the stress state in this region (e.g., ISHII et al., 1983).

Figure 1 shows the FEM model with the region size of 200×660 km. We use triangular elements. The total numbers of nodes and elements are 815 and 1354, respectively. The model consists of the elastic upper crust labeled by the numeral 1, the viscoelastic lower crust labeled by 2 and 4, the viscoelastic mantle wedge labeled by 3, 5 and 6, the elastic subducting Pacific plate labeled by 7, and the underlying asthenosphere labeled by 8. For the viscoelastic constitutive relation, we use that of a standard linear solid, which is similar to a Maxwell solid. We use this viscoelastic property in shear and an elastic solid in bulk. As described later, for numerical stabilization after applying gravitation, we use a standard linear solid in which an elastic spring element with small stiffness is added in parallel with a Maxwell element as shown in Figure 2. The stiffness of the added spring is assigned to be only 5 percent of the stiffness of the Maxwell element. Table 1 gives the material properties.

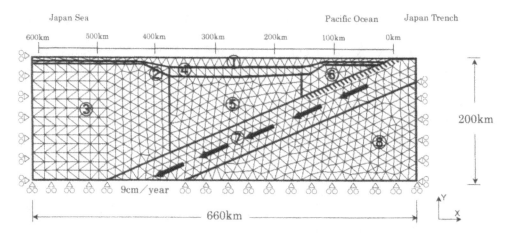

Figure 1
FEM mesh for 2-D model in northeast Japan (200 × 660 km) and boundary conditions. The numbered regions have different properties; the elastic upper crust ①, the viscoelastic lower crust ② and ④, the mantle wedge ③, ⑤ and ⑥, the elastic subducting Pacific plate ⑦, and the underlying asthenosphere ⑧. The corresponding properties are given in Table 1. The black arrows indicate the displacement rate of 9 cm/yr assigned at the central nodes of the plate. The upper plate boundary with hatches to a depth of 57 km indicates the locked portion where frictional coefficient drops to cause coseismic and postseismic slip. On the other portions of the upper and lower plate boundaries, roller conditions along the boundaries are imposed.

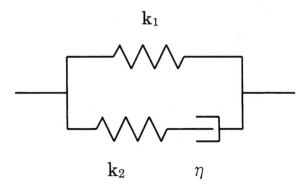

Figure 2
Standard linear solid used in this study. k_1 and k_2 are elastic stiffness and η is viscosity.

The boundary conditions along the four outer boundaries are also indicated in Figure 1. The top boundary is a free surface and the displacements normal to the other three boundary planes are fixed to zero, except for the portion corresponding to the subduction plate. The slab is not cut as shown in Figure 1, but has its extended upper and lower portions. Since the displacement rate is given at the central node of the slab, we give free boundary condition at the upper and lower end of the slab.

Table 1

Material properties of the model regions

No.	Rigidity (G) (10^{11} Pa)	Poisson Ratio	Viscosity (η) (10^{19} Pa s)	Relaxation Time (η/G) (years)	Density (ρ) (g/cm^3)
1	0.330	0.226	Elastic		2.6
2	0.481	0.258	11.44	75	2.9
3	0.695	0.273	11.16	51	3.37
4	0.427	0.253	10.12	75	2.9
5	0.589	0.273	2.38	13	3.37
6	0.735	0.245	11.54	50	3.37
7	1.908	0.258	Elastic		3.44
8	0.695	0.273	100.	456	3.37

2.2. Introduction of a Plate Interface into the FEM Model

FEM usually analyzes a continuous medium, therefore we need special techniques for introducing displacement discontinuities in the medium, such as a plate interface or faulting surface due to an earthquake. There have to date been several techniques proposed. In seismology, the "split-node technique" (MELOSH and RAEFSKY, 1981) has frequently been used for representing earthquake faulting. In this technique, fictitious double couples of forces equivalent to a given amount of displacement discontinuity are applied on the nodes adjacent to the split node as well as the split node itself. WANG (1995) employs the FEM code "TEKTON" to simulate fault slip, in which code the fault locking and unlocking is implemented, via the "slippery-node" technique of MELOSH and WILLIAMS (1989). MIYASHITA (1987) employs a thin layer element with low rigidity and viscosity. TAYLOR et al. (1996) apply the transformational strain method to thin fault elements. There are other techniques which allow an actual discontinuity in displacement, such as the joint element method using thin laminar material and joint elements (e.g., ZIENKIEWICZ, 1977).

Among these techniques, the "master-slave method" is most suitable for representing the evolving slip on the plate interface, on which considerable slip is accumulated due to subduction of the plate over many cycles. In the master-slave method, the pairing of master and slave nodes can change during the evolution of slip and it allows substantial slip on the interface.

There are several general purpose FEM codes available at present, in which the master-slave method is implemented. We use the FEM code "ABAQUS" in this study because it has been frequently used in engineering problems of frictional contact and allows a fault to be defined by the master-slave method. On the plate interface, we define master elements in the subducting plate, and the slave elements in the overlying mantle wedge and the underlying asthenosphere.

Along the upper and the lower plate interfaces except for the locked portion, the free slip condition is imposed. On the locked portion of the upper plate interface to a depth of 57 km, the changing friction is assigned to realize the locked and unlocked state as described below.

2.3. Implementation of Gravity

Though there are several strategies for treating the state of self-gravitation, we apply gravity in the following way. First, we set free slip conditions along the upper and lower plate interfaces, and fix the central nodes of the inclined plate to prevent the distortion of the plate due to gravitation. Under this condition, we apply gravity during a period of 15,000 years to secure a stable state. Viscoelastic flow due to gravity occurs following the elastic deformation in response to the sudden application of gravity. If we use a Maxwell solid, the deformation under gravitation continues and it takes considerably longer to reach a stable state. Therefore, in order to reach a stable state in a reasonably finite time, we use a standard linear solid instead of a Maxwell solid to represent inelasticity. Since the stiffness of the added elastic spring element is only 5 percent of that of the Maxwell solid as mentioned previously, the behavior during a few thousand years of earthquake cycles is not different from that for the Maxwell solid. The deformation at the stable state is shown in Figure 3. In the stable state, the surface is almost uniformly depressed by 3 km, and the mantle wedge slides by 2.5 km on the plate interface.

As seen in Figure 3, the underlying asthenosphere is partly detached from the plate because the central nodes at the plate are fixed and the free slip condition allows the underlying asthenosphere to descend. This is physically unreal, and we ignore the underlying upper mantle portion at present, though we need to consider the effect of this portion generating future friction. Now we use a free boundary condition on the lower boundary throughout the experiments and the underlying upper mantle portion produces no affects on the results.

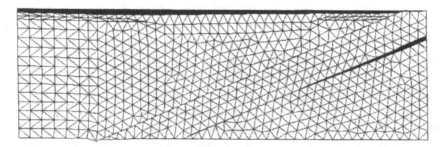

Figure 3
Deformation at the stable state in 15,000 years after the application of gravity. The deformation is exaggerated twice. Black portions at the ground surface and around the lower plate boundary represent depression at the surface and the portion of asthenosphere detached from the overlying plate.

2.4. Simple Earthquake Cycle with Steady Subduction of the Pacific Plate

After reaching a stable state under gravitation, we start the steady subduction of the Pacific plate by imposing the displacement rate of 9 cm/yr at the central nodes of the plate (arrows in Figure 1). As stated above, because partial detachment of the underlying asthenosphere and free slip are imposed on the lower plate interface, the asthenosphere is not coupled with the upper portion and has no contribution to the earthquake cycle in this simulation.

We define the following earthquake cycle with locked and unlocked states. In one earthquake cycle, the friction coefficient of the locked portion on the upper plate interface to a depth of 57 km, indicated by hatches in Figure 1, has a finite value during the one-hundred-year interseismic period and zero during the one-year coseismic and postseismic period. HEKI et al. (1997) reported that the afterslip following the 1994 Sanriku-oki earthquake lasted one year. Accordingly, we tentatively assign one year for the postseismic period here. With the steady subduction of the Pacific plate under gravitation, the earthquake cycle is periodically repeated to calculate the resultant earthquake fault slip and the associated deformation in the 2-D viscoelastic medium.

We performed several simulations by changing the interseismic value of friction coefficient μ. The frictional shear stress τ on the plate interface is calculated as

$$\tau = \mu(\sigma - P) \tag{1}$$

$$P = \rho_w g h \tag{2}$$

where σ and P are the normal stress and the hydrostatic pore pressure on the plate interface at a depth of h, and ρ_w and g are the density of water and gravity acceleration. Here, since we discuss the displacement field for earthquake cycles in comparison with the experiment of WANG (1995), we assign such a large friction value of 0.5 that the earthquake fault zone is mostly locked in interseismic periods.

3. Simulation Result

3.1. Slip along the Plate Interface during Earthquake Cycles

Figure 4 shows the development of slips at different depths along the plate interface during the first three earthquake cycles. For the locked portion to a depth of 57 km, the amount of earthquake fault slip at the first earthquake cycle is only 2.5 m, which is 26 percent of the total accumulated slip of 9 m during the one-hundred-year interseismic period. This result agrees well with that of WANG (1995). As stated in WANG (1995), this small rebound occurs because the elastic shear stress in the mantle wedge induced by the earthquake resists the elastic

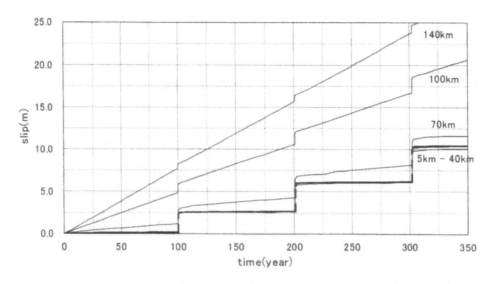

Figure 4
Evolution of slip at the selected depths along the plate interface at the first three cycles.

rebound of the overriding plate. Further, he showed that the larger the region of strain accumulation is compared to the earthquake rupture, the smaller is the earthquake fault slip, and WANG considered that such a delayed response in the viscoelastic mantle wedge is one of the factors explaining the observed small seismic coupling.

However, as can be clearly seen in Figure 4, the amount of earthquake fault slip increases cycle by cycle. Figure 5 enlarges the portion of each 4 years including coseismic and postseismic stages, and compares the amount of coseismic and postseismic slip at a depth of 5 km during the 31 cycles. This figure displays more clearly the development of slip amount cycle by cycle than Figure 4. The locked portion has negligible slip during the interseismic period, and coseismic slip followed by postseismic slip for one year.

Figure 6 shows the development of slip along the plate interface during 31 earthquake cycles. The downwarping shape of the slip curves for the locked portion at the first 15 cycles means that earthquake fault slip is rapidly growing during these initial earthquake cycles. After about the 15th cycle, slip curves on the locked portion come to be parallel to those at great depths, indicating that the growth rate of earthquake fault slip gradually decreases for later cycles.

3.2. Slip Distribution along the Plate Interface at Earthquake Cycle 31

Figure 7 displays the slip distribution along the plate interface for the interseismic and the coseismic and postseismic periods, respectively, at earthquake cycle 31.

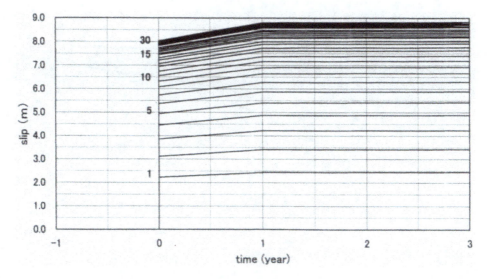

Figure 5
Comparison of the amount of coseismic and postseismic slip at a depth of 5 km on the plate interface during each 31 cycles. The time is aligned at the earthquake onset for each cycle, and slip for each cycle is plotted for only 4 years including coseismic and postseismic periods. Inserted numerals indicate earthquake cycle numbers.

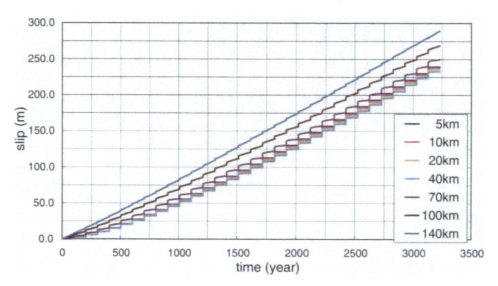

Figure 6
Development of slips at the selected depths along the plate interface during 31 cycles.

During the one-hundred-year interseismic period, the locked portion with friction coefficient of 0.5 actually has a small displacement. Slip increases from the surface to the deepest locked portion and reaches 1 m. This small slip on the deep locked

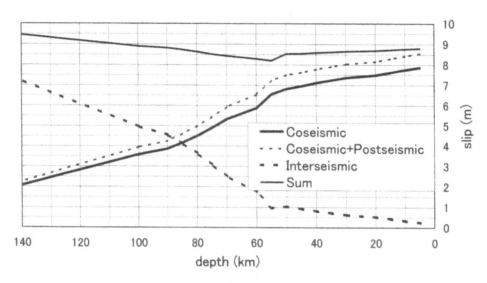

Figure 7
Distribution of slip change on the plate interface during the interseismic, coseismic, and postseismic periods at earthquake cycle 31. Dotted and solid lines indicate the slips during the interseismic and coseismic periods, respectively. Long-dotted and thin lines represent the sum of coseismic and postseismic slip and the sum of all slips at earthquake cycle 31.

portion for the interseismic period occurs because we impose completely free slip on the deeper portion of the plate interface. On the deeper free slip portion, slip increases downward from the deepest locked portion and reaches the total amount of relative plate motion at depths.

The shallow locked portion with almost no slip during the interseismic period has coseismic slip of 8 m followed by postseismic slip of larger than 0.5 m. The coseismic slip is 89% of the total relative plate motion and the total fault slip including the postseismic one reaches 95%. The coseismic slip is 8–7 m followed by 0.5 m of postseismic slip on the locked portion, after which the slip decreases on the free slip lower portion, in contrast to behavior during the interseismic period. Accordingly, the total slip during one earthquake cycle including interseismic, coseismic and postseismic periods is approximately 9 m, which is the total relative plate motion. At earthquake cycle 31, therefore, the interplate earthquake releases most of the displacement supplied by the plate motion during the interseismic period, and no displacement is accumulated during one earthquake cycle in contrast to those at early cycles.

The coseismic slip is not confined to the locked portion but extends to the deeper free slip portions. This is easily understood from the assigned free slip condition in the deeper portions. However, it has been observed that the actual fault region of great interplate earthquakes is confined to the locked portion at subduction zones. This point is discussed in the next section.

4. Discussion

We discuss a physical mechanism for the increase of earthquake fault slip with cycles. First, for this purpose, we introduce a simple experiment. Then, we examine the increase of fault slip in comparison with this simple experiment. Third, we discuss the extension of the earthquake slip to the deeper portions, which is different from observations. Finally, we give an implication for new earthquake simulation with a nonlinear friction in a viscoelastic medium.

4.1. A Simple One-degree-of-freedom Experiment

First, to gain physical insight on the increase of fault slip, we introduce a simple one-degree-of-freedom experiment as shown in Figure 8, which produces fundamentally the same increase of slip as in our viscoelastic FEM modeling. In this numerical experiment we define a simple earthquake cycle, in which we pull the one end of a standard linear solid element with a constant velocity of v_0 during a locked interseismic period of T and release the end during unlocked coseismic and postseismic periods of ΔT, while the other end is fixed. This earthquake cycle is iterated so that we can secure the increase of rebound slip during each of the unlocked coseismic and postseismic periods. In the standard linear solid element, an elastic spring with the elastic constant k_1 and a Maxwell solid, are connected in parallel whereas an elastic spring with the elastic constant k_2 and a dashpot with the viscosity η are connected in series. We image that the elastic spring and the Maxwell solid element correspond to the crust and the mantle wedge in our FEM modeling.

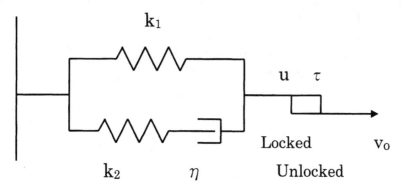

Figure 8

A simple one-degree-of-freedom spring and dashpot model. We consider a standard linear solid element, in which a spring with the elastic constant k_1 and a Maxwell solid with a spring of elastic constant k_2 and a dashpot of viscosity η are connected in parallel. The element is fixed at one end and another end is connected with a moving element of a constant velocity v_0 in an interseismic locked period. In unlocked coseismic and postseismic periods, the one end is disconnected with the moving end so that no force is applied to the element. From an initial state with no displacement of each spring and dashpot, earthquake cycles with locked interseismic and unlocked coseismic and postseismic periods are iterated. u and τ represent displacement and stress at the one end of the standard linear solid element, respectively.

For this simple earthquake cycle we can derive analytical expressions for the displacement u and the stress τ at the one end of the standard linear solid element as follows:

(1) Locked interseismic period $(t_0 < t < t_0 + T)$

During the locked interseismic period, the end is connected with the moving element and is elongated with the same velocity of v_0. Assuming the initial displacement is u_0 and the initial stress is zero at the starting time of an earthquake cycle $t = t_0$, we have the displacement u and the stress τ,

$$u(t) = v_0(t - t_0) + u_0 \tag{3}$$

$$\begin{aligned}\tau(t) &= k_1 u(t) + \eta v_0 - (\eta v_0 + k_1 u_0) \exp(-(t - t_0)/\tau_m) \\ &= k_1 v_0(t - t_0) + (\eta v_0 + k_1 u_0)(1 - \exp(-(t - t_0)/\tau_m))\end{aligned} \tag{4}$$

where τ_m is the Maxwell time and is defined by

$$\tau_m = \eta/k_2. \tag{5}$$

Equation (4) indicates that the accumulated stress caused by a moving element has two components. One is the stress in proportion to the accumulated displacement of spring 1 during the interseismic period. The second component is growing to the viscous stress and the initial stress of spring 1 with the characteristic time of the Maxwell time. Thus, if the Maxwell time is considerably less than the time of one earthquake cycle, the stress accumulation during an interseismic period is rapid at the early time of the Maxwell time and then becomes linear in time as shown later.

(2) Unlocked coseismic period $(t = t_0 + T)$

Since one end is detached from the moving element, the stress applied to the standard linear solid element is completely released to be zero at the coseismic period $t = t_0 + T$. Only the elastic springs respond to this coseismic instantaneous stress change and produce the coseismic slip. Then the displacement is reduced by the rebound slip Δu,

$$\Delta u = \tau(t_0 + T)/(k_1 + k_2). \tag{6}$$

Here the stress is calculated in eq. (4).

(3) Unlocked postseismic period $(t_0 + T < t < t_0 + T + \Delta T)$

Also during the postseismic period, since one end of the standard linear element is detached from the moving element, the stress is zero. Accordingly, the dashpot element is relaxed and the displacement is viscoelastically reduced as

$$u(t) = (u(t_0 + T) - \Delta u) \exp(-(t - (t_0 + T))/\tau_r) \tag{7}$$

where the τ_r is the relaxation time of the system and is written as

$$\tau_r = \eta(1/k_1 + 1/k_2). \tag{8}$$

It is noted that during the postseismic period the displacement changes with the relaxation time of the system, which is different from the Maxwell time effective during the interseismic period.

(4) Again locked $(t = t_0 + T + \Delta T)$

At the end time of the earthquake cycle $t = t_0 + T + \Delta T$, the end of the standard linear solid element is connected again with the moving element. The displacement at the time is

$$u(t_0 + T + \Delta T) = (u(t_0 + T) - \Delta u) \exp(-\Delta T/\tau_r). \tag{9}$$

This displacement is the initial one for the next cycle, and the next cycle for (1)–(4) is iterated.

We start the earthquake cycle from the state with no displacements of any springs and of the dashpot. In this experiment we use the parameters as follows:

$$T = 100 \text{ years}, \quad \Delta T = 1 \text{ year}, \quad v_0 = 10 \text{ cm/year}, \quad k_1 = k_2 = 6.34 \times 10^{10} \text{ Pa},$$

$$\eta = 10^{19} \text{ and } 10^{20} \text{ Pa} \cdot \text{s}, \quad \tau_m = 5 \text{ and } 50 \text{ years corresponding to } \eta.$$

Since we use equal values of k_1 and k_2, we have $\tau_r = 2\tau_m$ from eqs. (5) and (8). Thus, in case of prolonged Maxwell time, the system relaxation time is 100 years comparable to the time of an earthquake cycle.

4.2. Simulation Results in the Experiment and their Implication for a Physical Mechanism for the Increase of Earthquake Fault Slip with Cycles

For two cases with different Maxwell times, Figures 9 and 10 show the stress and the displacement histories, $\tau(t)$ and $u(t)$, at the end of the standard linear solid element, respectively. At the first cycle, the initial displacement u_0 is zero in eq. (4). However, the displacement accumulated during the first interseismic period is partially rebound and the residual displacement remains. Accordingly, following eq. (4) the stress accumulated during the locked interseismic period is slight at the first cycle, and then increases at later cycles. Coseismically, the accumulated stress is completely relaxed, and the coseismic slip is proportional to the accumulated stress following eq. (6). Therefore, the coseismic slip increases in proportion to the accumulated stress cycle by cycle. Figure 10 shows that, from the zero displacement point, the end point is moving to a equilibrium point cycle by cycle, and the accumulated stress is correspondingly increasing to a value as shown in Figure 9. Thus the increase of slip amount is a transient phenomenon which is seen in the system changing from a relaxed initial state to a stressed equilibrium state due to intermittent stressing by pulling.

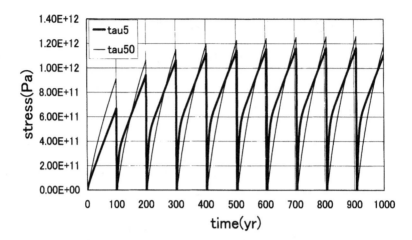

Figure 9
Stress time history at one end of the element during cycles in the experiment. Thick and thin lines indicate the stress histories for two cases of the Maxwell times of 5 and 50 years, respectively. Symbols of tau5 and tau50 mean the stresses for the Maxwell times of 5 and 50 years.

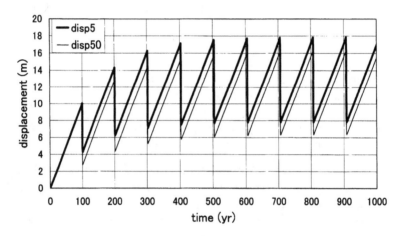

Figure 10
Displacement time history at one end of element during cycles in the experiment. Symbols are the same as in Figure 9, but this figure shows the displacement. Symbols of disp5 and disp50 mean displacements for the Maxwell times of 5 and 50 years.

For different Maxwell times we change the value of viscosity, keeping the values of elastic constant. Therefore, the case with the shorter Maxwell time has a smaller viscosity, and the reduced stress is accumulated during an interseismic period as understood from eq. (4). Accordingly, the coseismic slip is also smaller in that case. This small rebound at early cycles in the shorter Maxwell time case leads to larger equilibrium displacement than in the longer Maxwell time case.

It is noteworthy that the rate of stress accumulation is larger at the early stage of the interseismic period than at the later stage, as can be seen in the case of lesser Maxwell time of 5 years in Figure 9. This pattern of stress accumulation is easily understood from eq. (4) as stated previously. Usually, prescribing a constant subduction rate in an elastic medium, it is assumed that the stress is accumulated constantly in time during the interseismic period. As seen in this experiment, in a viscoelastic medium, the stress accumulation during an interseismic period is significantly different from this elastic view.

We artificially define a postseismic period of one year following a coseismic period. As understood from eq. (7), the characteristic time in this period is the system relaxation time, which is more considerable than the Maxwell time effective in the interseismic period. Since the postseismic period is very short, it is difficult to distinguish the small postseismic slip from the coseismic one in Figure 10. To see more details, we show the amount of coseismic slip and the total slip of coseismic and postseismic ones at cycles in Figure 11. In the case of the smaller system relaxation or Maxwell time, the coseismic slip is smaller and the postseismic slip is larger than those in the case of the extended time. Both slips in the case of the smaller relaxation time are increasing cycle by cycle to reach certain values, respectively. Interestingly, the converged amount of coseismic slip slightly exceeds 9 m. In this case, all of the accumulated displacement during the interseismic period is not released coseismically. Strictly speaking, in this case, the degree of seismic coupling in the final equilibrium sate is about 0.9. However, when including the postseismic slip, the total slip measures up to the accumulated slip of 10 m during the interseismic period, and

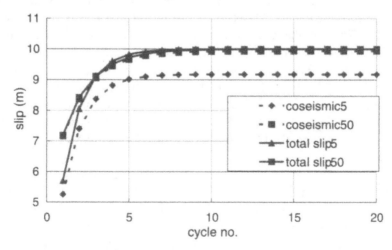

Figure 11
Coseismic and postseismic slips for respective cycles in the experiment. Dotted and thick lines represent the coseismic slip and the total slip of coseismic and postseismic slips, respectively, at cycles. The different symbols correspond to the case for different Maxwell times of 5 and 50 years.

the total rebound occurs during the coseismic and postseismic periods in the final equilibrium state. This is very similar to the development of the coseismic and postseismic slips in our FEM modeling, which can be seen in Figures 5 and 7. In the case of the large relaxation time, the postseismic slip is very minor and the total rebound occurs coseismically in the final equilibrium state. In this case of long relaxation time, the seismic coupling becomes 1.0. Thus, we have a very interesting finding. Namely, the seismic coupling is dependent on the relaxation time. The shorter the relaxation time becomes, the smaller the seismic coupling is.

From this experiment we can conclude that the mechanism for the increase of slip to a certain amount cycle by cycle is fundamentally that the overriding crust and mantle wedge are changing from the initial relaxed state with only gravitational stress, to the final stressed equilibrium state with deformation due to intermittent plate loading. In comparison with this simple experiment and the single earthquake simulation by WANG (1995), we discuss our results of FEM modeling in the following.

At the end of the first cycle, the amount of earthquake fault slip is only about 30 percent of the total relative plate motion, which has been already noted by WANG (1995). In the gravitational prestress state of the mantle wedge, the plate starts to subduct at the beginning of the first earthquake cycle in the simulation. At the first interseismic period, most elastic stress is stored in the upper crust due to the locking of the subducting plate, while the stress caused by the constant steady loading is minor in the viscoelastic medium. However, the elastic stress in the lower crust and the mantle wedge induced by the earthquake due to a sudden drop of frictional stress on the locked portion resists the elastic rebound of the upper crust. WANG (1995) applied an amount of slip to the locked fault and then unlocked it after a certain time; the stress is completely relaxed in the mantle wedge before the occurrence of the earthquake. Though the plate with the locked fault section is subducting in the interseismic period in our simulation, the stress in the viscoelastic lower crust and the mantle wedge before the occurrence of earthquake at the end of the first cycle is similar to that in WANG's (1995) calculation. If frictional stress continues to be zero during a period longer than the relaxation time of the system, fault slip grows to an amount equal to the total relative plate motion (WANG, 1995). In that case, complete rebound is realized and the stress in the crust and the mantle wedge is completely relaxed.

However, the frictional stress is zero only for one year after the occurrence of an earthquake in our simulation. Before the rebound response has been completed during the one-year postseismic period, locking of the subducting plate interface resumes. In our case, only partial rebound occurs during the one-year postseismic period and the displacement still remains in the crust and the mantle wedge above the plate. This remained displacement and the accumulated one due to the locking plate are released at the occurrence of the next earthquake, which causes the resultant earthquake fault slip to become larger than at previous cycles.

Finally, we discuss viscoelasticity and seismic coupling. Slight fault slip at the first earthquake cycle was considered to be a factor explaining moderate seismic coupling observed at actual subduction zones (WANG, 1995). Repetition of earthquake cycles in a viscoelastic medium, however, produces growing earthquake fault slip, and coseismic and postseismic fault slips reach certain amounts. The final amount of coseismic slip is dependent on the relaxation time. In the case of small relaxation time, the final coseismic fault slip is still smaller than the total relative plate motion, and the seismic coupling is less than 1.0. In this sense, WANG'S (1995) statement might still be correct. However, even in the case of the considerably short Maxwell time of 5 years or our FEM modeling, we can have the seismic coupling of 0.9 or larger, and it is quite difficult to explain such a small seismic coupling of 0.3 as observed in northeast Japan (e.g., PACHECO *et al.*, 1993). Accordingly, the viscoelasticity is not the main factor in seismic coupling.

4.3. Extension of Earthquake Slip to the Deeper Portions beyond the Locked one in our FEM Modeling

In our FEM modeling, the earthquake fault slip extends to the deeper portions beyond the locked one, which is different from the observations. WANG (1995) confines the free slip to a restricted portion and prevents the earthquake slip in the deeper portions.

We perform another simulation to prevent the deep earthquake slip. In the simulation, we lock the deeper portions on the plate boundary in the coseismic and postseismic periods, where the free slip is applied in the interseismic period. We assume that some manner of viscous coupling, which prevents a rapid motion, is applied to the deeper boundary. Subsequently the earthquake fault slip is confined to the locked region in the interseismic period. The slip is smaller than in the previous simulation, and it takes longer to reach a final equilibrium state. However, other features are the same as those in the previous simulation, though we show no results here. Therefore, our result on the increase of fault slip cycle by cycle holds for the case with the confined earthquake fault zone.

4.4. Technical Implication for New Earthquake Cycle Simulation with a Nonlinear Friction Law in a Viscoelastic Medium

Recent numerical experiments based on laboratory derived friction laws have successfully simulated earthquake cycles. Because these experiments have been performed on a single fault or on a single straight plate interface in an elastic medium, the sliding behavior becomes periodic and the steady state of earthquake cycle is obtained after only a few cycles from an initial state (e.g., KATO and HIRASAWA, 1997).

The simulation with a simple friction change in the present study indicates that the steady state of earthquake cycle in a viscoelastic medium seems to be obtained only after a few tens of earthquake cycles. If a nonlinear friction law such as the rate- and state-dependent friction law is implemented, the more cycles would be needed to accomplish the equilibrium state because of the numerically stronger unstable nature at the initial stage of simulation than that using such a simple friction law as in the present study. Thus, it takes considerably more time to perform the new earthquake cycle simulation with a nonlinear friction law in a viscoelastic medium than in an elastic medium. Therefore it would be necessary to develop numerical techniques to produce a final equilibrium stressed state as quickly as possible, say, by approximating the calculation in the transient state somehow.

5. Conclusions

Earthquake cycles in a laterally heterogeneous viscoelastic medium under gravitation at a subduction zone are simulated using a 2-D FEM model. We define a simple earthquake cycle with a steadily subducting plate, in which the locked portion of the plate interface has a finite friction coefficient during the interseismic period and has zero friction during coseismic and postseismic periods, and free slip is imposed on the lower portion of the plate interface. Resulting earthquake fault slip is smaller than the total relative plate motion at the first earthquake cycle, as previously noted by WANG (1995). The magnitude of coseismic fault slip, however, grows with a number of earthquake cycles, and reaches a limit value. The limit value is dependent on the relaxation time, and the seismic coupling can be less than 1.0 in the case of small relaxation time. However, only viscoelasticity cannot explain such a small seismic coupling of 0.3 observed at a subduction zone. These findings indicate that viscoelasticity of the earth is not the main factor in explaining the observed small seismic coupling. In comparison with a simple experiment, we show that the increase of earthquake fault slip cycle by cycle occurs in the transitional crust and mantle wedge which are changing from the initial relaxed state to the final stressed equilibrium state due to intermittent plate loading. New numerical models with viscoelasticity and nonlinear fault friction laws require a number of earthquake cycles to reach a limit cycle, and the computation will be substantially more formidable in a viscoelastic medium than in an elastic medium.

Acknowledgments

I would like to thank H. Nagasaka, CTI (Computer Technology Information) for his support in the calculations using ABAQUS and preparing figures. Support of H. Nakamura, RIST is acknowledged. I am also grateful to T. Miyatake and Z. Wu for their constructive review. The calculations were partly performed using SX-4 in GSI (Geographical Survey Institute), and support of S. Ozawa and T. Sagiya, GSI for using SX-4 is also acknowledged. This research was supported by grants from "Nankai Trough Project" and "Earth Simulator Project" of the Science and Technology Agency.

REFERENCES

HEKI, K., MIYAZAKI, S., and TSUJI, H. (1997), *Silent Fault Slip Following an Interplate Thrust Earthquake at the Japan Trench*, Nature *386*, 595–598.

HIROSE, H., HIRAHARA, K., KIMATA, F., FUJII, N., and MIYAZAKI, S. (1999), *A Slow Thrust Slip Event Following the Two 1996 Hyuganada Earthquakes beneath the Bungo Channel*, Geophys. Res. Lett. *26*, 3237–3240.

ISHII, H., SATO, T., TACHIBANA, K., HASHIMOTO, K., MURAKAMI, E., MISHINA, M., MIURA, S., SATO, K., and TAKAGI, A. (1983), *Crustal Strain, Crustal Stress and Microearthquake Activity in the Northeastern Japan arc*, Tectnophysics *97*, 217–230.

KATO, N. and HIRASAWA, T. (1997), *A Numerical Study on Seismic Coupling along Subduction Zones Using a Laboratory-derived Friction Law*, Phys. Earth Planet. Int. *102*, 51–68.

KAWASAKI, I., ASAI, Y., and TAMURA, Y. (1998), *Interplate Moment Release in Seismic and Seismogeodetic Bands and the Seismogeodetic Coupling in the Sanriku-Oki Region along the Japan Trench – A Basis for Mid-term/Long-term Prediction*, Zisin *50*, 292–307 (in Japanese).

MATSU'URA, M. and SATO, T. (1989), *A Dislocation Model for the Earthquake Cycle at Convergent Plate Boundaries*, Geophys. J. Int. *96*, 23–32.

MELOSH, H. J. and RAEFSKY, A. (1981), *A Simple and Efficient Method for Introducing Faults into Finite Element Computations*, Bull. Seismol. Soc. Am. *71*, 1391–1400.

MELOSH, H. J. and WILLIAMS, C. A. (1989), *Mechanics of Graben Formation in Crustal Rocks: A Finite Element Analysis*, J. Geophys. Res. *94*, 13,961–13,973.

MIYASHITA, K. (1987), *A Model of Plate Convergence in Southwest Japan, Inferred from Levelling Data Associated with the 1946 Nankaido Earthquake*, J. Phys. Earth *35*, 449–467.

PACHECO, J. F., SYKES, L. R., and SCHOLZ, C.H. (1993), *Nature of Seismic Coupling along Simple Plate Boundaries of the Subduction Type*, J. Geophys. Res. *98*, 14,133–14,159.

RICE, J. R. (1993), *Spatio-temporal Complexity of Slip on a Fault*, J. Geophys. Res. *98*, 9885–9907.

RUINA, A. (1983), *Slip Instability and State Variable Friction Laws*, J .Geophys. Res. *88*, 10,359–10,370.

SAVAGE, J. C. (1983), *A Dislocation Model of Strain Accumulation and Release at a Subduction Zone*, J. Geophys. Res. *88*, 4984–4996.

SCHOLZ, C. H., *The Mechanics of Earthquakes and Faulting* (Cambridge University Press, 1990).

STUART, W. D. (1988), *Forecast Model for Great Earthquakes at the Nankai Trough Subduction Zone*, Pure Appl. Geophys. *126*, 619–641.

TAYLOR, M. A. J, ZHENG, G., RICE, J. R., STUART, W. D., and DMOWSKA, R. D. (1996), *Cyclic Stressing and Seismicity at Strongly Coupled Subduction Zones*, J. Geophys. Res. *101*, 8363–8381.

THATCHER, W. and RUNDLE, J. B. (1984), *A Viscoelastic Coupling Model for the Cyclic Deformation due to Periodically Repeated Earthquakes at Subduction Zones*, J. Geophys. Res. *89*, 7631–7640.

THATCHER, W., MATSUDA, T., KATO, T., and RUNDLE, J. B. (1980), *Lithospheric Loading by the 1896 Riku-u Earthquake, Northern Japan: Implications for Plate Flexure and Asthenospheric Rheology*, J. Geophys. Res. *85*, 6424–6435.

TSE, S. T. and RICE, J. R. (1986), *Crustal Earthquake Instability in Relation to the Depth Variation of Frictional Slip Properties*, J. Geophys. Res. *91*, 9452–9472.

WANG, K. (1995), *Coupling of Tectonic Loading and Earthquake Fault Slips at Subduction Zones*, Pure Appl. Geophys. *145*, 537–559.

WANG, K., DRAGERT, H., and MELOSH, H. J. (1994), *Finite Element Study of Uplift and Strain across Vancouver Island*, Can. J. Earth Sci. *31*, 1510–1522.

ZIENKIEWICZ, O. C., *The Finite Element Method, Third Edition* (Macgraw-Hill Book Co., 1977).

(Received February 20, 2001, revised June 11, 2001, accepted June 15, 2001)

 To access this journal online:
http://www.birkhauser.ch

Pure appl. geophys. 159 (2002) 2221–2237
0033–4553/02/102221–17 $ 1.50 + 0.20/0

⌐Pure and Applied Geophysics

Development of a Finite Element Simulator for Crustal Deformation with Large Fault Slipping

ZHI SHEN WU,[1] YUN GAO,[2] and YUTAKA MURAKAMI[3]

Abstract — In this paper, a finite element model for simulating long-term crustal deformation with large slipping along fault interface is developed, where a rate- and state-dependent frictional law is introduced to represent the faulting processes and frictional behaviors of fault interface. Moreover, viscous and plastic material properties are used to simulate pressure solution creep and cataclasis, respectively. Throughout the simulations on a structural model of fault-bend folds, the distributions of the stress invariants, equivalent viscous plastic strain, and the traction on the fault interface are investigated. The sequence of deformation mechanisms during movement over a ramp is discussed. It is also found that this kind of frictional model is suitable to represent the rate-dependent behavior of fault slipping due to the movement over a ramp and the tractions on fault interface for treating the low frictional problem of fault-bend folds.

Key words: Fault-bend folds, viscous plasticity, frictional law, fault interface, master-slave method.

1. Introduction

The finite element method for numerically solving problems in continuum mechanics is finding increasingly wide applications to geological and geophysical problems. Though the general analysis of the finite elements is described in a number of excellent textbooks, there are numerous situations peculiar to geology for which no simple and efficient techniques have been devised such as the state of fault. When a fault develops in the interior of rock mass, previously contiguous portions of the mass are displaced relative to each other across a surface of discontinuity, accompanied with the large deformation and slipping.

A need for an efficient method of treating this situation has arisen and several schemes for treating faults and cracks in solids have been proposed. The first method

[1] Department of Urban and Civil Engineering, Ibaraki University, Hitachi 316-8511, Japan. E-mail: zswu@ipc.ibaraki.ac.jp
[2] Department of Urban and Civil Engineering, Ibaraki University, Hitachi 316-8511, Japan. E-mail: gaoyun@hcs.ibaraki.ac.jp
[3] Geophysics Department, Geological Survey of Japan, Tsukuba 305-8567, Japan. E-mail: murakami@gsj.go.jp
Corresponding author: Yun Gao

was devised by GOODMAN *et al.* (1968), in which a four-node element with zero thickness is introduced parallel to the fault plane. However, this technique is not well adapted to the description of fault slip, and low values of shear stiffness may result in unstable and inaccurate solutions if some of the eigenvalues of the stiffness matrix become small. Another approach was devised by SMITH (1974). In his method, the fault plane is defined by a series of pairs of adjacent nodes. The relative displacement of the nodes in each pair is specified by a boundary condition, while their average displacement is allowed to respond normally to the other forces and displacements applied to the finite element grid. However, such operations require complex bookkeeping, and the equations must be renumbered. The scheme of split node technique is introduced by JUNGELS and FRAZIER (1973). Displacements of nodes which are shared by different elements are normally the same for each element. Because the splitting can be introduced into the load vector, changes in the stiffness matrix are not required. However, since it is on the global level, special bookkeeping is required and it may lead to large unbalanced forces and moments action on the fault plane.

To solve the major slipping problem where the separation and sliding of fault interfaces with finite amplitude and arbitrary rotation may occur, the master-slave method is considered to be useful. With this approach, both sides of fault interface can be modeled as the master surface and the slave surface, respectively. Moreover, an appropriate constitutive relation which represents precisely the progressive behavior of fault interface, can be easily introduced to simulate the interfacial fracture behavior of complex fault. Existing research often used a simple frictional law, e.g., constant frictional coefficient, to represent the faulting processes (e.g., ERICKSON and JAMISON, 1995), which cannot describe the dependence of the friction force on slip history and offer complicated motion of faulting processes as observed from experiments of natural rock deformation. The authors (GAO *et al.*, 2000) introduced an interfacial viscous-plastic model to represent the nonlinear behavior of the fault interface with a ramp. However, the rate-dependent fractional behavior due to the fault movement over a ramp cannot be well represented by this kind of viscous-plastic model. For a fault-bend folding during movement over a ramp, the local slip rate is considerable, although the velocity of natural thrust sheet motion is rather moderate and whose slip propagation shows quasi-static behavior. Recently, due to numerous efforts regarding the experimental investigations and theoretical studies on rock faults, certain kinds of general physical models such as rate- and state-dependent frictional models (e.g., RUINA, 1983; KATO, and HIRASAWA, 1997) have been proposed. It is expected that this kind of experimentally motivated constitutive law can simulate the slip history and the complicated fault motion.

Based on the above reviews, regarding both numerical analysis and material model of fault, this paper is devoted to constructing an advanced finite element model to solve the active fault problem with significant slipping, in which the formulation of finite elements including a fault interface with nonlinear material behavior is derived based on the extended variational statement and finite deformation formulation, and the master-slave method is introduced to represent the large slipping of fault interface. In order to model precisely the nonlinear properties of fault interface, a rate- and state-dependent frictional constitutive relation is considered to simulate the natural deformation of faulting processes. Through these efforts, finally, a parametric study is carried out to simulate the long-term crustal deformation of fault-bend folding structures.

2. Finite Element Formulation

2.1 Extended Variational Statement

Consider a structural system in which the reference configuration V of a body exhibits large slipping along a fault interface S_c. The material behavior of the total structural system can be characterized by two kinds of constitutive relations on continuous domain and internal fault interface. One is a volumetric constitutive equation that relates stress and strain, while the other is a frictional interface constitutive relation between the traction and displacement jumps. Thus, there are two contributions to the internal virtual works: A volumetric contribution and an interfacial contribution. Based on a Lagrange description and the finite strain kinematics, the principle of virtual work is written as eq. (1) with considerable body forces, boundary forces and traction on internal fault interface and confining quasi-static deformations.

$$\int_V \{\delta\varepsilon\}^T \{s\}\, dV = \int_{S_0} \{\delta u\}^T \{f_0\}dS\} + \int_{S_c} \{\delta u\}^T \{f_{c0}\}dS + \int_V \{\delta u\}^T \{r_0\}dV. \quad (1)$$

Here, $\{s\}$ is the Lagrange stress vector, $\{\varepsilon\}$ the Lagrange strain vector, and $\{u\}$ the displacement vector. The body is subjected to a body force field $\{r_0\}$ in V, prescribed external traction $\{f_0\}$ on force boundary S_0 and an internal traction $\{f_{c0}\}$ caused by relative displacement jump on the fault interface S_c.

For the case of update Lagrange formulation, at a time step $m+1$ the stress vector $\{s\}^{m+1}$ can be expressed as the sum of Cauchy stress $\{\tau\}^m$ at the time step m and the incremental stress $\{\Delta s\}$. The load vectors can also be shown by an incremental formulation. Therefore, the incremental formulation of the principle of virtual work at the time step $m+1$ can be obtained as follows.

$$\int_V \{\delta\varepsilon\}^T \{\Delta s\}dV - \int_{S_0} \{\delta u\}^T \{\Delta f_0\}dS - \int_{S_c} \{\delta u\}^T \{\Delta f_{c0}\}dS - \int_V \{\delta u\}^T \{\Delta r_0\}dV$$

$$= -\int_{V^m} \{\delta\varepsilon\}^T \{\tau\}^m dV + \int_{S_0^m} \{\delta u\}^T \{f_0\}^m dS + \int_{S_c^m} \{\delta u\}^T \{f_{c0}\}^m dS$$

$$+ \int_{V^m} \{\delta u\}^T \{r_0\}^m dV \,. \quad (2)$$

2.2 Constitutive Equations in Continuous Domain

A strain hardening viscous-plastic material model is adopted to represent the material properties which include elastic-plastic behavior and a viscous creep behavior in the continuous domain, based on a mixed method in which a Mohr-Coulomb yielding function F and a Drucker-Prager viscous-plastic potential function Q^{vp} as shown in eq. (3) and (4),

$$F = \frac{\tau_1 - \tau_3}{2} - \sigma_Y \cos\phi - \frac{\tau_1 + \tau_3}{2}\sin\phi \tag{3}$$

$$Q^{vp} = \alpha I_1 + \sqrt{J_2} - K \; , \tag{4}$$

where τ_1 and τ_3 are the first and the third principal stresses, σ_Y the yielding stress, ϕ the internal-friction angle of the material, I_1 the first invariant of stress, and J_2 the second invariant of deviatoric stress. α and K can be expressed as the functions of σ_Y and ϕ.

When $Q^{vp} > 0$, considering a viscoplastic strain rate proposed by PERZYNA (1966), the stress-strain relation can be described as follows:

$$\{d\tau_{ij}\} = [D_e]\{d\varepsilon_{ij}\} - [D_e]\{d\varepsilon_{ij}^{vp}\} = [D_e]\{d\varepsilon_{ij}\} - [D_e]\gamma \left\langle \left(\frac{F}{F_0}\right)^{c_r} \right\rangle \left\{\frac{\partial Q^{vp}}{\partial \tau}\right\} , \tag{5}$$

where $[D_e]$, γ and c_r are the elastic coefficient matrix, the fluid coefficient and the creep coefficient, respectively. F_0 is generally given with a value of the yielding stress. As a measure of the viscous-plastic deformation, the equivalent viscous-plastic strain $\bar{\varepsilon}^{vp}$ is introduced.

$$\bar{\varepsilon}^{vp} = \sum d\bar{\varepsilon}^{vp} = \sum \sqrt{(2/3)d\varepsilon_{ij}^{vp}\,d\varepsilon_{ij}^{vp}} \; . \tag{6}$$

2.3 Master-slave Method and Frictional Law Model

As shown in Figure 1, the master-slave method is used to treat large slip problems of the deformation mechanism. Consider the contact between a node x_1 on the slave surface and a segment of the master surface described by the nodes x_2, x_3 and x_4. On this segment of the master surface in which global coordinates are used, the point x on the master surface corresponding to the point x_1 on the slave surface, whose position is defined completely by the interpolation function N_i of the node x_i on the segment, can be determined. Moreover, the relative displacements of the node x_1 after slipping can be obtained by

$$\{\Delta u_c\}^{m+1} = \{\Delta(\Delta u)\} + \{\Delta u_c\}^m = \{\Delta u(x_1) - \Delta u(x)\} + \{\Delta u_c\}^m \; , \tag{7}$$

where $\{\Delta u\}$ is the displacement increment, $\{\Delta(\Delta u)\}$ is the increment of the relative slip displacement between node x_1 and node x, and $\{\Delta u_c\}^m$ is the relative displacement of the node x_1 referring to the master surface at a time step m.

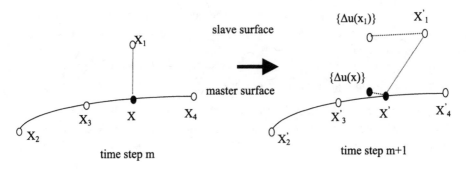

Figure 1
Master-slave slide line segment.

In order to investigate the complex frictional features of fault interface, many efforts have been made to establish a general friction law, in which the explicit expression of the relations between the tangent stress and the normal stress on fault is considered and identified by many experiments. For sliding at fixed normal stress τ_n, the shear stress τ_t resisting unidirectional slip is regarded as being a function of both the slip rate V and the state of the interface, where the latter evolves with unsteady slip, and it is assumed to obey the rate- and state-dependent frictional law proposed by RUINA (1983):

$$\tau_t = \mu \tau_n \quad \mu = \mu_0 + \theta + a \ln(V/V^*) \quad \frac{d\theta}{dt} = -\frac{V}{L}[\theta + b \ln(V/V^*)] \ , \tag{8}$$

where μ is the friction coefficient, V the sliding velocity, and V^* a reference velocity given arbitrarily. The constants μ_0, a, b and L characterize the frictional properties. The steady-state friction coefficient μ_{ss} is defined by the value of μ at $d\theta/dt = 0$ as $\mu_{ss} = \mu_0 + (a - b) \ln(V/V^*)$.

RUINA (1983) theoretically investigated the sliding behavior of a rigid block dragged through an elastic spring, where the frictional traction was assumed to act on the base of the rigid block. He also proposed an equation

$$r = a - b = \partial \mu / \partial(\ln V) < 0 \ , \tag{9}$$

which is a necessary condition for the occurrence of unstable slip. Under this condition, the sliding tends to be more unstable when $|a - b|$ takes a larger value or parameter L takes a smaller value. The traction increment caused by frictional interface can be described by an interfacial stiffness matrix $[K_c]$, the increment of the relative slip displacement $\{\Delta(\Delta u)\}$ and the frictional stress as follows

$$\{\Delta f_c\} = [K_c]\{\Delta(\Delta u)\} - \int_{S_c} [B]^T \left\{ \begin{matrix} \Delta \tau_t \\ 0 \end{matrix} \right\} ds \ , \tag{10}$$

where $[B]$ is the interfacial strain matrix.

It is observed that this model represents the rate- and state-dependent behavior of faulting processes in a relatively explicit manner and can be calibrated with the experimental data easily.

2.4 Time Integration and Iterative Procedure

For update Lagrange formulation, at a time step $m + 1$, the Kirchhoff stresses $\{\tau_k\} = [J]\{\tau\}$ in the continuous domain of structural system can be expressed by the following relation

$$\{\tau_k\}^{(n+1,m+1,k+1)} = \{\tau_k\}^{(n+1,m)} + \{\Delta\tau_k\}^{(n+1,m+1,k+1)} , \tag{11}$$

where $n + 1$ indicates the current load step and $k + 1$ indicates the current iterative step.

Symmetric Cauchy stress tensor $[\Delta\tau]$ of Cauchy stress vector $\{\Delta\tau\}$ is the composition of the Jaumann differential tensor $[\Delta\tau_k^*]$ and rotation tensor $\lfloor\dot{\Omega}\rfloor$.

$$[\Delta\tau] = [\Delta\tau_k] = \lfloor\Delta\tau_k^*\rfloor - \lfloor\dot{\Omega}\rfloor[\tau_k] - [\tau_k]\lfloor\dot{\Omega}\rfloor . \tag{12}$$

Based on the above equations, the incremental Lagrange stress vector $\{\Delta s\}$ can be written as

$$\{\Delta s\}^{(n+1,m+1,k+1)} = \{\Delta\tau_k^*\}^{(n+1,m+1,k+1)} + [D_G^*]^{(n+1,m+1)}\{\Delta\eta\}^{(n+1,m+1)}$$

$$\{\Delta\tau_k^*\}^{(n+1,m+1,k+1)} = [D_e]\{\Delta e\}^{(n+1,m+1,k+1)} - \{\Delta f^{v_p}\}^{(n+1,m+1,k)} \tag{13}$$

$$\{\Delta f^{v_p}\} = [D_e]\{\Delta\varepsilon^{v_p}\} ,$$

where $[D_G^*]$ is a modified stiffness matrix for large deformation, $\{\Delta e\}$ and $\{\Delta\eta\}$ are the first and second order terms of incremental strain, respectively. $\{\Delta\varepsilon^{v_p}\}$ and $\{\Delta f^{v_p}\}$ are the viscous plastic strain and the corresponding traction.

2.5 Finite Element Formulation

The formulation concerning time integration and iteration for the equation of the principle of virtual work can be obtained from eq. (2)

$$-\int_{V^{(n+1,m)}} \left(\{\delta e\}^T[D_e]\{\Delta e\}^{(n+1,m+1,k+1)} + \{\delta\eta\}^T[D_{Gt}^*]\{\Delta\eta\}^{(n+1,m+1,k+1)}\right)dV$$

$$-\int_{S_C^{(n+1,m)}} \{\delta(\Delta u)\}^T[K_c]^{(n+1,m)}\{\Delta(\Delta u)\}^{(n+1,m+1,k+1)}dS$$

$$= \int_{V^{(n+1,m)}} \{\delta e\}^T[D_e]\{\Delta f^{v_p}\}^{(n+1,m+1,k+1)}dV$$

$$-\int_{S_C^{(n+1,m)}} \{\delta(\Delta u)\}^T\left([B]^T\left\{\begin{matrix}\Delta\tau_t\\0\end{matrix}\right\}\right)^{(n+1,m+1,k+1)}dS$$

$$- \left(\int_{V^{(n+1,m)}} \{\delta e\}^T \{\tau_k\}^{(n+1,m)} dV - \int_{S_0^{(n+1,m)}} \{\delta u\}^T \{f_0\}^{(n+1,m)} dS \right.$$

$$\left. - \int_{S_c^{(n+1,m)}} \{\delta(\Delta u)\}^T \{f_c\}^{(n+1,m)} dS - \int_{V^{(n+1,m)}} \{\delta u\}^T \{r\}^{(n+1,m)} dV \right) \quad (14)$$

Upon finite element discretization with respect to the incremental nodal variables $\{\Delta a\}$, the finite element formulation can be derived as

$$[K^*]^{(n+1,m+1)} \{\Delta a\}^{(n+1,m+1,k+1)} = \{\Delta P^{v_p}\}^{(n+1,m,k+1)} - \{\Delta P_c\}^{(n+1,m,k+1)} - \{\varphi\}^{(n+1,m)} .$$
$$(15)$$

Here, m, n and k indicate the time step, the load step and the iteration step, respectively. The total stiffness matrix $[K^*]$ consists of incremental elastic stiffness, geometrical stiffness and interfacial stiffness. $\{\Delta P^{v_p}\}$ is the nodal load vector produced by the viscous-plastic behavior in the continuous domain and $\{\Delta P_c\}$ is the nodal load vector produced by the frictional stress. $\{\varphi\}$ is the residual vector and the subscript 'c' refers to fault interface.

3. Numerical Simulation

3.1 Structural Model

A fault-bend folding structure is used to demonstrate the feasibility of the proposed numerical technique and to investigate the long-term crustal deformation behavior due to the large fault slipping. Figure 2 shows such a 2-D structural model which is similar to the reference written by ERICKSON and JAMISON (1995) who investigated the structure of the thrust sheet in Pine Mountain using a finite element

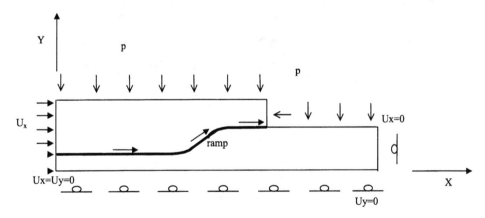

Figure 2
Structural model of fault-bend folds.

model by considering low and constant frictional behavior of the fault interface. In this paper we focus on investigating the effects of rate- and state-dependent frictional behavior along the fault interface.

The structural model consists of a hanging-wall block and a footwall block. Figure 3 shows the FEM mesh of the structural model. The hanging wall is 4500 m long and the top side is 1500 m high; the foot-wall is 7000 m long and the left side is 333 m high. Moreover, the ramp is 500 m high and 1000 m long, producing a ramp angle of 26.5°. A surface pressure of 75 MPa is applied to the top and the right side

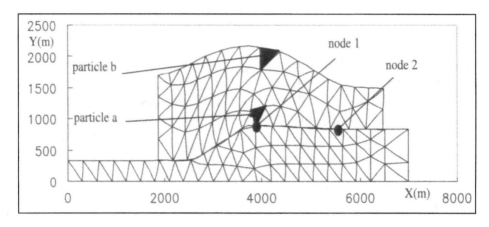

Figure 3
Mesh of structural model.

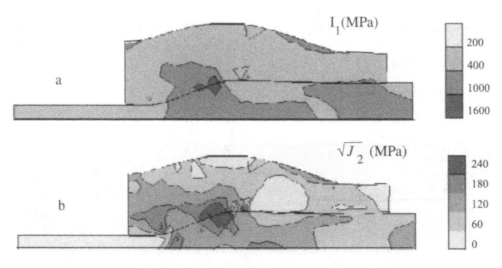

Figure 4
Contours of (a) I_1 and (b) $\sqrt{J_2}$ at displacement 2 km ($r = 0.005$).

of the hanging-wall block. A zero displacement boundary condition, $U_x = U_y = 0$, is used along the left side of the footwall block, $U_y = 0$ along the base of the model and $U_x = 0$ along the right side of the footwall block. A displacement rate of 25 m per 2500 y time step is imposed on the left side of the hanging wall, in which the velocity (1 cm y^{-1}) is consistent with the estimates of natural thrust sheet motion (ERICKSON and JAMISON, 1995). The models are run to a displacement of the left side of the model of 2.5 km in 100 time steps. Identical to the work of ERICKSON and JAMISON (1995), low friction coefficient is used to minimize internal deformation relative to fault slip for the ramp angle used.

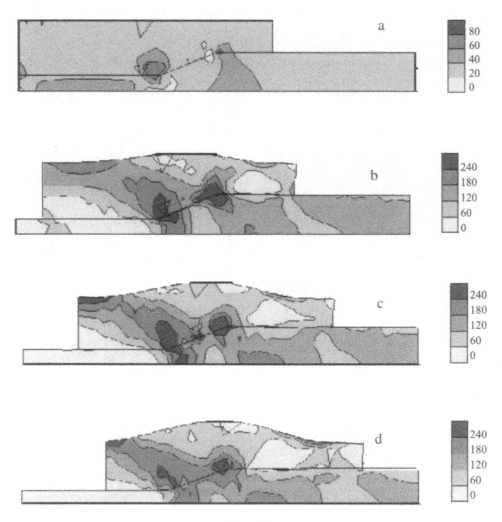

Figure 5
Contours of $\sqrt{J_2}$ during different slipping states (Unit: MPa).

The materials in the continuous domain have density 2500 kg m^{-3}, Young's modulus 3×10^4 MPa, Poisson's ratio 0.25, yield strength 27 MPa, stress-strain slope 2300 MPa, viscous-plastic fluid coefficient 5×10^{-14} s^{-1}, and creep coefficient 1. On the fault interface the frictional coefficient $\mu_0 = 0.01$ is considered.

3.2 Results and Discussions

When the hanging-wall block slides forward to the position as shown in Figure 3, the highest I_1 occurs above the upper ramp and the foreland of the footwall block, whereas the highest $\sqrt{J_2}$ is close to the ramp hinges as shown in Figure 4. Figure 5 shows $\sqrt{J_2}$ distribution during the slipping development. In the hanging-wall block, the highest $\sqrt{J_2}$ occurs from the area near the two ramp hinges to the upper

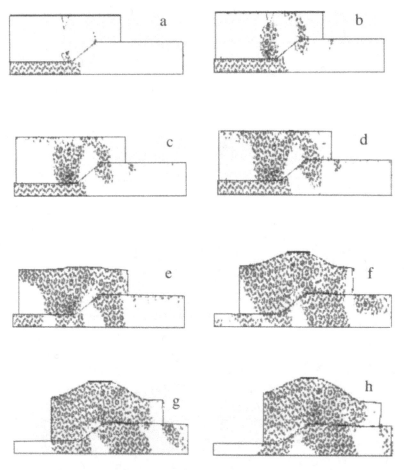

Figure 6
Yielding states in development of slip.

hinterland stage, and in the footwall block it occurs from the area near the two ramp hinges to the foreland stage.

As for the development of the yielding status as shown in Figure 6, the nodes from the area near the lower and upper ramp hinges to the upper plat of hanging-wall block yield initially. With the development of slip, the number of yielding nodes increases to the hinterland and the ramp, and at last most of the nodes are under yielding state. In the footwall block, the nodes of the lower plat and in proximity of the upper ramp hinge are yielding initially. From the area near the lower and upper ramp-hinges, the number of yielding nodes increases to the foreland. But at last, the status of the nodes in the lower flat becomes unyielding.

By following individual particles, a and b marked in Figure 3, the stress paths in J_2 stress space and the equivalent viscous-plastic strain histories can be tracked. The viscous-plastic strain accumulates only when the stress status is under yielding. Because a strain-hardening plastic material is used, the yielding surface moves

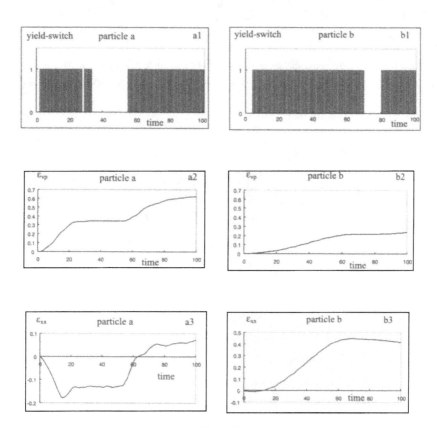

Figure 7
Development of yield state, ε_{v_p} and ε_{xx} (each time step represents 2500 years).

upward in J_2 stress space as the viscous-plastic strain accumulates. As shown in Figure 7, the particle a in the lower hanging-wall block is under yielding state when the value of 'yield-switch' is not zero. It undergoes viscous-plastic horizontal shortening over the lower flat because its stress state is on the yield surface. As the particle a passes over the lower ramp hinge, it continues to undergo horizontal shortening. When the particle is above the ramp, the stress state falls below the yielding surface because of increasing I_1 and decreasing $\sqrt{J_2}$, and thus, further accumulation of viscous-plastic strains temporarily emerge. The incremental strain changes from horizontal shortening to extension approximately midway up the ramp. Over the upper ramp hinge, I_1 decreases and $\sqrt{J_2}$ increases, and the stress state returns to the yielding surface, resulting in an additional viscous-plastic strain. This second phase of viscous-plastic deformation is a horizontal extension, which is superposed on the earlier phase of horizontal shortening. As the particle a moves over the upper flat, $\sqrt{J_2}$ decreases and the stress state falls below the yielding surface. The particle b in the upper hanging-wall block displays a similar deformational history, however the stress path of J_2-space fluctuates. The magnitude of viscous-plastic strain in the hanging-wall block decreases with distance from the fault. All of these results are similar to the ones of the reference of ERICKSON and JAMISON (1995) when the low frictional behaviors are adopted.

When the hanging wall slides forward to the position as shown in Figure 3, the highest viscous-plastic strain occurs from the area above the upper ramp and near

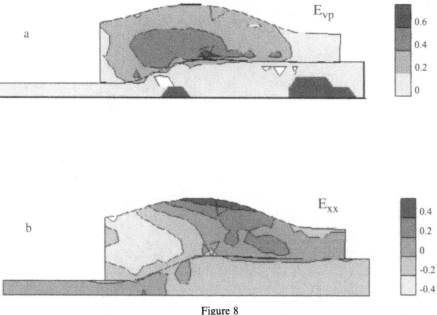

Figure 8
Contours of (a) equivalent viscous-plastic strain ε_{v_p} and (b) horizontal component of strain ε_{xx}.

the upper ramp hinge with an ellipse shape in the hanging-wall block. In the syncline of the upper ramp hinge, the horizontal component strains ε_{xx} are negative where there are horizontal shortenings. Corresponding to this, in the anticline of the upper ramp hinge, the strains ε_{xx} are positive where there are horizontal extensions (Fig. 8).

As for the effect of rate- and state-dependent behavior of the fault surface on the stress distributions, the developments of $\sqrt{J_2}$ of both particles are very similar regarding $r = 0$ and $r = 0.005$. However, by comparing the case of $r = 0.005$ and $r = 0.05$, the results of the particle are obviously different even though the results of the particle b only slightly change as shown in Figure 9. Therefore, it can be realized that a node nearer the fault interface has more sensitivity on stress distribution due to the considerable local slip rate of faulting action. Also it can be observed that the large values of parameter r hinder the advancement of the hanging wall. From Figure 10 the difference of traction due to a response of

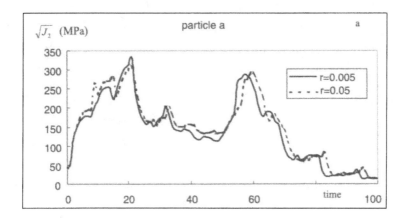

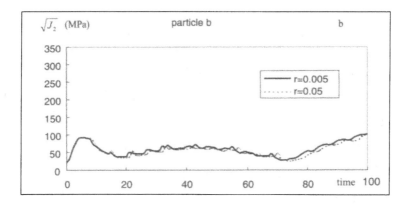

Figure 9
$\sqrt{J_2}$ Development with time due to different values of parameter r (each time step represents 2500 years).

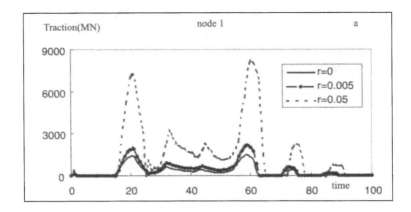

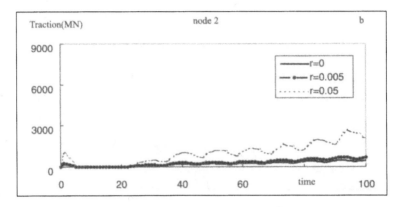

Figure 10
Development of tractions with time due to different values of parameter r (each time step represents 2500 years).

parameter r is large, whether the material on the fault interface passes over the ramp or not. If a node is close to the lower ramp hinge, its traction will increase with time and a larger value will appear when it arrives at the position of the lower ramp hinge. And from this time it will decrease to a very small value in about 10,000 years. In the following process of advancement, it will repeat this kind of phenomenon after each 20,000 years with different amplitudes. As it passes over the upper ramp hinge the largest value appears, and this phenomenon is the most evident of the largest value of $r = 0.05$ among the three discussed cases. Because of the existence of the ramp, the value of the variation on $\ln(V/V^*)$ becomes large, therefore the considerable effect of the rate-dependence is induced. However, our discussion is limited as regards the low frictional behavior, and the case of high cohesive and frictional behavior of fault interface is expected to be discussed in the near future.

3.3 Integration Error

Viscous-plastic problems are often solved numerically by an integration approach of the viscous-plastic strain. Therefore, the integration tolerance of the viscous-plastic strain must be chosen to control the convergence, and iteration is required. Measuring the error as

$$\max\left(\dot{\varepsilon}^{vp}|_{t+\Delta t} - \dot{\varepsilon}^{vp}|_t\right)\Delta t \leq \Delta\varepsilon^{\text{err}} , \tag{16}$$

where $\Delta\varepsilon^{\text{err}}$ is the tolerance which is defined by choosing an acceptable error tolerance, it is recognized that it is important to select a very conservative value of $\Delta\varepsilon^{\text{err}}$, therefore an acceptable solution can usually be obtained with higher values.

From Figure 11, when the advancing velocity is 3.0×10^{-10}m/s with $\Delta\varepsilon^{\text{err}} = 0.0001$, the $\sqrt{J_2}$ development of both particles a and b tends to be stable. And for this fault structure model, the variety of the velocity is also important and it

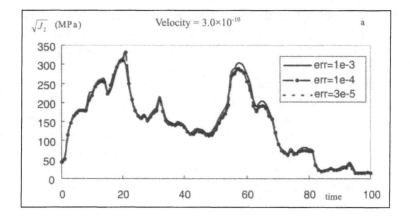

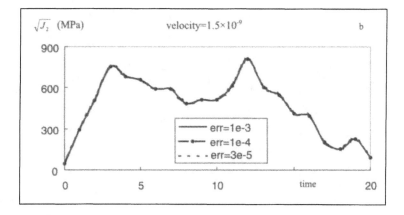

Figure 11
Integration error (a) time step represents 2500 years and (b) time step represents 12500 years

can enormously effect the total system. As the advancing velocity is promoted 1.5×10^{-9} m/s while $\Delta\varepsilon^{err}$ is only equal to 0.001, the $\sqrt{J_2}$ development of both particles a and b may tend to be stable.

4. Conclusions

In this paper a general finite element model for crustal deformation with substantial slipping on a fault interface, where a rate- and state-dependent interfacial constitutive relation is introduced, is developed. By using the developed model, the long-term viscous-plastic deformations of fault-bend folds including the rate- and state-dependent behavior due to faulting slipping over a ramp are simulated, with the detailed investigations of the distributions of stress invariants and the yielding processes. Furthermore, the structural simulation procedure will tend to be stable when tolerance is defined by choosing an acceptable error tolerance. From these results, it is determined that the developed computational model can be used to simulate the deformation mechanism which can be compared with the natural phenomenon of the faulting-processes of fault-bend folding structures. Moreover, this kind of frictional law model is suitable to represent the rate-dependent behavior of fault slipping due to the movement over a ramp and the tractions on fault interface for treating the low-frictional problem of fault-bend folds as a quasi-static problem.

Although the proposed finite element model is based on a general 3-D formulation, in order to demonstrate clearly the feasibility of the proposed algorithm and make a clear parametric analysis, only the 2-D problems are considered in this study. And due to the reason that the 2-D model is too simple to represent a realistic crustal deformation, a three-dimensional analysis with the proposed model is under consideration to be realized in the near future.

REFERENCES

DVORKIN, E. N. (1990), *Finite Elements with Displacement Interpolated Embedded Localization Lines Insensitive to Mesh Size and Distortions*, J. Numerical Methods in Engin. *30*, 541–564.

ERICKSON, G. and JAMISON, W. R. (1995), *Viscous-plastic Finite-element Models of Fault-bend Folds*, Struct. Geol. *17*(4), 561–573.

GAO, Y., WU, Z. S., and MURAKAMI, Y. (2000), *Viscous-plastic Analysis of Crustal Deformation of Fault-bend Folds*, J. Appl. Mech. *3*, 585–594.

GOODMAN, R. E., TAYLOR, R. L., and BREKKE, T. L. (1968), *A Model for the Mechanics of Jointed Rock*, J. Soil Mech. and Found. Proc. ASCE *94*, 637–658.

HIRAHARA (1999), *Quadratic Finite Element Simulation of Large Earthquake with Friction Law*, Earth *24*, 155–160.

JUNGELS, P. H. and FRAZIER, G. A. (1973), *Finite Element Analysis of the Residual Displacements for an Earthquake Rupture: Source Parameters for the San Fernando Earthquake*, J. Geophys. Res. *78*, 5062–5083.

KATO, N. and HIRASAWA, T. (1997), *A Numerical Study on Seismic Coupling along Subduction Zones Using a Laboratory-derived Friction Law*, Phys. Earth Planet. *102*, 51–68.

MELOSH, H. J. and RAEFSKY, A. (1981), *A Simple and Efficient Method for Introducing Faults into Finite Element Computations, 71*(5), 1391–1400.

RUINA, A. L. (1983), *Slip Instability and State Variable Friction Laws*, J. Geophys. Res. *88*, 10,359–10,370.

SMITH, A. T. (1974), *Time-dependent Strain Accumulation and Release at Island Arcs: Implications for the 1946 Nankaido Earthquake*, Ph.D. Thesis, MIT, Cambridge.

WU, Z. S. and YIN, J. (1998), *Analysis of mixed-mode brittle fractures by mixed finite element with internal displacement discontinuities*, Proceed. Conf. *Computational Engineering and Science*, pp. 231–236.

(Received February 20, 2001, revised June 22, 2001, accepted June 25, 2001)

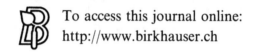

To access this journal online:
http://www.birkhauser.ch

Pure appl. geophys. 159 (2002) 2239–2259
0033–4553/02/102239–21 $ 1.50 + 0.20/0

❙ Pure and Applied Geophysics

3-D Viscoelastic FEM Modeling of Crustal Deformation in Northeast Japan

HISASHI SUITO,[1*] MIKIO IIZUKA,[2] and KAZURO HIRAHARA[1*]

Abstract — As a first step toward establishing a standard earthquake cycle model in Japan, we simulate the crustal deformation during the past 100 years in northeast Japan, using a 3-D FEM based on the kinematic model. Then, we compare the computed results with the observed long-term leveling data and the recent GPS data. On the whole, although the effect of the subducting PAC is dominant, coseismic deformation of the interplate earthquakes can be clearly seen in the inland. Moreover, the postseismic deformation of the earthquakes due to the viscoelastic upper mantle seriously affects the inland movements, and continues for a few decades. Our modeling, including the effects of the interplate earthquakes and the three-dimensional viscoelastic inhomogeneity, reasonably explains the observed movement. Finally, we stress that the viscoelastic effect should be taken into consideration in the analyses, even if no earthquakes occur in the analyzed period.

Key words: 3-D viscoelastic analysis, postseismic deformation, interplate earthquake, FEM, northeast Japan, crustal deformation.

1. Introduction

It is well known that large thrust-type earthquakes have occurred along the plate boundary around the Japan Islands with a recurrence interval of 100 to 150 years. For long-term earthquake prediction, it is necessary to construct an adequate earthquake cycle model along with continuous observations of seismic activity, crustal deformation and the investigation of the historical documents. As a first step toward realizing a standard earthquake cycle model at subduction zones, this study demonstrates the crustal deformation since 1900 in northeast Japan.

A number of researchers have attempted to model the crustal deformation with various time scales using finite element method (FEM), for examples from Japan (SHIMAZAKI, 1974; HASHIMOTO, 1984; MIYASHITA, 1987; YOSHIOKA *et al.*, 1989; HUANG *et al.*, 1998), the United States (WANG *et al.*, 1994; LUNDGREN *et al.*, 1995)

---

[1] Department of Earth and Planetary Sciences, Graduate School of Science, Nagoya University, Furo-cho, Chikusa-ku, Nagoya, 464-8602, Japan.
E-mails: suito@eps.nagoya-u.ac.jp, hirahara@eps.nagoya-u.ac.jp
[2] Research Organization for Information Science and Technology, 1-18-16, Hamamatsu-cho, Minato-ku, Tokyo, 105-0013, Japan. E-mail: iizuka@tokyo.rist.or.jp
*Now at Graduate School of Environmental Studies, Nagoya University

and Andes (WDOWINSKI and BOCK, 1994; LIU *et al.*, 2000). Concerning northeast Japan, most previous studies of the crustal deformation using FEM ignored the three-dimensional inhomogeneity or viscoelastic effects because their analyses were two-dimensional or elastic analyses (KATO, 1979; SATO *et al.*, 1981; SATO, 1988; MIURA *et al.*, 1989).

In the present paper we construct a three-dimensional viscoelastic finite element model with realistic plate and crust structures, based on the seismic velocity structure and the spatial distribution of the micro-earthquakes which were obtained by previous studies in northeast Japan. Then we compute the displacements using a 3-D viscoelastic FEM code, GeoFEM (IIZUKA *et al.*, 1999) under more realistic assumptions, and finally compare the computed results with the observed ones to evaluate our modeling.

2. Tectonic Setting in Northeast Japan

Northeast Japan has attracted the interest of many earth scientists because of its tectonic importance as a typical island arc region (e.g., YOSHII, 1979; HASEGAWA *et al.*, 1983). The region is considered to be subject to the interaction of three plates, the Pacific plate (PAC), the North American (NAM) or the Okhotsk (OKH) plate and the Amurian (AM) or the Eurasian (EUR) plate (Fig. 1). The PAC is subducting with a rate of 8 cm/yr beneath the NAM (or OKH) (SENO *et al.*, 1996). A number of great thrust-type interplate earthquakes have reoccurred continuously filling up the non-rupture regions along this plate boundary as shown in Figure 1. Moreover, significant afterslip and slow earthquakes were observed here (HEKI *et al.*, 1997; KAWASAKI *et al.*, 1995). It has been considered that the seismic coupling in this region is about 0.3 (PACHEO *et al.*, 1993). But recently, coupling ratio in this region was estimated to be 0.8–1.0 deduced from inversion analyses using recent geodetic data (EL-FIKY and KATO, 1999; ITO *et al.*, 2000; MAZZOTTI *et al.*, 2000).

On the other hand, subduction at the eastern margin of the Japan Sea was proposed by NAKAMURA (1983) and KOBAYASHI (1983). The eastward subduction was supported by the eastward dipping fault geometry of the 1983 Nihonkai-Chubu earthquake. OHTAKE (1995) studied the seismicity around this region, and reexamined this plate boundary. The AM (or EUR) subducts eastward beneath the NAM (or OKH) with a rate of about 1 cm/yr (SENO *et al.*, 1996; HEKI *et al.*, 1999).

Geodetic observations in Japan have been studied by many authors with a variety of time scales, for examples from the whole Japan islands (DAMBARA, 1971; KATO *et al.*, 1998; ISHIKAWA and HASHIMOTO, 1999; SAGIYA *et al.*, 2000), northeast Japan (KATO, 1979; MIURA *et al.*, 1989; EL-FIKY and KATO, 1999), and southwest Japan (EL-FIKY *et al.*, 1999).

Figure 2 shows the present horizontal displacement rate in northeast Japan during the period from February 1996 to June 1999, obtained from GPS observations

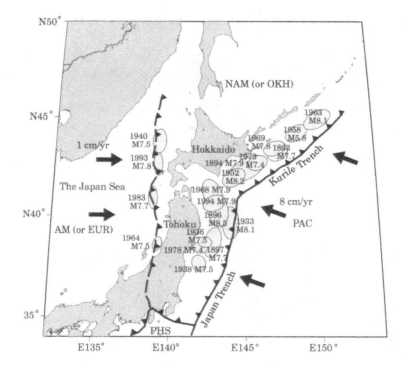

Figure 1

Major plate boundaries and source regions of large interplate earthquakes around the Northeast Japan. AM, EUR, NAM, OKH, PHS and PAC represent the Amurian, Eurasian, North American, Okhotsk, Philippine Sea and Pacific plates, respectively.

by Geographical Survey Institute, Japan (GSI) (MIYAZAKI, 1999, personal communication). The reference point is Niigata which is indicated by a white circle. In Figure 2, we find a large horizontal displacement rate along the Pacific coast, reaching 3 cm/yr, and the amount of the displacement rate decreases in the inland area toward the west. However, the small displacement rate compared with those in the other regions can be seen in the northern part of the Tohoku district. This small rate has been interpreted to be due to a significant afterslip after the 1994 Sanriku Haruka-Oki earthquake (HEKI et al., 1997). Although the observed data here are those from 1996 to 1999, the small displacement rate there may imply the existence of postseismic effects (SAGIYA et al., 2000). The eastward movement in the central part of the Tohoku district is caused by the seismic activity around the Mount Iwate.

Long term leveling surveys in the Tohoku district have been carried out by GSI since 1900 (KATO, 1979). These survey data summarized in Figure 3 reveal characteristic features. The most fundamental mode of vertical movement is the steady tilt of the Tohoku district toward the Pacific coast. This deformation is typically observed at the subduction zone and is caused by the subduction of the Pacific plate beneath northern Japan. Another predominant feature is the uplift in

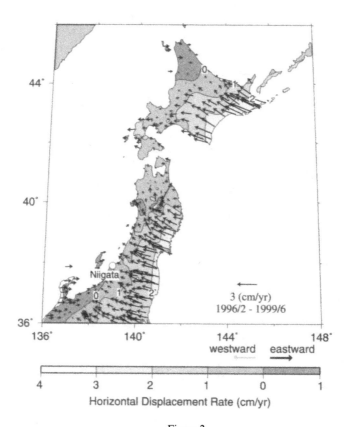

Figure 2
The observed horizontal displacement rate in northeast Japan during the period from February 1996 to June 1999 obtained from GPS observations (MIYAZAKI, 1999 personal communication). The white circle represents the reference point.

the southern part of the district. This uplift has been interpreted as "backbone" uplift of the island arc that extends farther to the south of the area (DAMBARA, 1971). More locally, there is observed an area of subsidence in the central area of the district. This subsidence is considered to be the postseismic deformation caused by the 1896 Riku-u earthquake (THATCHER *et al.*, 1980; SUITO and HIRAHARA, 1999).

3. Finite Element Modeling

We use a finite element method (FEM) to compute the displacements in the region around northeast Japan. In FEM, complex boundary shapes and internal variations of material properties can be easily handled. Therefore, this method is suitable to simulate the problems where heterogeneities and complex geometries play an important role. There have been numerous studies applying this method to solid

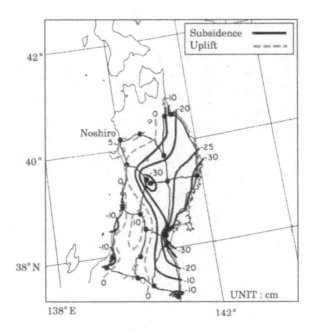

Figure 3

Accumulated vertical movement observed in the Tohoku district during the period 1900–1975 (modified after KATO, 1979). The vertical movement is given in centimeters, together with leveling routes. It is assumed that the benchmark at Noshiro did not move during the period.

earth problems (e.g., JUNGELS and FRAZIER, 1973; SHIMAZAKI, 1974; KOSLOFF, 1977; RICHARDSON, 1978; YOSHIOKA et al., 1989).

Figures 4a and b give the finite element meshes in the horizontal view and the vertical section along the line A-A', respectively. The model space is 200 km deep, 1250 km wide and 1150 km long. The total numbers of nodes and elements are 23,520 and 21,080, respectively. Boundary conditions are also shown in Figure 4. Taking the X, Y and Z coordinates as indicated in Figure 4, we assume that the model surface is a free surface and the other five outer boundaries normal to the X, Y or Z axes can have slip components only in Y-Z, X-Z or X-Y planes. The barbed lines and the dashed thick lines in Figure 4a represent the plate boundaries and the isodepth contours of the upper boundary of the subducting PAC, respectively. We construct the plate boundary at the eastern margin of the Japan Sea and the configuration of the subducting PAC, referring to OHTAKE (1995) and HASEGAWA et al. (1983), respectively. Letters A-G and I-O show the approximate horizontal projections of fault zone of the interplate earthquakes considered in this study. Table 1 gives the occurrence time and slip amounts of the interplate earthquakes. The explanation for the assumed earthquakes is given in the next section.

Since the earthquakes and the subduction considered in this study are a displacement discontinuity, efficacious devices must be introduced into FEM

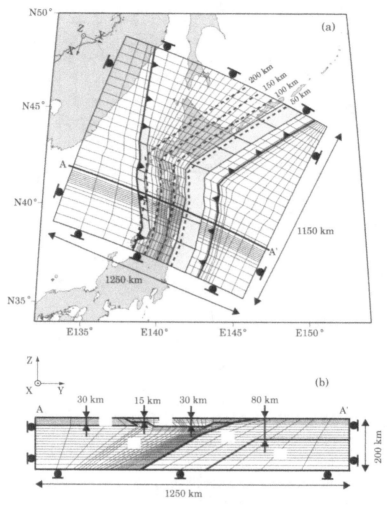

Figure 4

FEM meshes used in this study. (a) Horizontal projection of the finite element mesh. The barbed lines mark the plate boundaries. Rectangles are horizontal projections of the fault planes for interplate earthquakes considered in this study. A white line in the central Tohoku district represents the fault of the 1896 Riku-u earthquake. Thick dashed lines indicate the isodepth contours of the upper boundary of the PAC subducting beneath northeast Japan assumed in this study. For boundary conditions, see the text. (b) Vertical projection of the finite element mesh along A-A'. The portions (1), (2) and (3) represent the crust, the upper mantle and the plate, respectively.

computation. Several schemes for treating faults have been proposed. Among them, MELOSH and RAEFSKY (1981) have developed the "split node technique", which seems to be most suitable for representing the fault. This technique has several advantages over the others, one of which is that no net forces or moments are induced on finite element grid. Therefore, we apply this technique to compute the displacement due to the kinematic earthquake cycle proposed by SAVAGE (1983).

Table 1

Space-time distribution and slip amount of interplate earthquakes considered in this study. The letters A–G and I–O correspond to the regions shown in Fig. 4a. The amounts of slip are quoted from: (1) SATO (1988); (2) TANIOKA and SATAKE (1996); (3) KATO (1979); (4) NISHIMURA et al. (1998); (5) HASHIMOTO (1994); (6) See text

| Along the Japan Sea | | | | | | Along the Japan and Kurile Trench | | | | | | | |
O	N	M	L	K	J	I	G	F	E	D	C	B	A
								1987 M7.7 3.5 m[1]	1896 M8.5 5.7 m[2]	←1894→ M7.9 6.6 m[6]		1893 M7.7 6.0 m[6]	1958 M8.1 5.1 m[1]
						1940 M7.5 1.1 m[1]	1938 M7.5 2.3 m[1]	1936 M7.5 1.6 m[3]					
								1978 M7.4 2.0 m[1]	1968 M7.9 4.1 m[1]	1952 M8.2 4.0 m[1]	1973 M7.4 1.6 m[1]	1969 M7.8 2.9 m[1]	
					1993 M7.8 4.0 m[5]				1994 M7.9 1.5 m[4]				
			1983 M7.7 4.5 m[1]										
1964 M7.5 4.0 m[1]													

Table 2

Material properties used in this study. These values are quoted from SATO et al. (1981) and SUITO and HIRAHARA (1999). The numerals in the first column correspond to the portions described in Figure 4b

No.	Rigidity μ $(\times 10^{10}\ \text{Pa})$	Poisson's ratio	Viscosity η $(\times 10^{19}\ \text{Pa}\cdot\text{S})$	Relaxation time $\tau\ (=\eta/\mu)$ (years)	Material type
(1) Crust	3.30	0.226	–	–	Elastic
(2) Upper Mantle	5.89	0.273	0.93	5	Maxwell
(3) Plate	19.1	0.258	–	–	Elastic

Our model is composed of the elastic crust and plate, and the viscoelastic upper mantle wedge with the Maxwell time of 5 years. Table 2 tabulates the material properties assigned to each region shown in Figure 4b. These values are taken from SATO (1981) and SUITO and HIRAHARA (1999). As described in Figure 4b, the thickness of the crust, PAC and AM (or EUR) are assumed to be 15–30 km, 80 km and 30 km, respectively. Lateral variations in the crustal thickness are deduced from the seismic velocity structure determined by explosion and tomographic studies (RGES, 1977; ZHAO *et al.*, 1992).

4. Model Assumption

4.1 Subduction

It is widely accepted that the degree of interplate coupling varies with depth in the contact zone along a plate boundary surface (RUFF and KANAMORI, 1983; PETERSON and SENO, 1984). Following SAVAGE (1983), we consider that the subducting oceanic plate and the overlying continental plate are coupled at intermediate depths and decoupled at the shallower or deeper parts of the plate boundary. Figure 5 illustrates the dislocation model of an earthquake cycle. During an interseismic period, a region at intermediate depth remains locked, while shallower and deeper portions are decoupled, and a steady slip proceeds there. Then as a result of the steady slip at shallower and deeper portions, the tectonic stress accumulates in the locked region. Such a state can be expressed as the superposition of a uniform steady slip over the whole plate boundary and a back slip in the locked region. According to MATSU'URA and SATO (1989), the deformation produced by the uniform steady slip contributes to the surface deformation. However, in our modeling, we express the subduction as a back slip in the locked region, because our modeling focuses on the deformation at the continental plate. More consideration of this problem is discussed in Section 6.1.

We assumed the rate of back slip in the Japan trench to be 8 cm/yr (full coupling) and 1 cm/yr in the eastern margin of the Japan Sea. The coupled zone, where the

Earthquake Cycle = (a) Interseismic + (b) Coseismic

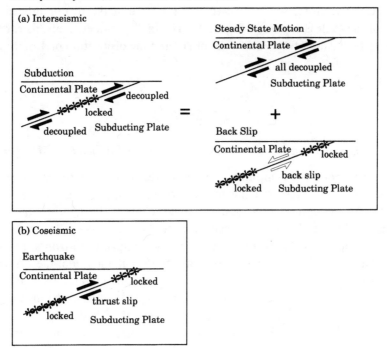

Figure 5

Schematic diagram of the kinematic earthquake cycle model after SAVAGE (1983).

back slip is assigned, is assumed to be at depths ranging from 15 to 60 km in the Japan trench, and 0 to 30 km at the eastern margin of the Japan Sea.

4.2 Earthquakes

The interplate earthquakes considered in this study are listed in Table 1. Letters A–G and I–O correspond to the respective fault regions described in Figure 4a. The slip amount of each interplate earthquake is mainly quoted from SATO (1988). Assuming the time predictable model proposed by SHIMAZAKI and NAKATA (1980), we determined the slip amount of the 1893 and the 1894 earthquakes, though it is not certain that this model is applicable to this region. Fault parameters assumed in this study are the dip of 20°W and the rake of 90° for PAC, and the dip of 30°E and the rake of 90° in the eastern margin of the Japan Sea. Although the 1964 Niigata and the 1993 Hokkaido Nansei-Oki earthquakes have western dip faults, we would like to represent primarily the eastward motion of the Amurian plate. Therefore, we simply assumed the eastward dip fault for all earthquakes at the eastern margin of the Japan Sea. As reported by THATCHER et al. (1980) and SUITO and HIRAHARA (1999), the 1896 Riku-u earthquake in the central Tohoku district has affected the crustal deformation in the Tohoku district, and its postseismic deformation continued over a few decades. Therefore, we include the Riku-u earthquake in our modeling. The position of the fault is shown in Figure 4a. The fault parameters of this earthquake assumed to be the dip of 45°E, the rake of 90° and the dislocation of 4 m (SUITO and HIRAHARA, 1999).

5. Result

5.1. Crustal Deformation Due to the Subduction of the PAC and the AM (or EUR)

We derived from trial computation of the subduction of the PAC and the AM (or EUR) that the delayed response due to the viscoelastic upper mantle disappears after about 10^5 years in our model. We assumed that the rate of the deformation due to the back slip is calculated by the displacement in this steady state, following eq. (10) in MATSU'URA and SATO (1989). Figure 6 shows the computed horizontal and vertical displacement rate in this steady state only due to the subduction of the PAC and the AM (or EUR). The westward movements are dominant, and their magnitudes decrease from east to west continuously (Fig. 6a). On the other hand, there is the steady tilt of the island toward the Pacific coast over the island (Fig. 6b). The effect of the AM (or EUR) subduction does not clearly appear in inland. These deformations are the basic features in northeast Japan in our modeling. In the following sections we show several patterns of the deformation since 1900, including the interplate earthquakes which actually occurred in northeast Japan. We compute the sequence of the interplate earthquakes independently and calculate the displacement by linear combination of the subduction and the sequence of the interplate earthquakes.

5.2. Crustal Deformation since 1900 in Northeast Japan

5.2.1. Horizontal displacement rate

Figure 7 indicates the horizontal displacement rate vectors every ten years, superposed on the contour maps of the rate with an interval of 1 cm/yr. It is noted that the rate is the average one for the displacement accumulated since 1900.

Before the period of 1950, westward movements are dominant at the Pacific coast and their magnitudes decrease from east to west throughout the periods except for certain areas and periods. These features are typical in the interseismic displacement field as shown in Figure 6a. Eastward movements in the north Tohoku district and southwest Hokkaido at the first period are due to the postseismic effects of the previous interplate earthquakes which occurred in 1890s. These movements gradually turn westward in the following period and are accelerated with time. These deformations indicate that the postseismic effects gradually decrease, and the effect of the subduction proceeds there.

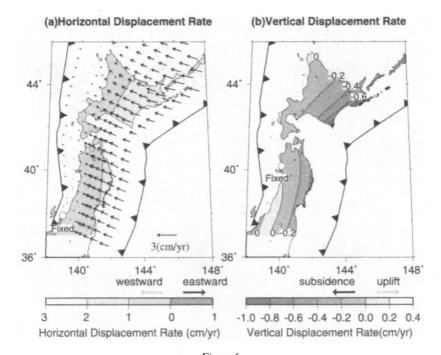

Figure 6

Computed crustal movement due to the subduction of the PAC and the AM (or EUR). The barbed lines mark the plate boundaries. The white circles represent the reference point. (a) Horizontal displacement rate vectors superposed on the contour map of the rate with an interval of 1 cm/yr. (b) Contour map of vertical movement rate with an interval of 0.2 cm/yr.

After the year 1940, eastward movements appear after the occurrence of the interplate earthquakes, which are displayed in red rectangles in each figure. These are the coseismic and postseismic effects of each interplate earthquake. It seems that these postseismic movements continue for a few decades.

There is a local westward area in the southern region of the Japan Sea coast in the 1900–1970 periods. These westward displacements are coseismic ones due to the 1964 Niigata earthquake. Although other earthquakes occur in 1940 and 1983 along the Japan Sea coast, their effects do not appear clearly in inland.

5.2.2. Vertical movement

Figure 8 displays the contour maps of accumulated vertical movement at the surface every ten years. These figures show the movement relative to that at the reference point indicated by a red circle in each figure, and this point is fixed to be 0 movement in each period.

The most fundamental mode of the vertical movement is a steady tilt of the island toward the Pacific coast. This feature does not change with time. Another predominant feature is steady and continuous uplift in the southwestern part of the

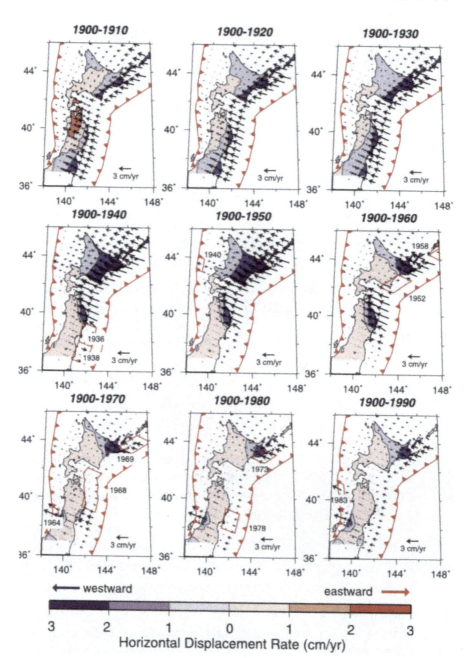

Figure 7

Accumulated horizontal displacement rate vectors every ten years superposed on the contour map of the rate with an interval of 1 cm/yr in northeastern Japan. Red color region represents the eastward movement and the blue one is the westward movement. The barbed lines mark the plate boundaries. The red rectangular regions correspond to the horizontal projection of the faults for the interplate earthquakes which occurred in the respective period.

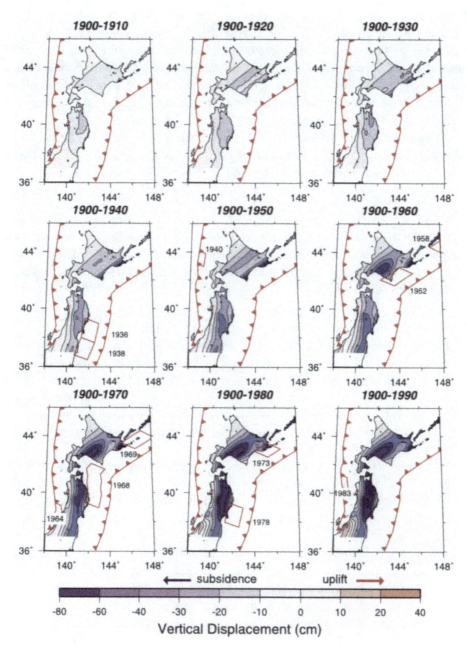

Figure 8

Contour map of accumulated vertical movement every ten years; the contour interval being 10 cm in northeastern Japan. Red color represents the uplift region and blue is the subsidence region. The red circle indicates the reference point. Other symbols are the same as in Figure 7.

Tohoku district and the northern part of the Hokkaido. While the subducting PAC produces the steady tilt of the island towards the Pacific coast, as described in Figure 6b, we find the subsidence region in inland and the uplift along the Pacific coast after the occurrence of the interplate earthquake, which is displayed by red rectangles in each figure. These are the coseismic and postseismic deformations. Moreover, these deformations continue for a few decades. Therefore, the deformation in the first period contains the postseismic effects which occurred in 1890s.

A local subsidence appears in the central part of the Tohoku district. This subsidence is caused by a postseismic deformation due to the 1896 Riku-u earthquake, as suggested by THATCHER *et al.* (1980) and SUITO and HIRAHARA (1999).

5.3. Comparison with the Observed Data

In order to give a quantitative evaluation of our modeling, we compare the computed results with the vertical movement obtained from the leveling survey, which was studied by KATO (1979), and the horizontal displacement rate obtained from the GPS survey by GSI.

5.3.1. Long-term vertical movement in the Tohoku district

Figure 9 shows the computed result of accumulated vertical movement during the period of 1900–1975 with a contour interval of 10 cm. The observed one shown in Figure 3 agrees moderately with the computed result except for some areas in the northern part of the district. Since there are few observation points in the northern part of the district, we do not discuss this area. The magnitude of the subsidence is large in our modeling compared with the observed one. Discussions regarding this discrepancy are referred later.

It has been considered that observed land subsidence along the Pacific coast is due to the subduction of the PAC, and is released by interplate earthquakes (KATO, 1979). However, our interpretation based on this modeling is the following: Subducting PAC produces the steady tilt of the island towards the Pacific coast (see Fig. 6b). The inland subsidence is mainly produced by the interplate earthquakes, and this subsidence continues for approximately a few decades due to the relaxation of the viscoelastic upper mantle. Therefore, the observed land subsidence in most of the inland is produced by the 1936 Kinkazan-Oki and the 1968 Tokachi-Oki earthquakes.

The local subsidence observed in the central part of the district is caused by the postseismic effect of the 1896 Riku-u earthquake, as suggested by THATCHER *et al.* (1980) and SUITO and HIRAHARA (1999). However its amount of subsidence is large in our computation, considering the subducting PAC and the occurrence of the interplate earthquakes. The parameters related to this earthquake may need to be reexamined.

On the other hand, the uplift observed in the southern part of the district has been interpreted as "backbone" uplift of the island arc that extends farther to the south of the area (DAMBARA, 1971). KATO (1979) and EL-FIKY and KATO (1999) also followed

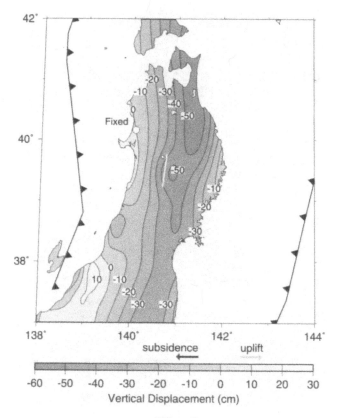

Figure 9
Contour map of the accumulated vertical movement spanning 75 years (1900–1975), with a contour interval of 10 cm in the Tohoku district. The white circle indicates the reference point. The white line represents the fault of the 1896 Riku-u earthquake.

DAMBARA's (1971) interpretation. However, our modeling suggests that this uplift would be explained by the subduction of the PAC. If we change the coupling region in the southern Tohoku region, the uplift region in the southwest Tohoku district may move eastward.

5.3.2. Recent horizontal displacement rate in northeast Japan

In Figure 10, we show the horizontal displacement rate during the period of 1996–1999. The reference point is shown in a white circle. The computed result is fairly consistent with the observed one shown in Figure 2 except for some regions. In north Hokkaido, the direction of the observed rate vectors is eastward, but that direction in our results is westward. If the back slip rate of the AM (or EUR) increases, these observed eastward movements might be explained. However, since this region is closer to the pole of relative motion between NAM and AM (HEKI et al., 1999), the back-slip rate should be less than in the southern parts. Accordingly,

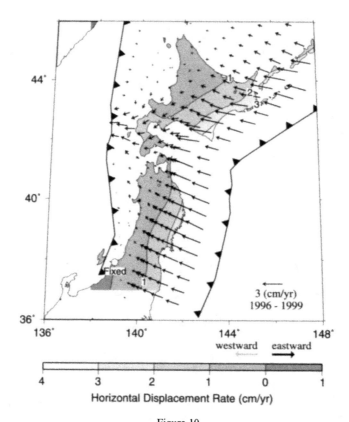

Figure 10
Horizontal displacement rate vectors superposed on the contour map with an interval of 1 cm/yr during the period of 1996–1999 in northeast Japan. The white circle represents the reference point.

to explain the observed results, we need to move the boundary landward. Actually, the boundary around this region contains uncertainty in this region.

The large displacement rate appears at southwest Hokkaido both in the observed and computed results. This movement is the effect of the postseismic deformation due to the 1993 Hokkaido Nansei-Oki earthquake. In the northern part of the Tohoku district, on the other hand, the small displacement rate appears in the observed results (Fig. 2). Nonetheless the computed rate is not small. The significant afterslip occurred in this region following the 1994 Sanriku Haruka-Oki earthquake, and slow earthquakes have occurred in 1989 and 1992 around this plate boundary (HEKI *et al.*, 1997; KAWASAKI *et al.*, 1995, 1998). However, our modeling does not consider the afterslip and slow earthquakes. Hence, the discrepancy of the observed results and computed ones in this region arises from the existence of the afterslip or slow earthquakes.

Though no earthquake occurs in this period, the magnitude of the rate is quite different in southwest Hokkaido and the Pacific coast compared with Figure 6a. This difference results from the consideration for the postseismic deformation of the interplate earthquakes due to the viscoelastic upper mantle.

6. Discussion

6.1. Consideration of the Steady Slip

In this study we expressed the subduction as the back slip and fairly explained the observed long-term (75 years) movement. However, MATSU'URA and SATO (1989) pointed out that the uniform steady slip contributes to the surface deformation and cannot ignore for the long term (one earthquake cycle) deformation. We also computed the steady slip model and checked that the steady slip also contributes to the surface deformation in our modeling. Nevertheless we could not explain the observed movement in northeastern Japan if considering the steady slip. Hence, we feel there is difficulty if we consider the steady slip in discussing the deformation during one earthquake cycle or less at the subduction zone based on the dislocation model.

6.2. Coupling Depth and Ratio along the Japan Trench

In this study the coupling depth along the Japan trench is considered to be 60 km, which is deduced from the distribution of the microearthquake and the deepest extent of large interplate earthquakes (e.g., HASEGAWA et al., 1994). However, previous studies which endeavored to explain the observed vertical or horizontal movement have claimed that the coupling depth extends to a depth of about 100 km (KATO, 1979; SENO, 1979; ITO et al., 2000). Our computation, including the effects of the interplate earthquakes and viscoelastic delayed response, suggests that the coupling depth is 60 km to explain the observed movements. Previous studies assumed the elastic material, and did not consider the effect of the interplate earthquakes and their postseismic deformations.

On the other hand, previous studies (e.g., KATO, 1979; SENO, 1979; MIURA et al., 1989) have estimated coupling ratios to be about 0.3 in the Japan trench. Recently, KAWASAKI et al. (1998) proposed that the interplate coupling ratio was estimated to be 0.5–0.85 in the Japan trench when including the slow earthquakes. Moreover, a high coupling ratio of about 0.8 to 1.0 has been estimated by some researchers using recent GPS data (LE PICHON et al., 1998; ITO et al., 2000; MAZZOTTI et al., 2000). They interpreted the discrepancy between the previous studies and their results as follows: To estimate the coupling ratio, long-term geodetic data were used in the previous studies. However, there is a possibility that the deformations associated with coseismic slips and slow slips are not well removed from the long-term data. Therefore, the geodetic coupling ratio obtained in their studies may underestimate the coupling ratio. But our modeling shows that considering the interplate earthquakes and the viscoelastic effect, the coupling ratio along the Japan trench is estimated to be 1.0, in spite of using the long-term geodetic data. However, the subsidence computed by our modeling is large compared with the observed one, as mentioned before (see Figs. 3 and 9). According to our modeling, the subducting PAC produces a steady tilt of the island towards the Pacific coast (Fig. 6b), and the interplate earthquakes produce subsidence in the inland (Fig. 8). Hence, certain possibilities are considered. (1) The coupling ratio changed with time during this period. (2) The coupling region changes with a place along this plate boundary. (3) Slow or silent earthquakes occurred in this period, as suggested by KAWASAKI *et al.* (1998). (4) Significant afterslip occurred after the interplate earthquakes such as the 1994 Sanriku Haruka-Oki earthquake (HEKI *et al.*, 1997).

Finally, we stress some important aspects which our modeling implies. Although many researchers have investigated the crustal deformation assuming an elastic medium, viscoelastic effect should be considered in their analyses, which seriously affects the surface deformation, as pointed out by MATSU'URA and SATO (1989). Even if earthquakes do not occur during the observed period, viscoelastic effects continue for a few decades and postseismic deformations caused by earthquakes occurring before the period are probably included in the observed data. Elastic analyses considering only the analyzed period, hence, lead to an erroneous interpretation of the interplate coupling.

7. Conclusion

The spatial distribution and the temporal change of the crustal deformation since 1900 in northeastern Japan have been investigated through FEM modeling, based on the kinematic earthquake cycle model. To evaluate our modeling, we compare the computed results with the long-term vertical movement during the period of 1900–1975 and the present horizontal displacement rate obtained by GPS observations during the period of 1996–1999. Our interpretations of the crustal deformation for the past 100 years in northeast Japan are as follows:

1. The subducting PAC produces the steady tilt of the island towards the Pacific coast. The effect of the slow subduction of the AM (or EUR) from the Japan Sea side does not clearly appear in the inland. Therefore, the westward movement is dominant and its magnitude decreases from east to west continuously due to the subduction of the PAC.

2. After the year 1940, coseismic and postseismic deformations of the interplate earthquakes can be clearly seen in the inland movement. Postseismic deformation of the interplate earthquakes due to the viscoelastic upper mantle continues for a few decades in the inland area.

3. The subsidence observed from the long-term survey in the central Tohoku district was previously interpreted to be due to the subducting PAC (KATO, 1979). However, our modeling indicates that this is mainly the coseismic and postseismic

deformation caused by the interplate earthquakes, since the subduction of the PAC produces the steady tilt of the island toward the Pacific coast.

4. Considering the effect of the interplate earthquakes and viscoelastic material, we can adequately explain the observed data, assuming the coupling ratio of 1.0 and the depth 60 km.

Acknowledgements

We are indebted to Prof. M. Hashimoto and an anonymous reviewer for their critical reviews. The authors used a software package, GMT (WESSEL and SMITH, 1991) to draw most of the maps used in this paper. This research was supported by grants from "Earth Simulator Project" of the Science and Technology Agency.

REFERENCES

DAMBARA, T. (1971), *Synthetic Vertical Crustal Movements in Japan During the Recent 70 Years*, J. Geod. Soc. Jpn. *17*, 100–108 (in Japanese with English abstract).

EL-FIKY, G. S. and KATO, T. (1999), *Interplate Coupling in the Tohoku District, Japan, Deduced from Geodetic Data Inversion*, J. Geophys. Res. *104*, 20,361–20,377.

EL-FIKY, G. S., KATO, T., and OWARE, E. N. (1999), *Crustal Deformation and Interplate Coupling in the Shikoku District, Japan, as Seen from Continuous GPS Observation*, Tectonophysics *314*, 387–399.

HASEGAWA, A., UMINO, N., TAKAGI, A., SUZUKI, S., MOTOYA, Y., KAMEYA, K., TANAKA, K., and SAWADA, Y. (1983), *Spatial Distribution of Earthquakes beneath Hokkaido and Northern Honshu, Japan*, J. Seismol. Soc. Jpn., Ser. 2, *36*, 129–150 (in Japanese with English abstract).

HASEGAWA, A., HORIUCHI, S., and UMINO, N. (1994), *Seismic Structure of the Northeastern Japan Convergent Margin: A Synthesis*, J. Geophys. Res. *99*, 22,295–22,311.

HASHIMOTO, M. (1984), *Finite Element Modeling of Deformations of the Lithosphere at an Arc-Arc Junction: The Hokkaido Corner, Japan*, J. Phys. Earth *32*, 373–398.

HASHIMOTO, M. (1994), *Crustal Movements Associated with the 1993 Southwestern off Hokkaido Earthquake*, J. Geographical Survey Institute *81*, 61–65 (in Japanese).

HEKI, K., MIYAZAKI, S., and TSUJI, H. (1997), *Silent Fault Slip Following an Interplate Thrust Earthquake at the Japan Trench*, Nature *386*, 595–598.

HEKI, K., MIYAZAKI, S., TAKAHASHI, H., KASAHARA, M., KIMATA, F., MIURA, S., VASILENKO, N. F., IVASHCHENKO, A., and AN, K. (1999), *The Amurian Plate Motion and Current Plate Kinematics in Eastern Asia*, J. Geophys. Res. *104*, 29,147–29,155.

HUANG, S., SACKS, I. S., and SNOKE, J. A. (1998), *Compressional Deformation of Island-arc Lithosphere in Northeastern Japan Resulting from Long-term Subduction-related Tectonic Forces: Finite-element Modeling*, Tectonophysics *287*, 43–58.

IIZUKA, M., GARATANI, K., NAKAJIMA, K., NAKAMURA, H., OKUDA, H. and YAGAWA, G. (1999), *GeoFEM: High-performance Parallel FEM Geophysical Applications*, ISHPC99, Second International Symposium Proceedings, High Performance Computing, Lecture Notes in Computer Science *1615*, 292–303.

ISHIKAWA, N. and HASHIMOTO, M. (1999), *Average Horizontal Crustal Strain Rates in Japan during Interseismic Period Deduced from Geodetic Surveys (Part 2)*, J. Seismol. Soc. Jpn., Ser. 2, *52*, 299–315 (in Japanese with English abstract).

ITO, T., YOSHIOKA, S. and MIYAZAKI, S. (2000), *Interplate Coupling in Northeast Japan Deduced from Inversion Analysis of GPS Data*, Earth Planet. Sci. Lett. *176*, 117–130.

JUNGELS, P. H. and FRAZIER, G. A. (1973), *Finite Element Analysis of the Residual Displacements for an Earthquake Rupture; Source Parameters for the San Fernando Earthquake*, J. Geophys. Res. *78*, 5062–5083.

KATO, T. (1979), *Crustal Movements in the Tohoku District, Japan, during the Period 1900–1975 and their Tectonic Implications*, Tectonophysics *60*, 141–167.

KATO, T., EL-FIKY, G. S., OWARE, E. N., and MIYAZAKI, S. (1998), *Crustal Strains in the Japanese Islands as Deduced from Dense GPS Array*, Geophys Res. Lett. *25*, 3445–3448.

KAWASAKI, I., ASAI, Y., TAMURA, Y., SAGIYA, T., MIKAMI, N., OKADA, Y., SAKATA, M., and KASAHARA, M. (1995), *The 1992 Sanriku-Oki, Japan, Ultra-slow Earthquake*, J. Phys. Earth *43*, 105–116.

KAWASAKI, I., ASAI, Y., and TAMURA, Y. (1998), *Interplate Moment Release in Seismic and Seismo-geodetic Bands and the Seismo-geodetic Coupling in the Sanriku-Oki Region along the Japan Trench – A Basis for Mid-term/Long-term Prediction-*, J. Seismol. Soc. Jpn., Ser. 2, *50*, supplement, 293–307 (in Japanese with English abstract).

KOBAYASHI, Y. (1983), *The initiation of Plate Subduction*, Monthly Earth *5*, 510–524 (in Japanese).

KOSLOFF, D. (1977), *Numerical Simulations of Tectonic Processes in Southern California*, Geophys. J. Roy. Astr. Soc. *51*, 487–501.

LE PICHON, X., MAZZOTTI, S., HENRY, P., and HASHIMOTO, M. (1998), *Deformation of the Japanese Islands and Seismic Coupling: An Interpretation Based on GSI Permanent GPS Observations*, Geophys. J. Int. *134*, 501–514.

LIU, M., YANG, Y., STEIN, S., ZHU, Y., and ENGELN, J. (2000), *Crustal Shortening in the Andes: Why do GPS Rates Differ from Geological Rates?* Geophys. Res. Lett. *27*, 3005–3008.

LUNDGREN, P., SAUCIER, F., PALMER, R., and LANGON, M. (1995), *Alaska Crustal Deformation: Finite Element Modeling Constrained by Geologic and Very Long Baseline Interferometry Data*, J. Geophys. Res. *100*, 22,033–22,045.

MATSU'URA, M. and SATO, T. (1989), *A Dislocation Model for the Earthquake Cycle at Convergent Plate Boundaries*, Geophys. J. Int. *96*, 23–32.

MAZZOTTI, S., LE PICHON, X., HENRY, P., and MIYAZAKI, S. (2000), *Full Interseismic Locking of the Nankai and Japan-west Kurile Subduction Zones: An Analysis of Uniform Elastic Strain Accumulation in Japan Constrained by Permanent GPS*, J. Geophys. Res. *105*, 13,159–13,177.

MELOSH, H. J. and RAEFSKY, A. (1981), *A Simple and Efficient Method for Introducing Faults into Finite Element Computations*, Bull. Seismol. Soc. Am. *71*, 1391–1400.

MIURA, S., ISHII, H., and TAKAGI, A. *Migration of vertical deformations and coupling of island arc plate and subducting Plate. In Slow Deformation and Transmision of Stress in the Earth*, Geophys. Monogr. Ser., 49 (eds.) S. C. COHEN and P. VANICEK, pp. 125–138, AGU, (Washington, D.C. 1989).

MIYASHITA, K. (1987), *A Model of Plate Convergence in Southwest Japan, Inferred from Leveling Data Associated with the 1946 Nankaido Earthquake*, J. Phys. Earth *35*, 449–467.

NAKAMURA, K. (1983), *Possible Nascent Trench along the Eastern Japan Sea as the Convergent Boundary between Eurasian and North American Plates*, Bull. Earthq. Res. Inst. *58*, 711–722 (in Japanese with English abstract).

NISHIMURA, T., MIURA, S., TACHIBANA, K., HASHIMOTO, K., SATO, T., HORI, S., MURAKAMI, E., KONO, T., NIDA, K., MISHINA, M., HIRASAWA, T., and MIYAZAKI, S. (1998), *Source Model of the Co- and Postseismic Deformation Associated with the 1994 far off Sanriku Earthquake (M7.5) Inferred from Strain and GPS Measurements*, Tohoku Geophys. J. (Sci. Rep. Tohoku Univ., Ser. 5) *35*, 15–32.

OHTAKE, M. (1995), *A Seismic Gap in the Eastern Margin of the Sea of Japan as Inferred from the Time-space Distribution of Past Seismicity*, The Island Arc *4*, 156–165.

PACHEO, L., SYKES, L. R., and SCHOLZ, C. H. (1993), *Nature of Seismic Coupling along Plate Boundaries of the Subduction Type*, J. Geophys. Res. *98*, 14,133–14,159.

PETERSON, E. T., and SENO, T. (1984), *Factors Affecting Seismic Moment Release Rates in Subduction Zones*, J. Geophys. Res. *89*, 10,233–10,248.

RESEARCH GROUP FOR EXPLOSION SEISMOLOGY (1977), *Regionality of the Upper Mantle around Northeastern Japan as Derived from Explosion Seismic Observations and its Seismological Implications*, Tectonophysics *37*, 117–130.

RICHARDSON, R. M. (1978), *Finite Element Modeling of Stress in the Nasca Plate: Driving Forces and Plate Boundary Earthquakes*, Tectonophysics *50*, 223–248.

RUFF, L. and KANAMORI, H. (1980), *Seismicity and the Subduction Process*, Phys. Earth Planet. Interiors. *23*, 240–254.

SAGIYA, T., MIYAZAKI, S., and TADA, T. (2000), *Continuous GPS Array and Present-day Crustal Deformation of Japan*, Pure Appl. Geophys. *157*, 2303–2322.

SATO, K., ISHII, H., and TAKAGI, A. (1981), *Characteristics of Crustal Stress and Crustal Movements in the Northeastern Japan Arc I: Based on the Computation Considering the Crustal Structure*, J. Seismol. Soc. Jpn., Ser. 2, *34*, 551–564 (in Japanese with English abstract).

SATO, K. (1988), *Stress and Displacement Fields in the Northeastern Japan Island Arc as Evaluated with Three-dimensional Finite Element Method and their Tectonic Interpretations*, Tohoku Geophys. J. (Sci. Rep. Tohoku Univ., Ser. 5) *31*, 57–99.

SATO, R., *Handbook of the Seismic Fault Parameters in Japan* (Kashima Press 1988) (in Japanese).

SAVAGE, J. C. (1983), *A Dislocation Model of Strain Accumulation and Release at a Subduction Zone*, J. Geophys. Res. *88*, 4984–4996.

SENO, T. (1979), *Intraplate Seismicity in Tohoku and Hokkaido, and Large Interplate Earthquake: A Possibility of a Large Interplate Earthquake off the Southern Sanriku Coast, Northern Japan*, J. Phys. Earth *27*, 21–51.

SENO, T., SAKURAI, T., and STEIN, S. (1996), *Can the Okhotsk Plate be Discriminated from the North American Plate?*, J. Geophys. Res. *101*, 11,305–11,315.

SHIMAZAKI, K. (1974), *Preseismic Crustal Deformation Caused by an Underthrusting Oceanic Plate, in Eastern Hokkaido, Japan*, Phys. Earth Planet. Inter. *8*, 148–157.

SHIMAZAKI, K. and NAKATA, T. (1980), *Time-predictable Recurrence Model for Large Earthquakes*, Geophys. Res. Lett. *7*, 279–282.

SUITO, H. and HIRAHARA, K. (1999), *Simulation of Postseismic Deformations Caused by the 1896 Riku-u Earthquake, Northeast Japan: Reevaluation of the Viscosity in the Upper Mantle*, Geophys. Res. Lett. *26*, 2561–2564.

TANIOKA, Y. and SATAKE, K. (1996), *Fault Parameters of the 1896 Sanriku Tsunami Earthquake Estimated from Tsunami Numerical Modeling*, Geophys. Res. Lett. *23*, 1549–1552.

THATCHER, W., MATSUDA, T., KATO, T., and RUNDLE, J. B. (1980), *Lithospheric Loading by the 1896 Riku-u Earthquake, Northern Japan: Implications for Plate Flexure and Asthenospheric Rheology*, J. Geophys. Res. *85*, 6429–6435.

WANG, K., DRAGERT, H., and MELOSH, H. J. (1994), *Finite Element Study of Surface Deformation in the Northern Cascadia Subduction Zone*, Can. J. Earth Sci. *31*, 1510–1522.

WDOWINSKI, S. and BOCK, Y. (1994), *The Evolution of Deformation and Topography of High Elevated Plateaus 2. Application to the Central Andes*, J. Geophys. Res. *99*, 7121–7130.

WESSLE, P. and SMITH, W. H. F. (1991), *Free Software Helps Map and Display Data*, Abstr. EOS Trans. Am. Geophys. Union *72*, 445–446.

YOSHII, T. (1979), *A Detailed Cross Section of the Deep Seismic Zone Beneath Northeastern Honshu, Japan*, Tectonophysics *55*, 349–360.

YOSHIOKA, S., HASHIMOTO, M., and HIRAHARA, K. (1989), *Displacement Fields due to the 1946 Nankaido Earthquake in a Laterally Inhomogeneous Structure with the Subducting Philippine Sea Plate – A Three-dimensional Finite Element Approach*, Tectonophysics *159*, 121–136.

ZHAO, D., HASEGAWA, A., and HORIUCHI, S. (1992), *Tomographic Imaging of P and S Wave Velocity Structure Beneath Northeastern Japan*, J. Geophys. Res. *97*, 19,909–19,928.

(Received February 20, 2001, revised June 11, 2001, accepted June 15, 2001)

To access this journal online:
http://www.birkhauser.ch

Pure appl. geophys. 159 (2002) 2261–2270
0033–4553/02/102261–10 $ 1.50 + 0.20/0

© Birkhäuser Verlag, Basel, 2002

▌Pure and Applied Geophysics

Combined GPS and InSAR Models of Postseismic Deformation from the Northridge Earthquake

ANDREA DONNELLAN,[1] JAY W. PARKER,[2] and GILLES PELTZER[3]

Abstract — Models of combined Global Positioning System (GPS) and Interferometric Synthetic Aperture Radar (InSAR) data collected in the region of the Northridge earthquake indicate that significant afterslip on the main fault occurred following the earthquake. Additional shallow deformation occurred to the west of the main rupture plane. Both data sets are consistent with logarithmic time-dependent behavior following the earthquake indicative of afterslip rather than postseismic relaxation. Aftershocks account for only about 10% of the postseismic motion. The two data sets are complimentary in determining the postseismic processes. Fault afterslip and shallow deformation dominate the deformation field in the two years following the earthquake. Lower crustal deformation may play an important role later in the earthquake cycle.

Key words: Earthquake, InSAR, GPS, afterslip, Northridge.

Introduction

California is well instrumented with GPS and seismic instruments and serves as an excellent laboratory for studying the complete earthquake cycle. The 1994 Northridge earthquake provides an excellent opportunity for determining postseismic processes for several reasons. The earthquake occurred within a GPS network being used to measure shortening across the Ventura basin (Fig. 1). InSAR data were collected less than two months before the Northridge earthquake and two years following the earthquake, making for an excellent comparison between GPS and InSAR observations. The Northridge earthquake occurred on a buried thrust fault resulting in vertical postseismic motions that make it possible to discriminate between fault afterslip and lower crustal relaxation.

_____

[1] Jet Propulsion Laboratory, California Institute of Technology, 4800 Oak Grove Drive, Pasadena, California, U.S.A. and University of Southern California, Los Angeles, California, U.S.A.
E-mail: donnellan@jpl.nasa.gov
[2] Jet Propulsion Laboratory, California Institute of Technology, 4800 Oak Grove Drive, Pasadena, California, U.S.A. E-mail: Jay.W.Parker@jpl.nasa.gov
[3] Jet Propulsion Laboratory, California Institute of Technology, 4800 Oak Grove Drive, Pasadena, California, U.S.A. and University of California, Los Angeles, California, U.S.A.
E-mails: Gilles.F.Peltzer@jpl.nasa.gov, mailto:peltzer@ess.ucla.edu

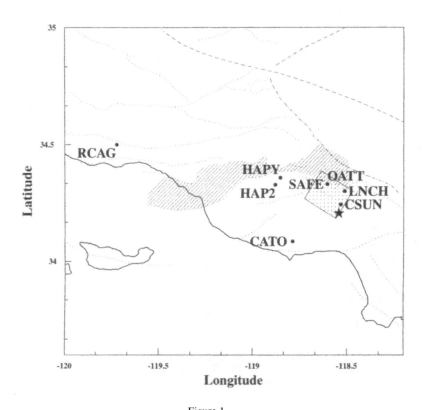

Figure 1

Location of the GPS stations. The coastline is marked by the heavy line. The Ventura basin is shaded. The epicenter of the Northridge earthquake is marked by a star and the shaded rectangle indicates the rupture area.

GPS analysis techniques have now improved to the point that daily absolute horizontal and vertical positions can be determined to 3 and 8 mm, respectively (ZUMBERGE *et al.*, 1997). Using continuous data, horizontal velocities accurate to 1 mm/yr can be achieved in 5 years (ARGUS and HEFLIN, 1995). Campaign style measurements can yield velocities accurate to 3–5 mm/yr over two years (DONNELLAN and LYZENGA, 1998) and to better than 2 mm/yr over longer timespans (SHEN *et al.*, 1996). Semi-continuous or frequent measurements collected following the earthquake indicate a time-dependent postseismic signal that decays within two years of the earthquake.

The technique of radar interferometry consists of combining Synthetic Aperture Radar (SAR) images of the same area acquired from repeated passes on a given orbit to extract the interferometric phase, which provides for each pixel of the scene a measure of the antenna-ground path length difference between the two images. After appropriate corrections for orbit configuration and topography, the interferometric phase depicts the line of sight component of the surface displacement that occurred during the time interval covered by the two images (GABRIEL et al., 1989). The technique has been successfully applied to map the surface displacement field related to earthquakes (e.g., MASSONET et al., 1993; ZEBKER et al., 1994; PELTZER et al., 1998) and surface strain related to post-seismic relaxation processes (e.g., PELTZER et al., 1996, 1998). The main error on the line of sight displacement estimate comes from variations in the phase propagation delay through the troposphere (e.g., GOLDSTEIN, 1995; ZEBKER et al., 1997). Such a signal does not generally exceed a phase cycle (28 mm of line of sight change) in the Los Angeles area.

GPS results from campaign and continuous SCIGN data indicate that a narrow band of shortening runs along the front of the Transverse Ranges through the Ventura and northern Los Angeles basins (DONNELLAN et al., 1993; ARGUS et al., 1999). The shortening rates are 7–10 mm/yr and 5–6 mm/yr for the Ventura and Los Angeles basins, respectively. Analysis of the data shows nearly pure shortening indicating thrust faulting environments.

Forward and inverse elastic modeling, when combined with geologic data, have been useful in estimating fault slip rate and geometry for the Ventura basin. The Northridge earthquake occurred along the southeastern portion of the basin on a fault similar to that defined by elastic forward models. While the slip rate and geometry of the faults can be well-described by elastic models the fault locking depths are shallower than the local earthquakes. Viscoelastic finite element models in which a ductile lower crust relaxes between earthquakes and the basin is composed of compliant sediments explains the concentrated strain rates and deep seismogenic depths (HAGER et al., 1999). In the following discussion elastic models seem to adequately describe the postseismic deformation occurring in the two years following the Northridge earthquake. Viscous deformation is likely to have a longer response time and probably dominates later.

Post Northridge Results

GPS data collected following the Northridge earthquake show significant postseismic deformation on the order of 30% of the deformation produced by the mainshock (DONNELLAN and LYZENGA, 1998). Aftershocks show a similar sense of motion as the GPS results, but only account for about 10% of the measured motions, suggesting that at least 90% of the postseismic motion has occurred aseismically (DONNELLAN and LYZENGA, 1998). The GPS data are consistent with afterslip on the main rupture plane, but also suggest shallow deformation to the west. InSAR measurements provide an independent measurement of postseismic deformation associated with the Northridge earthquake.

The Northridge earthquake occurred while the European Space Agency ERS-1 satellite was operating in a 3-day repeat mode, which was not providing global coverage of the surface of the earth at low latitudes. The first postseismic ERS data acquisitions of the area of Northridge were possible after March, 1995, when the satellite was placed again on a 35-day repeat cycle. The interferogram we use in this study covers the 11/8/1993–12/6/1995 time interval, thus including both the coseismic displacement signal and the signal associated with 2 years of postseismic deformation (Fig. 2a). The time period is close to the time period of the GPS data. The SAR data correlate well in the San Fernando Valley where much of the deformation from the Northridge earthquake took place. Decorrelation occurs in the hills to the north of the valley. SAR observations are present at the locations of the GPS stations, although an unwrapping error causes the value at CSUN to be suspect.

To analyze the postseismic deformation we removed from the interferogram a modeled phase corresponding to the coseismic signal (Fig. 2b). We used the variable slip distribution model determined by inversion of seismic and geodetic data by WALD *et al.* (1996). The residual phase (Fig. 2c) depicts, in principle, surface displacements related to postseismic processes but also includes incorrectly modeled

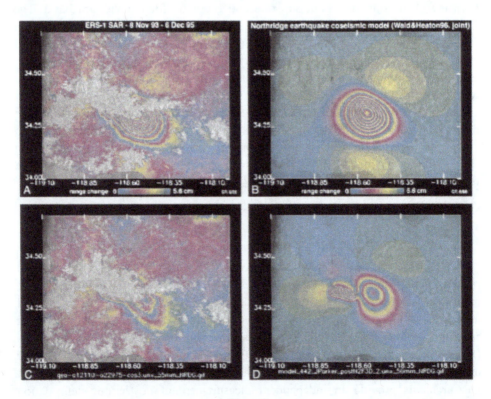

Figure 2
A) Observed interferogram for the time period November 1993 to December 1995. B) Modeled phase corresponding to the coseismic signal. C) Residual interferogram after removal of the coseismic signal. D) Modeled phase from the best fit postseismic inversion.

coseismic displacements. The results match qualitatively well where the InSAR signal does not decorrelate (Fig. 2d). To assess the validity of the approach, we first compared the displacement observed with InSAR data at GPS sites with the postseismic displacement measured with the GPS instruments.

We fit a linear trend to the GPS data, however, sites closer to the rupture plane show a clear nonlinear postseismic trend. The observable postseismic transients decay with time with most of the motion occurring within the first year after the earthquake. The horizontal and vertical postseismic motions, at stations where they are significant, can be fit by both exponential decay and logarithmic functions. Models of the GPS results indicate that afterslip, consistent with a logarithmic decay (MARONE et al., 1991), is the dominant mechanism in the two years following the earthquake (DONNELLAN and LYZENGA, 1998). Because the GPS data were sampled nonuniformly and infrequently and the InSAR data were only sampled twice a constant velocity fit to the data is the most reasonable way to model the dominant postseismic process using all of the available data. Additionally, for sites more than one fault dimension away, the errors are large enough that it is difficult to discriminate between linear and nonlinear trends in the data.

Since the InSAR and GPS data do not cover the exact same time intervals, we scaled the InSAR data to be compatible with the GPS velocities. We removed the coseismic signal (WALD et al., 1996) from the InSAR data, and then scaled the InSAR by the best exponential and logarithmic fit functions of the GPS data (Fig. 3). The first GPS observations were made about three days after the Northridge earthquake, and substantial afterslip may have occurred during that time. Coseismic offsets calculated from the GPS observations are consistently larger than those of the WALD et al. (1996) model, which is based largely on seismic observations, suggesting a fair amount of immediate postseismic deformation. We tested for the best-scaled fit of the InSAR data (e.g., with or without rapid postseismic deformation for different models) for the time intervals 1/17/94–12/6/95 and from 1/20/94–12/6/95. We correlated the line-of-sight (LOS) station displacements calculated from the GPS solutions with the LOS displacements observed with InSAR at those locations. The best correlation is for the logarithmic afterslip model scaled from three days after the mainshock to December 1995. This model takes into account rapid postseismic deformation. The InSAR displacement is relative and since the reference site RCAG is not in the radar frame, the InSAR measurements can be shifted arbitrarily. There is likely an unwrapping problem near site CSUN, which, if solved, may shift it by −2.8 cm or −5.6 cm. Afterslip is best fit by a logarithmic function, which is consistent with the inversions of just the GPS data that suggest that afterslip was the dominant mechanism in the first two years following the earthquake. Models of viscoelastic relaxation (exponential decay) or slip on the downdip extension of the fault do not fit well. A particular problem with exponential relaxation is that the sites over the rupture plane should show subsidence, but the observed results show about 12 cm of uplift, which is consistent with afterslip.

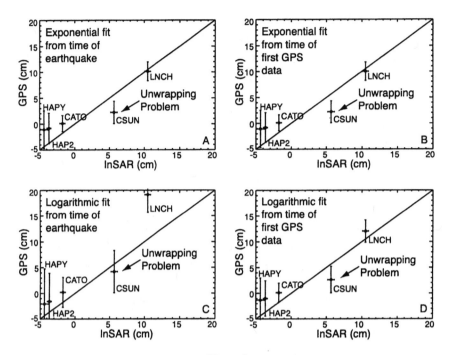

Figure 3
Correlation between GPS and InSAR for various assumptions on the style of postseismic motion.

Friction Rate Parameters

The calculated logarithmic function that fits the data can be used to estimate friction parameters for the Northridge fault. The initial coseismic slip rate for the thickness-averaged region undergoing afterslip is 174 mm/day, which is within the range of that observed for the Superstition Hills earthquake. The friction rate parameter is about 0.002 which matches laboratory values for poorly consolidated materials (C. Marone, written communication, June 1998).

Inversions

The GPS and postseismic InSAR observations are qualitatively similar. We see deformation over the rupture plane and to the west of the rupture. The magnitude of the deformation is also similar in both cases. Because both the correlation between the GPS and InSAR data and the inversions of the GPS data strongly point toward fault afterslip following the earthquake we used the InSAR data to improve the inversion, particularly west of the rupture plane (Fig. 2d). The inversion code is based on OKADA's (1985) methods for a dislocation in an isotropic elastic medium. The inversion model uses a residual-minimization procedure based on a downhill simplex simulated annealing algorithm (LYZENGA et al., 2000), for which nine fault parameters can be solved (location, depth, dip, length, width, slip).

The results for the main fault plane are nearly identical to those observed with the GPS only solution (Table 1), further indicating that afterslip was the dominant mechanism following the earthquake. By adding the InSAR data we were able to free every parameter for the auxiliary fault plane. In the combined solution the potency, or moment, is about a factor of 1.8 greater than in the GPS only solution. The size of the fault patch is about one third the size of the fault patch in the GPS only solution, while the amount of slip on the fault is substantially greater (Fig. 4). The depth to the top of the "fault" is slightly deeper. The amount of slip at depth on this auxiliary fault is about 50 cm, which while large is probably not unreasonable. A qualitative look at the InSAR results shows very localized deformation west of the mainshock rupture suggesting that displacement on a long planar structure in this region is likely. The residuals indicate that the model is consistent with the data, particularly for the GPS data (Fig. 5).

As previously reported, the auxiliary fault does not correspond to any mapped fault but may rather be indicative of general deformation of the upper crust as a result of the mainshock. This "fault," which is more likely representative of broad deformation in the upper crust, coincides with shallow aftershocks that are also interpreted as deformation of a quasi-elastic material (UNRUH et al., 1997).

Table 1

Fault parameters for combined and GPS only inversions. The models shown here represent typical inversions but are nonunique. The GPS and combined GPS/InSAR models are in good agreement for the fit to the main rupture plane. The greater sampling of the InSAR to the west of the rupture area improves the fit for the auxiliary plane. The combined model should be taken as a qualitative example fit. The errors are difficult to assess due to lack of a rigorous error model on the InSAR data, errors in the coseismic model, and varying sampling intervals of the InSAR data. The X and Y coordinates are east and north distances, respectively, of the northeast (upper right) corner of the plane from the epicenter of the Northridge earthquake (N34.20883, W118.54067). Depth is to the top of the fault plane. Positive strikeslip component indicates left-lateral motion, and positive dip-slip indicates thrust motion. Errors are 1σ and are not scaled by the χ^2/dof

Parameter	Combined Main Plane	Auxiliary Plane	GPS only Main Plane	Auxiliary Plane
X (km)	7.7 ± 0.5	−6.4 ± 0.7	8.8 ± 1.3	−5.6 ± 0.3
Y (km)	13.0 ± 1.0	12.7 ± 0.3	12.8 ± 1.9	13.3 ± 1.1
Strike (°)	300.0	286.6 ± 3.0	300.0	277.8
Dip (°)	40.0	54.0 ± 7.7	40.0	38.0
Depth (km)	7.3 ± 1.0	0.5 ± 0.6	5.3 ± 1.5	0.1 ± 0.7
Width (km)	18.8	5.2 ± 1.2	18.8	8.6
Length (km)	12.1	13.1 ± 1.1	12.1	23.0
Strike-slip (mm/yr)	25.5 ± 22.4	−308.9 ± 97.1	36.0 ± 30.2	−12.1 ± 22.2
Dip-slip (mm/yr)	276.0 ± 40.5	413.3 ± 107.0	203.3 ± 59.4	93.8 ± 53.4

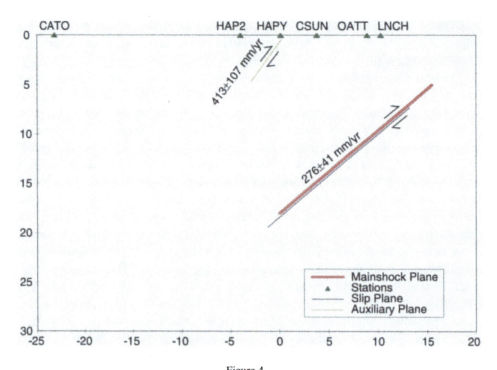

Figure 4

Cross section showing mainshock rupture plane and calculated faults.

The reported χ^2/dof for the combined inversion is 0.3. However, it is highly sensitive to the assumed SAR "uncertainty." That assumed uncertainty is partly arbitrary because the SAR data are scaled in such a way that the few GPS results have comparable weight in the inversion to the heavily sampled SAR data. The SAR uncertainties include largely systematics like model removal and flattening, and are far from white. The dominant source of that error is most certainly highly correlated systematics, therefore we do not scale the formal reported errors.

Conclusions

Both GPS and InSAR data collected in association with the Northridge earthquake indicate that a significant amount of afterslip occurred on the mainshock plane in the two years following the earthquake. In addition, a significant amount of localized deformation occurred to the west of the fault plane, which cannot be linked to a mapped fault. The results imply that the upper crust near Northridge is inhomogeneous and contains localizations of softer material, or bedding plane faults. Future observations will indicate if afterslip ceases and other

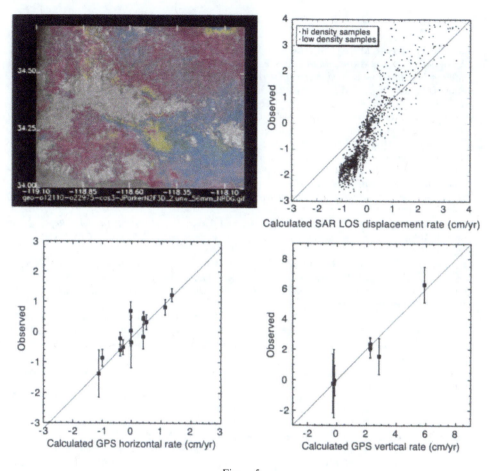

Figure 5

Residual interferogram of the observed minus the modeled data. Correlation between the SAR line of site observed displacement rates and the GPS rates. The GPS results fit better than the InSAR results.

mechanisms, such as viscoelastic relaxation of the lower crust, begin to dominate later in the earthquake cycle.

Acknowledgments

This work was carried out at the Jet Propulsion Laboratory, California Institute of Technology, under contract with NASA. Support was also provided by the U.S. Geological Survey and National Science Foundation through the National Earthquake Hazards Reduction Program (NEHRP) and the Southern California Earthquake Center (SCEC). SCEC is funded by NSF Cooperative Agreement EAR-8920136 and USGS Cooperative Agreements 14-08-0001-A0899 and 1434-HQ-97AG01718. The SCEC contribution number for this paper is 398.

REFERENCES

ARGUS, D.F., HEFLIN, M.B., DONNELLAN, A., WEBB, F.H., DONG, D., HURST, K.S., JEFFERSON, D.C., LYZENGA, G.A., WATKINS, M.M., and ZUMBERGE, J.F. (1999), *Shortening and Thickening of Metropolitan Los Angeles Measured and Inferred by Using Geodesy*, Geology 27, 703–706.

ARGUS, D.F. and HEFLIN, M.B. (1995) *Plate Motion and Crustal Deformation Estimated with Geodetic Data from the Global Positioning System*, Geophys. Res. Lett. *15*, 1973–1976.

DONNELLAN, A. and LYZENGA, G.A. (1998), *GPS Observations of Fault Afterslip and Upper Crustal Deformation Following the Northridge Earthquake*, J. Geophys. Res. *103*, 21,285–21,297.

DONNELLAN, A., HAGER, B.H., KING, R.W., and HERRING, T.A. (1993), *Geodetic Measurement of Deformation in the Ventura Basin Region, Southern California*, J. Geophys. Res. *98*, 21,727–21,739.

GABRIEL, A.G., GOLDSTEIN, R.M., and ZEBKER, H.A. (1989), *Mapping Small Elevation Changes Over Large Areas: Differential Radar Interferometry*, J. Geophys. Res. *94*, 9183–9191.

GOLDSTEIN, R. (1995), *Atmospheric Limitations to Repeat-Track Radar Interferometry*, Geophys. Res. Lett. *22*, 2517–2520.

HAGER, B.H., LYZENGA, G.A., DONNELLAN, A., and DONG, D. (1999), *Reconciling Rapid Strain Accumulation with Deep Seismogenic Fault Planes in the Ventura Basin, California*, J. Geophys. Res. *104*, 25,207–25,219.

LYZENGA, G. A., PANERO, W. R. and DONNELLAN, A. (2000) *The Influence of Anelastic Surface Layers on Postseismic Thrust Fault Deformation*, J. Geophys. Res. *105*, 3151–3157.

MARONE, C.J., SCHOLZ, C.H., and. BILHAM, R. (1991), *On the Mechanics of Earthquake Afterslip*, J. Geophys. Res. *96*, 8441–8452.

MASSONNET, D., ROSSI, M., CARMONA, C., ADRAGNA, F., PELTZER, G., FEIGL, K., and RABAUTE, T. (1993), *The Displacement Field of the Landers Earthquake Mapped by Radar Interferometry*, Nature *364*, 138–142.

OKADA, Y. (1985), *Surface deformation due to shear and tensile faults in a half-space*, Bull. Seimol, Soc. Am. *75*, 1135–1154.

PELTZER, G., ROSEN, P., ROGEZ, F., and HUDNUT, K. (1996), *Postseismic rebound in fault stepovers caused by pore fluid flow*, Science *273*, 1202–1204.

PELTZER, G., ROSEN, P., ROGEZ, F., and HUDNUT, K. (1998), *Poroelastic rebound along the Landers 1992 earthquake surface rupture*, J. Geophys. Res. *103*, 30,131–30,145.

SHEN, Z., JACKSON, D.D., and GE, B.X. (1996), *Crustal deformation across and beyond the Los Angeles basin from geodetic measurements*, J. Geophys. Res. *101*, 27,957–27,980.

UNRUH, J.R., TWISS, R.J., and HAUKSSON, E. (1997), *Kinematics of postseismic relaxation from aftershock focal mechanisms of the 1994 Northridge, California earthquake*, J. Geophys. Res. *102*, 24,589–24,603.

WALD, D.J., HEATON, T.H., and HUDNUT, K.W. (1996), *The slip history of the 1994 Northridge, California, earthquake determined from strong-motion, teleseismic, GPS, and leveling data*, Bull. Seimol. Soc. Am.*86*, S49–S70.

ZEBKER H.A., ROSEN, P.A., and HENSLEY, S. (1997), *Atmospheric effects in interferometric synthetic aperture radar surface deformation and topography maps*, J. Geophys. Res. *102*, 7547–7563.

ZEBKER, H. A., ROSEN, P., GOLDSTEIN, R.M., GABRIEL, A., and WERNER, C.L. (1994), *On the derivation of coseismic displacement fields using differential radar interferometry: The Landers earthquake*, J. Geophys. Res. *99*, 19,617–19,634.

ZUMBERGE, J.F., HEFLIN, M.B., JEFFERSON, D.C., WATKINS, M.M., and WEBB, F.H. (1997), *Precise point positioning for the efficient and robust analysis of GPS data from large networks*, J. Geophys. Res. *102*, 5005–5017.

(Received February 27, 2001, revised June 11, 2001, accepted June 25, 2001)

 To access this journal online:
http://www.birkhauser.ch

Pure appl. geophys. 159 (2002) 2271–2283
0033–4553/02/102271–13 $ 1.50 + 0.20/0

⌐ Pure and Applied Geophysics

A Hidden Markov Model Based Tool
for Geophysical Data Exploration

ROBERT GRANAT[1] and ANDREA DONNELLAN[1]

Abstract—Unsupervised learning techniques provide a way of investigating scientific data based on automated generation of statistical models. Because these techniques are not dependent on *a priori* information, they provide an unbiased method for separating data into distinct types. Thus they can be used as an objective method by which to identify data as belonging to previously known classes or to find previously unknown or rare classes and subclasses of data. Hidden Markov model based unsupervised learning methods are particularly applicable to geophysical systems because time relationships between classes, or states of the system, are included in the model. We have applied a modified version of hidden Markov models which employ a deterministic annealing technique to scientific analysis of seismicity and GPS data from the southern California region. Preliminary results indicate that the technique can isolate distinct classes of earthquakes from seismicity data.

Key words: HMM, geophysics, annealing, clustering, seismic, gps.

Introduction

In recent years, computerized analysis techniques have become increasingly popular for use in scientific data analysis (DUDA *et al.*, 1973; FAYYAD *et al.*, 1996; FAYYAD and SMYTH, 1999). Here we discuss the application of one such technique to data mining of large geophysical data sets. By data mining, we mean the process of extracting interesting information from the data in cases where (1) either there is so much data that analysis by hand would be impractical or even intractable, or (2) in cases where trends in the data may be subtle enough to evade the notice of a human analyst.

Clustering methods are one approach to data mining. As the end result of these unsupervised learning algorithms, the data are grouped into several classes such that data belonging to one class are similar to other data in that class but dissimilar to data in other classes. When data points are viewed as vectors in feature space, points in the same class will appear clustered together. Clustering is a useful technique for data mining because it does not require that any of the data be labeled ahead of time;

[1] Jet Propulsion Laboratory, Pasadena, California, U.S.A.
E-mails: granat@aig.jpl.nasa.gov; andrea@aig.jpl.nasa.gov

the algorithm is free to determine which data are related without human supervision. This means unexpected patterns can be discovered that might have otherwise been overlooked because of bias.

The hidden Markov model (HMM) (BAUM, 1972; BAUM and EGON, 1967; BAUM and PETRIC, 1966; BAUM et al., 1970; BAUM and SELL, 1968; RABINER, 1989) forms the basis of one type of unsupervised learning clustering method that is particularly applicable to geophysical systems because it includes in the model the time relationship between different classes, or states of the system. The method assumes that the data were generated by an unknown statistical process which at each point in time can be in any one of a given number of states. Each state is associated with a probability distribution of observable outputs and a set of transition probabilities. These transition probabilities determine the probability that the system will be in each of the possible states at the next point in time, given the current state. An iterative approach is used to find the system model and state sequence that best explain the observed data. The data can be clustered by using the optimal state sequence: each data vector is labeled according to the system state at the time the data were generated. The advantage of this approach is that the end result says more than which data are related to which other data: it also provides information about the confidence with which each class assignment is regarded, and about the relationship between classes in time. For this reason, an accurate HMM has the potential to provide valuable predictive information when employed in a real time context.

Hidden Markov Models

A hidden Markov model λ with N states $S_1, \ldots, S_N$ consists of the initial state distribution $\pi = (\pi_1, \ldots, \pi_N)$, the transition probabilities from each state i to each state $j, A = (a_{11}, a_{12}, \ldots, a_{NN})$, and the observation probability density functions associated with each state, $B = (b_1, \ldots, b_N)$.

We wish to answer three main questions when using an HMM:
1. What is the probability of the observed data, given the model?
2. What is the optimal state sequence, given the observed data and the model?
3. What are the optimal model parameters, given the observed data?

What is the Probability of the Model Given the Observed Data?

For the series of observations $O = O_1 O_2 \cdots O_T$, we consider the possible model state sequences $Q = q_1 q_2 \cdots q_T$ to which this series of observations could be assigned. For a given fixed states sequence Q, the probability of the observation sequence O is given by

$$P(O|Q, \lambda) = \prod_{t=1}^{T} P(O_t|q_t, \lambda). \tag{1}$$

Assuming statistical independence of observations,

$$P(O|Q, \lambda) = b_{q_1}(O_1)b_{q_2}(O_2)\cdots b_{q_T}(O_T). \tag{2}$$

The probability of the given state sequence Q is

$$P(Q|\lambda) = \pi_{q_1}a_{q_1 q_2}a_{q_2 q_3}\cdots a_{q_{T-1} q_T}. \tag{3}$$

The joint probability of O and Q is the product of the above, so that

$$P(O, Q|\lambda) = P(O|Q, \lambda)P(Q|\lambda), \tag{4}$$

so that the probability of O given the model is obtained by summing this joint probability over all possible state sequences Q:

$$P(O|\lambda) = \sum_{\text{all } Q=q_1 q_2\cdots q_T} \pi_{q_1}b_{q_1}(O_1)a_{q_1 q_2}b_{q_2}(O_2)\cdots a_{q_{T-1} q_T}b_{q_T}(O_T). \tag{5}$$

Unfortunately, calculating this quantity is an $O(N^T)$ computation. To avoid excessive computation cost, an iterative approach is used. This is the *forward-backward* procedure. Consider the forward variable $\alpha_t(i)$ defined as

$$\alpha_t(i) = P(O_1\cdots O_t, q_t = S_i|\lambda). \tag{6}$$

This is the probability of observing the partial sequence $O_1\cdots O_t$ and that the system is in state S_i at time t, given the model λ. We can solve for $\alpha_t(i)$ inductively as follows:
1. Initialization:

$$\alpha_t(i) = \pi_i b_i(O_1), \quad i = 1,\ldots,N. \tag{7}$$

2. Induction:

$$\alpha_{t+1}(j) = \left[\sum_{i=1}^{N}\alpha_t(i)a_{ij}\right]b_j(O_{t+1}), \quad t = 1,\ldots,T-1, \quad j = 1,\ldots,N. \tag{8}$$

3. Termination:

$$P(O|\lambda) = \sum_{i=1}^{N}\alpha_T(i). \tag{9}$$

This is an $O(N^2 T)$ computation.

This provides the answer to the first of the three questions. Nevertheless, for future use we also consider the backward variable $\beta_t(i)$ defined as

$$\beta_t(i) = P(O_{t+1}\cdots O_T|q_t = S_i, \lambda). \tag{10}$$

This is the probability of the partial observation sequence from $t+1$ to the end, given that the system is in state S_i at time t and the model λ. Once again we can solve for $\beta_t(i)$ inductively:
1. Initialization:

$$\beta_T(i) = 1, \quad i = 1,\ldots,N. \tag{11}$$

2. Induction:

$$\beta_t(i) = \sum_{j=1}^{N} a_{ij} b_j(O_{t+1}) \beta_{t+1}(j), \quad t = T-1, \ldots, 1, \quad i = 1, \ldots, N. \tag{12}$$

Again, this is an $O(N^2 T)$ computation.

What is the Optimal State Sequence Given the Observed Data and the Model?

The answers to this question are varied, depending on the definition of "optimal" given. As an example, we consider the optimality criterion that the states q_t are individually most likely. This maximizes the expected number of correct states. To implement a solution, we define a variable

$$\gamma_t(i) = P(q_t = S_i | O, \lambda). \tag{13}$$

This is the probability of being in state S_i at time t, given the observation sequence and the model. Since

$$P(O_1 \cdots O_t, q_t = S_i | \lambda) P(O_{t+1} \cdots O_T | q_t = S_i, \lambda)$$
$$= P(q_t = S_i | \lambda) = P(q_t = S_i | O, \lambda) P(O | \lambda), \tag{14}$$

we can express $\gamma_t(i)$ in terms of the forward-backward variables:

$$\gamma_t(i) = \frac{\alpha_t(i) \beta_t(i)}{P(O | \lambda)} = \frac{\alpha_t(i) \beta_t(i)}{\sum_{i=1}^{N} \alpha_t(i) \beta_t(i)}. \tag{15}$$

Using $\gamma_t(i)$ we can solve for the individually most likely state q_t at time t, as

$$q_t = \operatorname*{argmax}_{1 \leq i \leq N} [\gamma_t(i)], \quad t = 1, \ldots, T. \tag{16}$$

What are the Optimal Model Parameters Given the Observed Data?

There is no known way to analytically solve the model which maximizes the probability of the output sequence. However, we can find a model which is locally maximized by using an iterative procedure. In this section we will present a computational algorithm for doing so, based on the EM algorithm of statistics (DEMPSTER et al., 1977). A more detailed derivation of the method is presented below in a different section.

We define a variable $\xi_t(i, j)$, the probability of being in state S_i in time t and state S_j at time $t+1$, given the model and the observation sequence:

$$\xi_t(i, j) = P(q_t = S_i, q_{t+1} = S_j | O, \lambda). \tag{17}$$

Using our definitions of the forward-backward variables, we can write

$$\xi_t(i,j) = \frac{P(q_t = S_i, q_{t+1} = S_j, O|\lambda)}{P(O|\lambda)}$$

$$= \frac{\alpha_t(i)a_{ij}b_j(O_{t+1})\beta_{t+1}(j)}{\sum_{i=1}^{N}\sum_{j=1}^{N}\alpha_t(i)a_{ij}b_j(O_{t+1})\beta_{t+1}(j)}. \tag{18}$$

Note that we can relate $\gamma_t(i)$ to $\xi_t(i,j)$ by summing of j:

$$\gamma_t(i) = \sum_{j=1}^{N}\xi_t(i,j). \tag{19}$$

If we sum $\gamma_t(i)$ over the time index t, we get a quantity which can be interpreted as the expected number of times that state S_i is visited over time. Equivalently, this quantity can be interpreted as the expected number of transitions from the state S_i, if we exclude the time $t = T$ from the summation. Similarly, summation of $\xi_t(i,j)$ over $t = 1, \ldots, T - 1$ can be interpreted as the number of transitions from state S_i to state S_j.

Using the above ideas, we can arrive at a set of reasonable reestimation formulas for the model parameters:

$$\pi_i^* = \gamma_1(i), \tag{20}$$

$$a_{ij}^* = \frac{\sum_{t=1}^{T-1}\xi_t(i,j)}{\sum_{t=1}^{T-1}\gamma_t(i)}, \tag{21}$$

$$b_j^*(k) = \frac{\sum_{\substack{t=1 \\ O_t=v_k}}^{T}\gamma_t(j)}{\sum_{t=1}^{T}\gamma_t(j)}, \tag{22}$$

where the updated model is $\lambda^* = (A^*, B^*, \pi^*)$.

Algorithm

In order to determine the optimal model and state sequence for an HMM, it is relatively straightforward to employ an iterative procedure such as the expectation-maximization (EM) algorithm (DEMPSTER et al., 1977). However, this method suffers from a local maxima problem; that is, the quality of result can be dependent on the initial conditions. This problem is particularly acute for the geophysical data under consideration due to the complexity of the underlying system. Simultaneously, the demand for high-quality and consistent results is increased due to the necessity of commanding scientific credibility. We attempt to reduce the dependency of the method on the initial conditions by employing the deterministic annealing method of UEDA and NAKANO (1998). In that work, the deterministic annealing method was presented in the context of a finite mixture model (DUDA and HART, 1973;

FUKUNAGA, 1990), which assumes that each observed data vector was generated by one of a number of component density functions. The problem of model estimation and class assignment for this problem can also be solved via the EM algorithm. Therefore extension of deterministic annealing to the HMM optimization problem is therefore fairly straightfoward. Nevertheless, there remain some important distinctions between the two cases.

In the mixture model case, the unknown model Θ is estimated so that the log-likelihood of the mixture probability density function over the observed data vectors X is given by

$$L(\Theta) = \sum_x \log p(x|\Theta). \tag{23}$$

The probability distribution is represented as:

$$p(x|\Theta) = \sum_{i=1}^{C} \alpha_i p(x|w_i, \theta_i), \tag{24}$$

where $p(x|w_i, \theta_i)$ is the conditional probability density function corresponding to the component w_i in some proportion α_i. The model $\Theta = (\alpha_1, \ldots, \alpha_C, \theta_1, \ldots, \theta_C)$ consists of unknown parameters associated with the parametric forms adopted for the C component densities.

In this mixture density estimation problem, an observed sample x is incomplete because the information w_i which indicates the component from which the sample originates is unobservable. Therefore the EM algorithm employs the so-called complete data log likelihood, specified as:

$$L_{\text{complete}}(\Theta) = \log \alpha_i p(x|w_i, \theta_i). \tag{25}$$

The goal of the EM algorithm is to maximize the incomplete data log likelihood $L(\Theta)$ by using the complete data log-likelihood $L_{\text{complete}}(\Theta)$. It is straightforward to derive that, at the k-th iteration of the algorithm, the posterior probability that x belongs to the i-th component w_i is

$$P(w_i|x, \Theta_i^{(k)}) = \frac{\alpha_i^{(k)} p(x|w_i, \theta_i^{(k)})}{\sum_j \alpha_j^{(k)} p(x|w_j, \theta_j^{(k)})}. \tag{26}$$

This quantity is key to both the expectation and maximization steps of the EM algorithm. However, due to the iterative nature of the method, it is highly dependent on the original parameters $\alpha_i^{(0)}, \theta_i^{(0)}$, and so the performance of the method is sensitive to the initial model choice $\Theta^{(0)}$.

The deterministic annealing approach replaces the posterior probability $P(w_i|x, \Theta_i^{(k)})$ with $P(x \in w_i)$, the probability that x comes from component w_i. By applying the principle of maximum entropy to specify the probability, and by minimizing the free energy at each iteration of the method, the new calculation becomes on dependent on a parameter β:

$$P(x \in w_i) = \frac{(\alpha_i^{(k)} p(x|w_i, \theta_i^{(k)}))^\beta}{\sum_j (\alpha_j^{(k)} p(x|w_j, \theta_j^{(k)}))^\beta}. \tag{27}$$

By analogy to the simulated annealing approach, $1/\beta$ corresponds to the "computational temperature." The full annealing method then wraps this modified EM algorithm with an outer loop that varies the parameter β, starting from $\beta_{min} \ll 1$ and increasing towards some β_{max}; the model parameters are carried across different values of β.

The annealing feature helps to avoid local maxima in following way: The process begins at $\beta = 0$, at which point all the components of the model are identical. As β gradually increases, the influence of each observation x is gradually localized. Because each new global maximum is close to the old one as β increases, the algorithm is able to track the global maximum while increasing β. At the same time, finer structure, which is closer to the true model, gradually emerges.

In the hidden Markov model case, we wish to maximize the log of the probability of the observation sequence $O = O_1 O_2 \cdots O_T$ given the model λ: that is, $P(O|\lambda)$. We remind the reader that we can express

$$P(O|\lambda) = \sum_{\text{all } Q = q_1 q_2 \ldots q_T} \pi_{q_1} b_{q_1}(O_1) a_{q_1 q_2} b_{q_2}(O_2) \cdots a_{q_{T-1} q_T} b_{q_T}(O_T). \tag{28}$$

Treating the hidden Markov model similarly to the finite mixture model, we want to optimize incomplete data log likelihood $L(\lambda) = \log P(O|\lambda)$ by using the EM algorithm. The EM procedure once again rests on the key computation of the posterior probability of a given state sequence Q given the model at the k-th iteration and the observation series:

$$P(Q|O, \lambda^{(k)}) = \frac{P(O|Q, \lambda^{(k)}) P(Q|\lambda^{(k)})}{\sum_{\text{all } Q} P(O|Q, \lambda^{(k)}) P(Q|\lambda^{(k)})}. \tag{29}$$

By a parallel derivation, we are able to arrive at the deterministic annealing variant of the calculation by replacing the posterior probability with $P(O \in Q)$:

$$P(O \in Q) = \frac{(P(O|Q, \lambda^{(k)}) P(Q|\lambda^{(k)}))^\beta}{\sum_{\text{all } Q} (P(O|Q, \lambda^{(k)}) P(Q|\lambda^{(k)}))^\beta}. \tag{30}$$

Once again, this calculation is dependent on a "computational temperature" parameter β.

In order to derive a practical implementation of this deterministic annealing Hidden Markov model (DAHMM), we note that the above calculation can be rewritten

$$P(O \in Q) = \frac{(\pi_{q_1} b_{q_1}(O_1) a_{q_1 q_2} b_{q_2}(O_2) \cdots a_{q_{T-1} q_T} b_{q_T}(O_T))^\beta}{\sum_{\text{all } Q} (\pi_{q_1} b_{q_1}(O_1) a_{q_1 q_2} b_{q_2}(O_2) \cdots a_{q_{T-1} q_T} b_{q_T}(O_T))^\beta}. \tag{31}$$

This implies that we can implement an efficient calculation of the DAHMM via modified Baum-Welch and forward-backward (BAUM and EGON, 1967; BAUM and SELL, 1968) procedures by a simple replacement of π_i, a_i, and b_{ij} in all places with π_i^β, a_i^β, and b_{ij}^β, while wrapping the whole in an outer loop that varies the parameter β.

Straightforward as this may sound, practical implementation requires addressing several issues. It is clear that if $\beta_{min} = 0$, the EM algorithm will result in a model with unchanging parameters. In fact, even if β_{min} is "too small," the different components of the model will fail to diverge towards the true values. However, if β_{min} is "too big," excessive importance will once again be placed on the initial starting conditions. Careful choice of the increment of β is also necessary. If β is increased too rapidly, the new global maxima may be too far from the old one to track it accurately; if β is increased too slowly the increase computation time makes the method useless in a scientific tool. Our preliminary investigations indicate that these issues are of great importance in practice.

Results

In our preliminary experiments, we applied our DAHMM method to GPS seismicity data collected in the southern California region. Figure 1 shows the results of the DAHMM method applied to GPS data collected in the city of Claremont, California (with $\beta_{min} = 1.0e^{-12}$, $\beta_{max} = 1$, and increments of 0.01). The data used in the analysis had three components: east-west displacement, north-south displacement, and vertical displacement. Using a five state model, the method is able to separate the data into distinct classes that correspond to physical events. The classes are indicated by different shades and vertical lines. There is one instance of class 2 in the midst of class 3, corresponding to sharp north-south and vertical movements at that time sample, but otherwise the classes are sequential. The states before and after the Hector Mine quake of October 1999 are clearly separated, and distinct in turn from a period in 1998 in which well groundwater drainage caused displacement in the vertical direction.

Figures 2–4 show the results of the DAHMM method (with $\beta_{min} = 1.0e^{-12}$, $\beta_{max} = 1$, and increments of 0.01) applied to seismicity data taken from the SCEC catalog, including only events between January 1st, 1960 and December 31st, 1999. The data were composed of five components: latitude, longitude, depth, magnitude, and time until next event. Only events of magnitude greater than or equal to four were included. Circles indicate the location of earthquakes which are members of the depicted class; circle size corresponds to the magnitude of each event. Lines represent the major fault structure of southern California.

Figure 2 shows a class of earthquakes (that is, a set of earthquake observations most likely have been generated by the HMM when it is in a particular state) that includes several major events, among them the Hector Mine and Landers earthquakes.

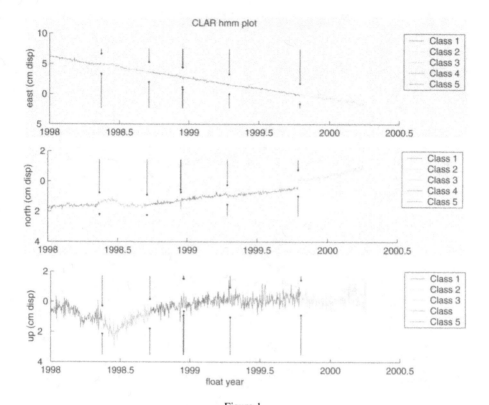

Figure 1

HMM analysis results for southern California GPS data. Classes associated with different regimes are indicated by line shading and indicator lines.

Figure 3 shows a different class of earthquakes of relatively large magnitude and a long time until next event.

Figure 4 shows the transition probabilities (a_{ij} values) for the class (HMM state) portrayed in Figure 3.

Note that the next event is most likely to belong to the class of large earthquakes portrayed in Figure 2, designated as class 16.

While the relationship between these two classes has not yet been fully investigated, this example demonstrates the potential of this type of analysis to distinguish between different classes of observed seismic events and to reveal previously unsuspected relationships between them. Analysis of this type has the potential to be useful for addressing many kinds of geophysical problems. Current methods for identifying aftershocks so that they can be separated from background earthquake activity (BRIGGS *et al.*, 1977; PRESS and ALLEN, 1995; REASENBERG, 1985) rely on simple and *ad hoc* models of earthquake activity; the HMM approach is a more objective and data-driven method for this task. Similarly, the HMM method provides a way of separating seismic "swarm" events (for example, those induced by

Class 16 means: lat=35 long=−117.4487 depth=7.1 mag=5.1 days to next event=0.052

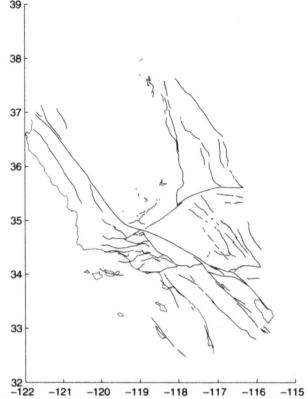

Class 16 sqrt(variances): lat=1.7 long=1.4 depth=4.7 magnitude=0.86 days to next event=0.083

Figure 2

HMM analysis results for southern California seismicity data. Earthquakes indicated by circles are all members of the same class, designated "16".

the actions of hydrothermal power plants in the Brawley seismic zone in southern California) from fault-related geophysical activity. (Preliminary results demonstrating this capability have already been obtained, but due to space considerations are not presented here.) As well, the time-dependent interaction of events that is extracted by the HMM lends itself well to identifying relationships between seismic events due to changes in stress within a fault system. For social and economic reasons, the potential to identify precursor events which signal large earthquakes is of considerable interest.

Future Work

Our preliminary experiment presented above was conducted assuming a fixed number of HMM states, thus ignoring the question of how many distinct classes of

Class 14 means: lat=36 long=−118.0364 depth=7.2 mag=5 days to next event=2.2

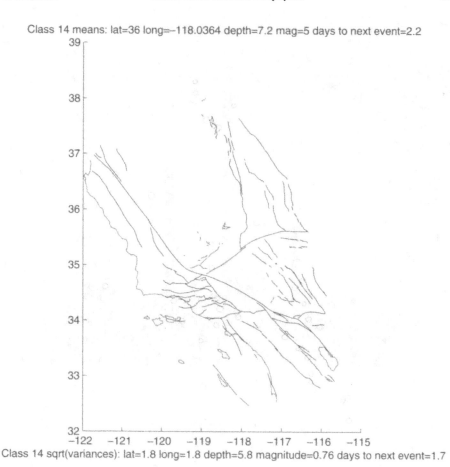

Class 14 sqrt(variances): lat=1.8 long=1.8 depth=5.8 magnitude=0.76 days to next event=1.7

Figure 3

HMM analysis results for southern California seismicity data. Earthquakes indicated by circles are all members of the same class, designated "14".

earthquakes exist in the data. Our current efforts focus on addressing this issue through development of automated methods for determining the number of classes, and thereby the HMM size.

However, other issues also need to be addressed. Computation time needs to be reduced in order to make the method usable as interactive exploratory tool. Currently it takes several hours to produce a result; we need results in a matter of minutes. A parallel implementation of the method may be one way to achieve this goal. As well, missing values cause trouble for the method; this is a problem which is particularly acute for GPS data. Intelligent interpolation will have to be used to fill in the missing values.

Future experiments include using geophysical data that include more detailed information about individual seismic events and more complete seismic catalogues

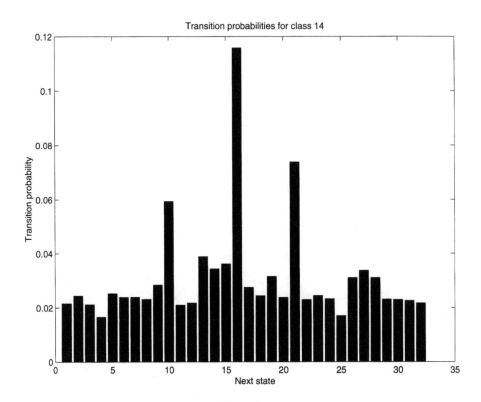

Figure 4
HMM analysis results: transition probabilities for southern California seismicity data, from the class
designated "14" to all classes.

(i.e., all low magnitude events are included). We intend to integrate the method into a
real-time data stream of seismic events, updating the model with each event. This will
allow for real-time predictive capability.

REFERENCES

BAUM, L. E. (1972), *An Inequality and Associated Maximization Technique in Statistical Estimation for Probabilistic Functions of Markov Processes*, Inequalities *3*, 1–8.

BAUM, L. E. and EGON, J. A. (1967), *An Inequality with Applications to Statistical Estimation for Probabilistic Functions of a Markov Process and to a Model for Ecology*, Bull. Am. Meteorol. Soc. *73*, 360–363.

BAUM, L. E. and PETRIC, T. (1966), *Statistical Inference for Probabilistic Functions of Finite State Markov Chains*, Ann. Math. Stat. *37*, 1554–1563.

BAUM, L. E. PETRIE, T., SOULES, G., and WEISS, H. (1970), *A Maximization Technique Occurring in the Statistical Analysis of Probabilistic Functions of Markov Chains*, Ann. Math. Soc. *41*, 164–171.

BAUM, L. E. and SELL, G. R. (1968), *Growth Functions for Transformations on Manifolds*, Pac. J. Math. *27*, 211–227.

BRIGGS, P., PRESS, F., and GUBERMAN, S. A. (1977), *Pattern Recognition Applied to Earthquake Epicenters in California and Nevada*, Geol. Soc. Am. Bull. *88*, 161–173.

DEMPSTER, A. P., LAIRD, N. M., and RUBIN, D. B. (1977), *Maximum Likelihood from Incomplete Data via the EM Algorithms*, J. Roy. Stat. Soc. *39*, 1–38.

DUDA, R. O. and HART, P. E., *Pattern Classification and Scene Analysis* (John Wiley and Sons, New York, 1973).

FAYYAD, U. M., DJORGOVSKI, S. G., and WEIR, N. (1996), *From Digitized Images to Online Catalogs – Data Mining a Sky Survey*. AI Mag. *17*, 51–66.

FAYYAD, U. M. and SMYTH, P. (1999), *Cataloging and Mining Massive Datasets for Science Data Analysis*, J. Comput. Graph. Stat. *8*, 589–610.

FUKUNAGA, K., *Introduction to Statistical Pattern Recognition* (Academic Press, New York, 1990).

PRESS, F. and ALLEN, C. (1995), *Patterns of Seismic Release in the Southern California Region*, J. Geophys. Res. *100*, 6421–6430.

RABINER, L. R. (1989), *A Tutorial on Hidden Markov Models and Selected Applications in Speech Recognition*, P. IEEE *77*, 257–286.

REASENBERG, P. (1985), *Second-order Moment of California Seismicity, 1969–1982*, J. Geophys. Res. *90*, 5479–5496.

STOLORZ, P. and CHEESEMAN, P. (1998), *Onboard Science Data Analysis: Applying Data Mining to Science-directed Autonomy*, IEEE. Intell. Syst. App. *13*, 62–68.

UEDA, N. and NAKANO, R. (1998), *Deterministic Annealing EM Algorithm*, Neural Networks *11*, 271–282.

(Received February 26, 2001, revised June 11, 2001, accepted, June 25, 2001)

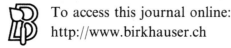

To access this journal online:
http://www.birkhauser.ch

Pure appl. geophys. 159 (2002) 2285–2309
0033–4553/02/102285–25 $ 1.50 + 0.20/0

© Birkhäuser Verlag, Basel, 2002

❙ Pure and Applied Geophysics

Geophysical Applications of Multidimensional Filtering with Wavelets

DAVID A. YUEN,[1] ALAIN P. VINCENT,[2]
MOTOYUKI KIDO,[3] and LUDEK VECSEY[4]

Abstract — We present imaging results in geophysics based on using multidimensional Gaussian wavelets as a filter in a 2-D Cartesian domain. Besides decomposing the field into various distinct lengthscales, we have also constructed the 2-D maps describing the spatial distributions of the maximum of the wavelet-transformed L_2-norm $E_{max}(x,y)$ and its corresponding local wavenumber $k_{max}(x,y)$, where x and y are the Cartesian coordinates. For geoid anomalies, using a wavelet filter extending to 90 degrees, we have discerned the distinct outlines of convergent and divergent tectonic zones and have conducted a quantitative comparison of the short-wavelength gravitational anomalies at those wavelengths between two different geographical locations. We have also compared the wavelet results with a nonlinear bandpass filter in the spectral domain where a Gaussian filter with the logarithm of the degree l acting as the argument has been employed. A wavelet solution, with a length-scale corresponding to 256 degrees, would need a filter with over 400 spherical harmonics centering around $l = 157$ for an optimal spatial fit. The computational effort with the bandpass filter technique greatly exceeds those associated with wavelets. We have also shown the ability of the wavelets to analyze the vastly different scales present in high Rayleigh number convection and the mixing of passive heterogeneities driven by thermal convection. Wavelets will be a useful tool for rapid analyzing of the large multidimensional fields to be captured in many other geophysical endeavors, such as the upcoming gravity satellite missions and satellite radar interferometry images.

Key words: Multidimensional wavelets, geoid, filtering.

1. Introduction

The past decade has witnessed the rapid development of wavelets or wavelet analysis, a new mathematical tool, which began in the area of seismology (GOUPILLAUD *et al.*, 1984) and analysis (GROSSMAN and MORLET, 1984) and has been quickly adopted by diverse fields in science and engineering. In a brief period of

---

[1] Department of Geology and Geophysics and Minnesota Supercomputing Institute, University of Minnesota, Minneapolis, MN 55455-0219, U.S.A. E-mail: davey@krissy.geo.umn.edu
[2] Department Physique, Universite de Montreal, H3C 3J4, Montreal, Quebec, Canada.
E-mail: vincent@cerca.umontreal.ca
[3] Frontier Research Program in Subduction Dynamics, JAMSTEC, 2-15 Natsushima-cho, Yokosuka, 237-0061, Japan. E-mail: kido@jamstec.go.jp
[4] Department of Geophysics, Charles University, Prague 8, Czech Republic.
E-mail: vecsey@karel.troja.mff.cuni.cz

15 years it has reached a certain level of maturity as a well-defined mathematical subject with a strong interdisciplinary character, which has certainly begun to make an impact in the geophysical community (e.g., KUMAR and FOUFOLA-GEORGIOU, 1997). Basically wavelets are mathematical transformations, which allow one to tackle problems of multiple-scale character in time and space both in visualization (MALLAT, 1998), analysis and also in the solution of nonlinear partial differential equations with multiple scales (VASILYEV *et al.*, 1997).

Our interest in wavelets stems from image processing in geophysical applications. Wavelet analysis is a relatively novel mathematical technique in this respect. From its name, a wavelet transform uses a small wave of zero mean, which is scaled and translated for scanning over the space of voluminous data. For a fixed scale, however small, the zero-mean constraint makes sure that all scales substantially larger than the given scale are eliminated. Thus features of a given scale are isolated and emphasized. This feature is responsible for the ability of wavelets to act as filter-banks for sharpening blurred images (e.g., STRANG and NGUYEN, 1996) and for bringing out the essential features in complex data sets. Furthermore, wavelet has a nonuniform window and is a natural extension of the window Fourier transform. Hence it is ideal for analyzing data sets with localized and discontinuous character such as plume-like structures in tomography (BERGERON *et al.*, 1999, 2000a, b). Most works on wavelets in geophysics have been focused on time series (ALEXANDRESCU *et al.*, 1995; GIBERT *et al.*, 1998; VECSEY and MATYSKA, 2001). One-dimensional tracks involving a two-dimensional filter have been employed by SIMONS and HAGER (1997) on locating the sources of postglacial rebound from gravity anomalies. Similar work on identifying the sources of gravitational potential fields was conducted by MOREAU *et al.* (1997, 1999). However, slight work has been carried out in using multidimensional wavelets on geophysical imaging until our investigations on analyzing seismic tomography (BERGERON *et al.*, 1999, 2000 a, b) and correlation analysis between different layers in a high-resolution seismic tomographic model (PIROMALLO *et al.*, 2001).

Much of what has been published on wavelets is of considerable mathematical complexity (e.g., FREENEN and WINDHEUSER, 1997), and in this form cannot be easily applied by geophysicists. In this paper we will discuss and explain in very basic terms our wavelet imaging processing techniques using multidimensional wavelets, which are applied to geoid anomalies and other examples involving mantle convection and mixing. After seeing these examples, one should not be surprised that in the near future wavelets will be a very viable tool in many areas of geophysics.

2. Multidimensional Wavelet Filtering

In this section we will present the mathematical derivation of the formulas to be used in the 2-D wavelet filtering. We begin with the isotropic continuous wavelet transform (CWT) (e.g., FARGE, 1992), which is distinctly different from discrete wavelet transforms (DWT) (e.g., MALLAT, 1998). The CWT plays the same role as the Fourier transform and is used for analysis and feature detection in signals, whereas the DWT is the analogue of the discrete Fourier transform and is more appropriate for data compression, signal reconstruction, and inversion of seismic data (CHIAO and KUO, 2001).

The continuous wavelet transform of a discretized field variable $f(\mathbf{x})$, like the temperature field, where $\mathbf{x}$ is the independent spatial variable, is defined as the convolution of f with a scaled and translated version of the wavelet function ψ ($\mathbf{x}$, $\mathbf{b}$, a), where $\mathbf{b}$ is the position vector and a is the wavelet scale. The continuous wavelet transform of a 2-D field $f(x, y)$ is given by

$$\tilde{f}(a, \mathbf{b}) = \frac{1}{a} \iint f(\mathbf{x}) \psi^* \left(\frac{\mathbf{x} - \mathbf{b}}{a} \right) d^2 x, \tag{1}$$

where * indicates complex conjugation. By varying the wavelet length scale a and moving along $\mathbf{b}$, we can construct a picture showing both the amplitude of any features versus the characteristic scale a and how this amplitude with the scale varies along the position vector $\mathbf{b}$ over a two-dimensional domain. We have used the Gaussian function and its higher-order derivatives for constructing the two-dimensional wavelets. The two-dimensional Gaussian wavelets, called DGp, take the form

$$\psi(\mathbf{x}) = \frac{\partial^P}{\partial x_i^P} \exp \left(\frac{-|\mathbf{x}|^2}{2} \right), \tag{2}$$

where p is the order of the spatial derivative. We have tested DG2, DG4, and DG8 functions. In the main we used DG2 functions.

For accelerating the computation, the spatial convolution integral can be replaced by a considerably faster operation in the Fourier spectral domain. Thus we must transfer to the global wavenumber $\mathbf{k}$ space. The Fourier transform of the wavelet, eqn. (2) is

$$\hat{\psi}(a\mathbf{k}) = (2\pi)^{1/2} |a\mathbf{k}|^2 \exp \left(\frac{1}{2} |a\mathbf{k}|^2 \right) \tag{3}$$

where $\mathbf{k}$ is the global (Fourier) wavenumber. We denote the fast-Fourier transform (FFT) of $f(\mathbf{x})$ as $\hat{f}(\mathbf{k})$. Then the convolution in eqn. (1) becomes simply a product in the spectral space which is expressed by

$$B(\mathbf{k}, a) = a\hat{f}(\mathbf{k}) \hat{\psi} * (a\mathbf{k}). \tag{4}$$

The wavelet transform, eqn. (1), is the inverse fast Fourier transform (IFFT) of the product B from the convolution theorem or

$$\tilde{f}(a, \mathbf{b}) = \iint B(\mathbf{k}, a) e^{i\mathbf{k} \cdot \mathbf{b}} \, dk. \tag{5}$$

We note that the wavelet scale a is not necessarily equal to the Fourier wavelength. In this case the wavelet transform automatically interpolates at the desired local "scale-number," $(1/a)$ although this will not bring out more information, once the Nyquist wavelength has been reached. We have changed the scale a by altering a mode parameter m in the following empirical relationship, which allows us to dial in the proper scale a.

$$a = \frac{1}{2^{\beta m}}, \tag{6}$$

where m is a positive number and sweeps from a small to large value according to the range of the length-scales being examined and $\beta = 0.22$ has been used as the tuning parameter.

We have used the Gaussian wavelet as a filter in constructing $\tilde{f}(a, \mathbf{b})$. We have not tackled the issue of reconstruction of the Gaussian wavelet transform, which would be given by the reconstruction formula in 2-D (e.g., KAISER, 1994) as

$$f(\mathbf{x}) = \frac{1}{C_\psi} \iiint \frac{da}{a^2} d^2\mathbf{b} \, \psi_{a,b}(\mathbf{x}) \tilde{f}(a, \mathbf{b}) \tag{7}$$

where $\psi_{a,b}(\mathbf{x})$ is the mother wavelet and $\tilde{f}(a, \mathbf{b})$ is the wavelet transform, C_ψ is a constant originating from the admissibility condition which demands a zero mean. This topic will be deferred to a future study. We note that the natural measure on the space $(a, \mathbf{b})$ is $da \, d^2\mathbf{b} / a^2$. It is invariant for both translation and dilation.

There are other types of wavelets specifically suited for the geoid problem. They are taken from the Green's function of the Poisson equation and are called Poisson wavelets (HOLSCHNEIDER, 1996). A one-dimensional version of the Poisson wavelets has been presented by MOREAU *et al.* (1999). Two-dimensional Poisson wavelets can be constructed from the formulae provided in MOREAU *et al.* (1999). A comparison of the Poisson wavelets with the Gaussian wavelets for the problem of geoid filtering would lie beyond the scope of this paper.

There are several advantages of wavelets over a local Fourier filtering (e.g., SIMONS *et al.*, 2000). Foremost is the fact that wavelets are considerably faster, as you can process a complete coverage of the data with only one FFT operation. Second, it is not a bandpass filter because of the compact support nature of the shape of the Gaussian wavelets. Third, in window Fourier transform (WFT) the window is of the same length for all scales being examined. The WFT will miss scales greater and smaller than the window. WFT also causes oscillations from the Gibbs' phenomena. Finally, wavelets can interpolate at smaller scales than the Nyquist cut-off length scale. In this case, the user should state his hypothesis or use extra information from another data set.

One of the crucial issues raised by the use of multidimensional wavelets is the sea of data resulting from the extra dimension coming from the two parameters a and $\mathbf{b}$ in the wavelet transform. The wavelet transform associated with a one-dimensional time-series needs a two-dimensional map in time and period for depicting the local wavelet power spectrum (e.g., TORRENCE and CAMPO, 1998). The local spectral portrayal of isotropic two-dimensional wavelets would require 3 dimensions and for three-dimensional wavelets, 4 dimensions. This demonstrates the huge demand on memory and problems in visualization of multidimensional data structure. YUEN et al. (2000), demonstrated the tremendous complexity and difficulty involved in examining two-dimensional convection with 2-D wavelets. Very convoluted three-dimensional spectral surfaces over (a, **b**) space were generated, which could not be interpreted easily. Extending this to 3-D wavelet scalogram would be a very difficult task from a visualization point of view.

In order to address this problem of visualization of multidimensional wavelet spectra, we (BERGERON et al., 1999, 2000a) have developed a low-dimensional parameterization scheme based on the maximum of the L_2-norm of the wavelet transform at a given point (x_i, y_i), $E_{max}(x_i, y_i)$ and the associated local wavenumber, $k_{max}(x_i, y_i)$ of the local spectrum about this point. Certain features such as the maximum strength of a signal can be enhanced in the E_{max} map and the boundaries between regions with sharply varying properties are highlighted by the k_{max} map. This approach has been applied to 3-D seismic tomography (BERGERON et al., 1999, 2000b), where the outlines of mantle plumes are brought out in a clearer light. In Figure 1 we show a schematic diagram of how the E_{max} and k_{max} maps are constructed from the local spectral data. We note that sometimes there exist multiple stationary points in the $E(k)$ curve. In this case we choose the stationary point with the largest magnitude in the L_2-norm. This is a computationally intensive task, as a local spectrum must be determined at each grid point. However casting the wavelet transform in terms of E_{max} and k_{max} helps to compress the amount of information, thus making it easier to digest mentally.

To be sure, our definitions for E_{max} and k_{max} are not well founded from a mathematical point of view but we may understand this from a physical standpoint as places where extrema or singularities in the physics may occur, be it in the gravity field or in thermal convection where the boundary layers are sites with a strong concentration of thermal energy.

3. The Wavelet-filtered Geoid

Geoid anomalies are undulations of the earth's equipotential surface with respect to a reference ellipsoid and have been used to constrain the viscosity structure of the mantle over the past 15 years (HAGER, 1984; RICARD et al., 1984; FORTE and MITROVICA, 1996; KING, 1995; CADEK et al., 1995; KIDO and CADEK, 1997; CADEK and FLEITOUT, 1999; KIDO and YUEN, 2000). There are various mechanisms for producing geoid anomalies, depending on the wavelength, such as the rotation of the earth, which can cause sea-level fluctuations (SABADINI et al., 1990). Longer wavelength geoids originate the source in the lower mantle (CHASE, 1979) and shorter

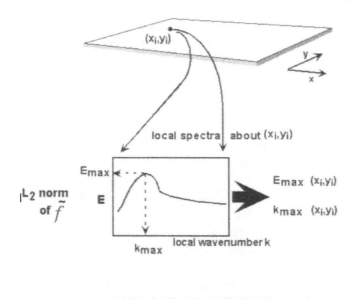

Figure 1

Schematic diagram showing the construction of maps for E_{max}, the maximum of the L_2-norm of the wavelet transform $\tilde{f}$ and k_{max}, the associated local wavenumber, from the local wavelet spectra concerning each point (x_i, y_i). E_{max} is extracted at each point along with the associated k_{max}.

wavelength anomalies are caused by heterogeneities in the lithosphere (HAGER, 1983; LE STUNFF and RICARD, 1995). Using spherical harmonic decomposition, one could use the geoid locally to study a particular area with short wavelength variations (CALMANT and CAZENAVE, 1987; BAUDRY and KROENKE, 1991; CAZENAVE *et al.*, 1995; HWANG *et al.*, 1998), intermediate wavelengths with emphasis on the transition zone (WESSEL *et al.*, 1994; KIDO and CADEK, 1997) and long wavelengths probing into the deep mantle (RICHARDS and HAGER, 1984; CAZENAVE *et al.*, 1989).

In this section the data set used (RAPP and PAULIS, 1990) corresponds to 90° in the spherical harmonics, which translates into roundly 200 km in the wavelet length scale. Because the Fourier transform is applied over the entire domain, we need to apply a window to the data set and we have selected the Parzen window for the entire Mercator projection. This window cuts out the signal at the poles although this is not a problem because the Mercator projected geoid is not usable for latitudes over about 45 degrees. This geoid is displayed in Figure 2a in the Mercator projection as described above. The horizontal axis is the longitude covering 360 degrees and the vertical axis is the latitude spanning 180 degrees. A map depicting the coastlines has been overdrawn on each panel of Figure 2. Strong maximum potential over Indonesia and strong minimum potential over India are obvious features. In this

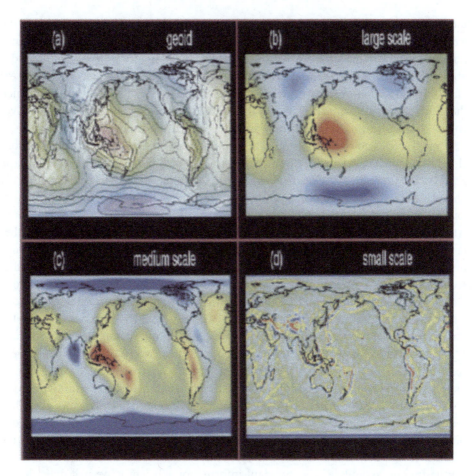

Figure 2
(a) Mercator projection of the geoid taken from RAPP and PAULIS (1990) but truncated at 90 degrees. Only the strong maximum potential over Ontong Java and strong minimum potential over southern India are distinctly visible features, in spite of a spatial resolution of about 200 km. (b) Data has been filtered with the second order Gaussian wavelet at scale a approximately 9000 km. (c) Geoid has been filtered with a Gaussian wavelet at scale a of approximately 4000 km. (d) Geoid has been filtered with a Gaussian wavelet at scale a of around 200 km.

picture, only the large scales are visible. Hereafter, the color map ranges from blue (minimum), green, yellow to red (maximum). The minima and maxima are indicated on each panel. Because we are only interested in qualitative features here, all the data displayed in the following figures have been normalized to kg m^2/sec^2.

We displayed the wavelet transformed geoids for large, intermediate and small scales respectively in Figures 2b, 2c and 2d. The wavelet filtered gravity potential (Fig. 2b) at a scale of approximately 9000 km would only show the largest scales of Figure 2a. More details are displayed when we wavelet filter the data at a wavelet length scale $a \sim 4000$ km (Fig. 2c). Strong depression of the gravity geoid appears over the southern tip of India and the west Atlantic, while strong maxima appear over the western and central Pacific and Andes. The presence of a maximum over Iceland is not relevant because of the strong distortion caused by the Mercator projection at latitudes above 45 degrees. These features are well known from previous studies (e.g., CROUGH and JURDY, 1980), however with wavelets we can selectively dial on the scale at which these features would appear. This is the important advantage held by wavelets over conventional truncated window analysis and spherical-harmonic expansion.

For a small enough resolution, *a* approximately 500 km, one can then discern clearly the subduction zones along the west coast of South America, Aleutian arc, East Pacific down in latitudes to Tonga (Fig. 2(b)) and mountain belts such as the Himalayas, and the Andes because they are formed by convergent plate processes. Over regions with extensional tectonics we can discern the African rift, the mid-Atlantic ridge, certain details of the Indian plate near Madagascar and other landmarks, such as the Hawaiian Island chain in the Pacific and the Atlas Mountains in northern Africa are also unveiled. The Australian coastline can be readily recognized on a high resolution gravity map (SANDWELL and SMITH, 1997). Europe has a more complicated geoid pattern, as there are many small tectonic provinces, which would require zooming in (see Fig. 4 below). Nearby one can discern the Zagros mountain belt, the Anatolian fault trace, and the outlines of the Arabian plate.

The local wavelet wavenumber k' is related to the inverse of the wavelet length scale a by $k' = 1/a$. The type of information produced by the local wavelet spectra analysis is indeed difficult to visualize, since at each geographical grid point we need to construct a whole spectrum. This operation (see Fig. 1) would demand at least a required memory of N^3 points, where N is the number of grid points along one direction. This approach becomes prohibitively large for a gravity data set with high resolution e.g., SMITH and SANDWELL (1997) and GRUBER *et al.* (2000). In order to synthesize and assimilate this abundant information efficiently, we will make use of $E_{\max}$ and $k_{\max}$, the two proxy quantities described above in Figure 1. Such a transformation can be viewed as a low-dimensional parameterization, a form of data compression. An additional piece of important information is the sign of the signal itself. Geoid anomaly can be negative or positive. Obviously this information is lost from using the L_2-norm. Consequently, we represent the local sign energy by using $E_{k'}$:

$$E_{k'} = \text{sign}(f(\mathbf{b}))|\tilde{f}(a, \mathbf{b})|^2. \tag{8}$$

The map for $k_{\max}$ represents the distribution of scales for which the absolute value of the gravity potential assumes a locally maximum value. This rendition can be viewed as a low-dimensional parameterization. If $k_{\max}$ were the smallest wavenumber for a given range in k, we would have a flat plateau on the $k_{\max}$ map. Variations of $k_{\max}$ take place along plate boundaries and other sites with sharp gradients and facilitate the detection of different tectonic provinces. Thus one would expect the $k_{\max}$ distribution to exhibit a sharp increase at any tectonic boundaries.

We have plotted the spatial distributions of E_{max} and k_{max} in Figures 3a,b. E_{max} has been normalized at each scale by the areal average of E_{max} for that scale. The scales range from $a = 600$ km down to $a = 200$ km. In the E_{max} map we can discern immediately a strong positive signal from the Andes, the Tibetan plateau and the Tethys-Alpine range, and a negative signal from southern India. One can also see the Hawaiian Islands and the Azores. All the plates and ridges as in Figure 3(a) are present as well on the k_{max} map. Fracture zones on the three major oceans can be picked out, as well as the boundaries of the Pacific, Nazca and North American plates.

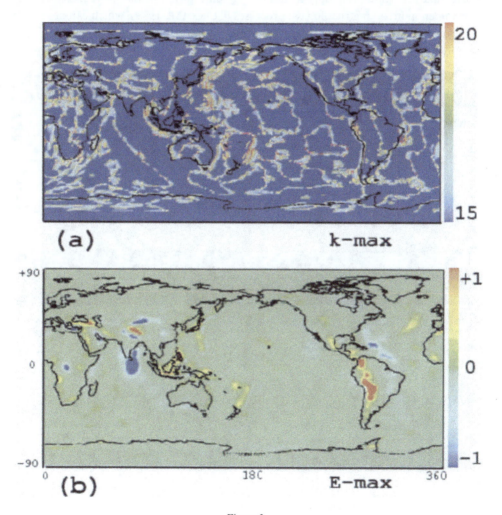

Figure 3
(a) Wavelet filtered k_{max} map of the data in the scale range ($a = 600 - 200$ km). The automatic local focusing allows that all the details of Figure 1b are present but also the plate boundaries in western U.S. and Mexico. Fracture zones on the Pacific, Atlantic and Indian Oceans can be discerned. (b) E_{max} map for scales between 200 and 600 km. Strong gradients in E_{max} are reflected in the k_{max} map below.

Also quite distinguishable are many convergent zones, such as South America, Indonesia, Japan, and Tonga (see zoom-in version in Figure 4). In fact, E_{max} gradients are related to small values in k_{max}. The boundaries of the zones associated with high E_{max} values are correlated with strong gradients and imply very small spatial scales resulting from the sharp front of the adjacent high and low gravity potential values along the ridges. This could be a criterion to limit the horizontal extent of the tectonic influence due to the oceanic ridge. The sudden appearance of coherent geoid structures at sufficiently short length scales of around 400 km, which are well correlated with plate boundaries, may be a manifestation of a self-organized critical phenomenon involving rheological processes (TURCOTTE, 1999), which can be related to plate boundary formation (e.g., BERCOVICI, 1996; REGENAUER-LIEB, and YUEN,

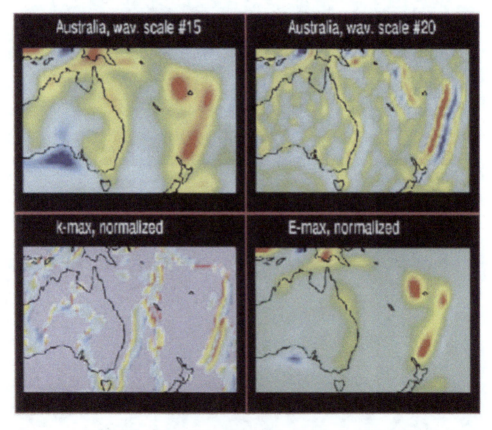

Figure 4

(a) Zoom-in of the Australian and Tongan region. Data have been filtered with the second-order Gaussian wavelet at scale a of about 600 km. Normalization is carried out with the regional maximum. (b) Zoom-in of the Australian and Tongan region. Data have been filtered with a second-order Gaussian wavelet at scale a of around 200 km. (c) Wavelet filtered k_{max} map of the zoomed-in Australian region in the scale range ($a = 600 - 200$ km). (d) Wavelet filtered E_{max} map of the zoomed-in Australian region in the scale range ($a = 600 - 200$ km).

2000). In this vein, TANIMOTO and OKAMOTO (2000) have shown that the rate of gravitational energy release from earthquakes in the last 20 years is also concentrated in narrow belts with characteristic length scales bounded by around 400 km.

Since wavelets have compact support, we can employ them to zoom into any region and renormalize the localized fields by using the regional maxima. This feature allows us to investigate regional geological questions, such as the different tectonic boundaries and mountain belts. In Figure 4, we have selected the Tonga-Australian region for this purpose of zooming-in. We have used the same scale range as in Figure 2 but with the modification for the regional renormalization. Figure 4 shows the regional geoid fields for two scales associated with 600 (mode 15) and 200 (mode 20) km. The regional E_{max} and k_{max} fields are displayed in the bottom row of Figure 4. The Tongan trench is clearly discerned. In the k_{max} map we can see the decomposition of the Australian region into three tectonic provinces which have been interpreted as continental roots from surface-wave tomography (SIMONS et al., 1999) and by isostatic response from multitaper spectral analysis (SIMONS et al., 2000). In the E_{max} map the Great Dividing Range in Eastern Australia is well delineated, as well as the Tonga subduction zone and the Ontong Java plateau.

In Figure 5 we compare regional geoid taken over Japan for spherical harmonics taken (1) between $l = 20$ and 90 (left panel) (2) a shorter window for modes between $l = 40$ and 90 (center panel) and one single wavelet pass (right panel) for a length-scale corresponding to $a = 400$ km ($m = 20$ according to eqn. (6)). Clearly the shorter spectral window (center panel) matches very closely with the geoid picture (right panel) obtained by the single wavelet. However the computational cost is considerably greater for the spectral window approach. A crucial problem in the window approach is finding the appropriate truncation level for viewing the small scales. This is indeed a tedious trial-and-error procedure, which can be extremely time-consuming. On the other hand, wavelet filtering requires a single pass and is much less painless than experimenting with different sizes of the window. This issue of windowed spectral transform versus continuous wavelet transform will be discussed in greater detail below in section 4.

Another interesting outcome from wavelet-transformed geoid is that we can compare the current local geoid values at a particular length scale for different tectonic sites around the world, in the same manner as one would employ a voltmeter to measure the relative potential difference at two different places. In Table 1 we compile a list of the relative geoidal values for some representative sites worldwide for $a = 400$ and 800 km. It is well known from global spectral analysis (KAULA, 1966) that the geoid associated with longer length scales has larger magnitudes. For $a = 800$ km, convergent belts, such as the Himalayas and Chilean Antiplano, have dominating values. What is surprising is the large negative value near Sri Lanka, which may have sublithospheric origins. At the shorter wavelet lengthscale, $a = 400$ km, active convergent zones, the Hellenic arc and the northern part of the Kermadec trench have the largest value, while the continental mountain belts, the Swiss Alps

Figure 5

Zoom-in views of the regional geoid over the Japanese region and comparison with a wavelet with a scale of 200 km. The leftmost panel consists of harmonic contributions from l − 20 to 90, the central panel contributions from $l = 40$ to 90, and the right panel contribution from one single wavelet transform with a scale a of around 200 km.

and the Rockies, have much smaller magnitudes as compared to the active convergent boundaries. Thus comparing local geoidal values by means of wavelet transforms is a new quantitative tool for measuring the relative intensity of contemporary tectonics at different locales.

4. Comparison of Wavelet Filtered Geoid with Nonlinear Spectral Filter

In this section we will advance one step further by determining the differences between a wavelet filtered geoid and a geoid filtered with a nonlinear filter in the

Table 1

Comparison of local geoidal values at different wavelet length scales

	$a = 1000$ km	$a = 400$ km
Caucus Mountains	0.31	0.033
Atlas Mountains	0.26	0.044
Carpathians	0.11	0.015
Hellenic Trench	−0.37	−0.076
Antiplano	0.54	0.10
Himalayas	0.40	0.063
Dead Sea	0.20	0.029
Zagros	0.21	0.035
Sri Lanka	−1.07	0.010
Colorado/Wyoming Rockies	...	0.025
N. Kermadec Trench	...	−0.057
S. Appenines	...	0.033
Somali Peninsula	...	0.026
Swiss Alps	...	0.022
S. Australian Rift	...	0.021

... means too small to be measured at that length scale.

spherical harmonic spectral domain. In this comparison we have employed a higher resolution geoid, EGM96, (LEMOINE *et al.*, 1997), which elevates to a spherical harmonic of 360 degrees. In Figure 6 we show the map of the EGM96 model, which is of a higher resolution than the geoid shown in Figure 2a. We can see several interesting features not evident in the previously shown geoid model. They include the high over the Antiplano region in South America and Indonesia, the negative anomalies off the east coast of India and over Hudson Bay in Canada.

The localization character of wavelets may be mimicked by the spectral method by using a moving weight function or a moving window to localize the regional features. Such an approach spatial-spectral localization of gravity and topographic data has been applied for analyzing postglacial rebound over Canada (SIMONS and HAGER, 1997) and the Venusian tectonics (SIMONS *et al.*, 1997). We have devised a nonlinear bandpass filter in the spectral domain, which is expanded in terms of spherical harmonies. This axisymmetric filter $P(l)$ given in terms of Legendre functions is portrayed in Figure 7 and is given explicitly by

$$P(l) = A(l) \exp\left(-\left(\frac{\log_2 l - \log_2 l_c}{w}\right)^2 \Big/ 2\right), \tag{9}$$

where $A(l)$ comes from the normalization constant of the Legendre function and is a montonically decaying function of l. The parameters l_c and w control the location and shape of the window function. We note that this window function is nonlinear because of its argument depending on $\log_2 l$. With this window function $P(l)$, we may write the spectrally filtered version of the geoid as

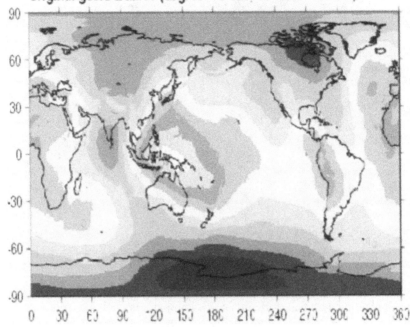

Figure 6
The geoid from the EGM96 model. This geoid model (LEMOINE *et al.*, 1997) contains contributions up to
360 degrees.

$$P(l) = \exp\left(-\left(\frac{\log l - \log l_c}{w}\right)^2 / 2\right)$$

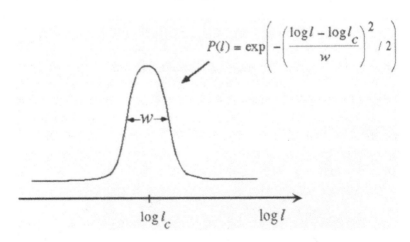

Figure 7
Nonlinear spectral filter of Legendre functions. Note that the argument of the filter is in log *l*. Hence it is
nonlinear. The center of the window is denoted by log l_c and the width of the window by *w*.

$$C(l, m) = P(l)G(l, m), \tag{10}$$

where $G(l, m)$ is the 2-D spectrum of the geoid, calculated from the EGM96 model (LEMOINE et al., 1997). From eqn. (9) we can then calculate the spatially filtered version of the geoid A_f by using the spectral coefficients C_{lm} in eqn. (9) and summing over the spherical harmonics up to the maximum degree L covered by the filter $P(l)$.

$$A_f(\theta, \phi) = \sum_{l,m}^{L} C_{lm} Y_{lm}(\theta, \phi), \tag{11}$$

where θ and ϕ are respectively the colatitude and longitude.

A similar set of formulas has also been derived by SIMONS et al. (1997), in the construction of the filtered gravity field in the spatial domain from a spectral window function, by making use of the selection rule in the triple product of the spherical harmonics in the surface integral. We have constructed the wavelet filtered geoid field $A_w(\theta, \phi)$ by using eqn. (5) and the scale a, corresponding to the spherical harmonic of degree l. In Figure 8 we compare the filtered geoid corresponding to maximum degree $L = 32$ for both $A_w(\theta, \phi)$ and $A_f(\theta, \phi)$ for which the center of the filter l_c is 19.7 and the width is 1.0. The values of the best fit between A_w and A_f are determined by the correlation between these two field quantities. As can be observed, the fit between spectrally filtered and the wavelet filtered geoids is fairly good, but A_f is required around 100 spherical harmonics because of the nature of the width w in the nonlinear spectral filter (see Fig. 7), while A_w is calculated with one wavelet function with a length scale corresponding to $32°$ of the spherical harmonic. The computational time is more than 50 times longer for the nonlinear spectral filter and grows with L.

For obtaining the best fitting parameters l_c and w, we have calculated the correlation value C between A_w and A_f as an inner product over the surface of the sphere. This operation takes the form

$$C = \frac{\int A_w(\Omega) A_f(\Omega) d\Omega}{\sqrt{\int A_f^2 d\Omega \int A_w^2 d\Omega}}, \tag{12}$$

where Ω is the solid angle of a sphere. The best fitting values of l_c and w are obtained by fitting the maximum of the value C over the l_c and w plane. An illustration showing the surface of C for the case of the central degree $l_w = 32$ is shown in Figure 9. We see here the optimal w (ordinate) is close to 1.0, while the optimal $\log_2 l_c$ is shifted downward by 0.6 along the logarithmic abscissa. The maximum in C is moderately localized and can be easily discerned from the landscape of the function.

One of the advantages of wavelet filtering is the ability to conduct local magnification readily in a particular region by using interpolation with the localization filter. In Figure 10 we present a comparison of the features using different scales of the wavelet and then compare these images with the corresponding

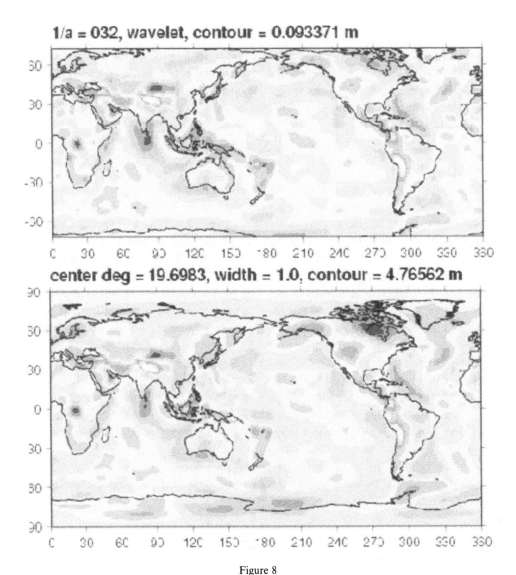

Figure 8
Comparison of the wavelet-filtered global geoid (left panel) with the spectrally filtered geoid (right panel) for a scale corresponding to $l = 32$ or a length scale of 1200 km. Contour levels are different because of different normalization between wavelets and spherical harmonics. Only the geoidal patterns can be compared.

best fit result obtained from using the nonlinear filter in the spectral domain. Because of different normalization conventions, we can only compare the geoidal patterns between wavelet-transformed and window-transformed geoids. We have surveyed scales from around 600 km down to 150 km. Inspection of Figure 10, reveals that the overall accurate shape in the geoid around Japan is a robust result, which can be

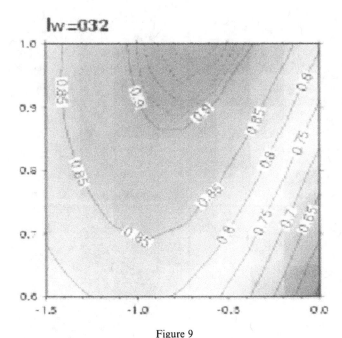

Figure 9

The sensitivity of the correlation between the wavelet-filtered geoid and the spectrally filtered geoid using eqn. (9). The ordinate and abscissa axes refer respectively to variations in w and l_c in Figure 7, depicting the filter.

captured with a length scale corresponding to $l = 64$, which would be approximately 600 km. We note here that wavelet filtered images are nearly free from Gibbs' oscillations. Conversely, bandpass filter produces oscillations for a window insufficiently long, as can be observed in the zoom-in panels associated with the spectral window (right column). This comparison of different scales also demonstrates, in a sense, the convergence of the wavelet solution for this range of scales. We also discover that it is possible to find a good fit with the nonlinear filter. However, many spherical harmonics are needed for the nonlinear filter $P(l)$. This number increases with smaller scale features. For a length scale corresponding to degree 256, one would need over 500 spherical harmonics, while for a degree of 128, one would need some 250 harmonics. In any case the computational efficiency of the wavelet transform over spectral filtering has been clearly established in this comparison, where for $l = 256$ the nonlinear spectral calculation is more than 200 times slower than the wavelet computation.

5. Other Applications of Wavelet Filtering

Thus far we have been dealing solely with the geoid dataset. Numerical modeling of dynamical processes in geophysics also produces an enormous amount of data,

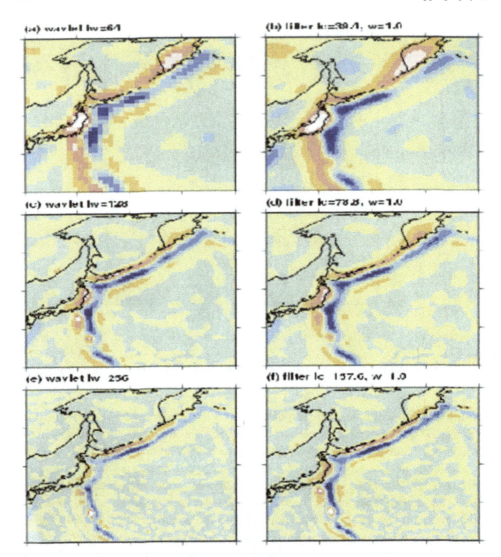

Figure 10

Comparison of the wavelet-filtered regional geoid (left column) with the corresponding spectrally filtered regional geoid over Japan (right column) for three different scales corresponding to $l = 64$ (first row), $l = 128$ (second row) and $l = 256$ (third row). The salient point dealing with the convergence in the wavelet filtered solution is illustrated by inspection of the left column in which the resolution is increased successively. Computational speed of the third row, involving panels (e) and (f), is some 200 times faster for the wavelet-filtered geoid.

sometimes 100 Gbytes for a long 3-D simulation (CSEREPES and YUEN, 2000), which must be interpreted visually and analyzed efficiently. There are other applications in geophysics in which wavelet filtering can play an important role in the analysis. They are mantle convection and mixing of passive scalar fields driven by convective flow.

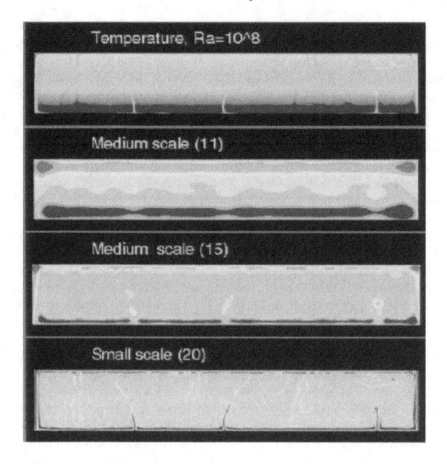

Figure 11

Temperature field (top panel) and wavelet-filtered temperature fields in thermal convection at Ra $= 10^8$. Calculations have been done in a axisymmetric spherical geometry (MOSER *et al.*, 1997) with a fourth-order finite-difference method. It has been mapped to a Cartesian geometry because the wavelets are in Cartesian geometry. The second, third and bottom panel respectively portray wavelet filtered temperature fields at scales corresponding to 1/3, 1/6 and 1/20 of the layer depth. The original temperature field has been calculated with 300 × 1500 finite difference grid points. This time-frame is taken after a quasi-equilibrium state has set in.

In Figure 11 we show the wavelet filtering of the temperature field produced in high Rayleigh number convection of an infinite Prandtl number fluid characteristic of the earth's mantle. The Rayleigh number is 10^8, which puts this in the hard turbulent convective regime (YUEN *et al.*, 1993). The number of grid points with fourth-order accuracy are 300 points in the vertical and 1500 points along the horizontal direction. Three distinct scales are portrayed, corresponding to the mode number $m = 11$, 15 and 20 given in equation (6). In this high Rayleigh number regime the maximum variations in the temperature are concentrated in both the vertical and horizontal boundary layers and they are brought out clearly by the small scale wavelet, corresponding to mode $m = 20$. The original temperature field is given in the top panel. For higher Rayleigh numbers than 10^8, wavelets with higher mode numbers are required to pick up the thin tendrils from the thermal boundary layers. There is a distinct transition in the morphology of the thermal fields from mode 11 to 15 (second to third panel) where the boundary-layer character is brought out. Knowledge of the spatial distribution of the large-scale component of the thermal fields is important for fundamental issues in geodynamics, such as global heat-transfer and geoid anomalies of the lower mantle. In short, a picture of convection painted with the panels involving the dynamics at different length scales can shed more interesting light than looking at all the scales simultaneously as in a single snapshot of the entire field or statistical analysis using correlation lengths (PUSTER et al., 1995).

Mixing of passive scalar fields driven by thermal convection (e.g., TEN et al., 1996, 1997) is another appropriate application for wavelet filtering. Wavelet transform has already been applied to two-dimensional dye concentration stirred up in turbulent jets (EVERSON et al., 1990). In that pioneering study the details of the inner structure of the jet were revealed by two-dimensional Gaussian wavelets. In Figure 12 we show the decomposition of a mixed passive scalar field driven by convection in temperature-dependent viscosity fluid (TEN et al., 1997) at a volumet-rically averaged Rayleigh number of 2×10^6. Because of downward cascade in the mixing process, the scalar field exhibits turbulent character (TEN et al., 1997). We have used 1000×1500 bicubic spline points in solving the hyperbolic equation governing the advection of passive scalar field by the velocity field taken from the convective flow. The original mixing frame is located at the top left panel, with the large scale lying in the top right panel and the next two smaller scales on the bottom panel. Again we can see the smallest scales capture the dynamics of the turbulent-like mixing at this stage of the process. There are still large clumps of heterogeneities persisting up to this time, as well as traces of intermediate scale heterogeneities. Wavelets are well suited to revealing the self-similar structures due to fractal features (TEN et al., 1997) and other inner structures of mixing. Looking from the largest to the smallest scales we see basically two types of structures present, the large-scale clumps and the small scale filaments, which act like streak lines for the small-scale flow. We note that the E_{max} and k_{max} maps of mixing can provide complementary information on the style of mixing at different scales, which have important ramifications in relating geochemical isotopic heterogeneities to length scales of convective mixing in the earth's upper mantle (e.g., GRAHAM et al., 2001).

6. Concluding Remarks

Unquestionably wavelet analysis is useful for problems in many applied disciplines as well as within mathematics itself (e.g., BURKE-HUBBARD, 1998), which

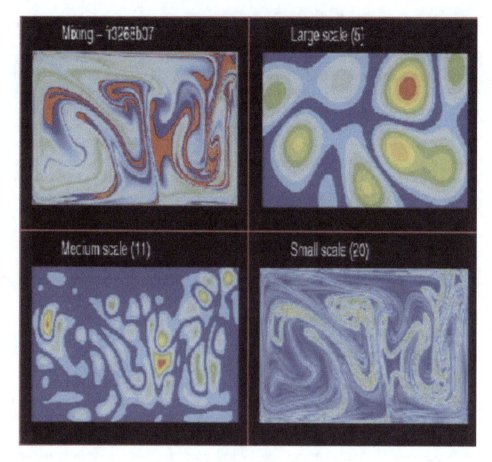

Figure 12
Wavelet-filtered mixing fields. The original mixing field is in the top left panel, while the wavelet-filtered fields with successively smaller length scales are portrayed respectively in the top right, bottom left and bottom right panels. The three scales would correspond to 1/3, 1/6 and 1/20 of the layer depth. The passive scalar field is driven by thermal convection in a temperature-dependent viscosity at a volumetrically averaged Rayleigh number of about 2×10^6. The original field of mixing has been calculated by solving the hyperbolic advection equation governing the evolution a passive scalar field using 1000×1500 bicubic spline points (TEN et al., 1996).

tells us that it possesses a special quality. Wavelet analysis provides a systematic new way to represent and analyze multiscale structures found in natural phenomena. Geophysics offers no exceptional barrier to the flow of new ideas generated from the use of wavelets.

In this paper we have presented our work carried out over the past year relative to using wavelets as a filtering tool in geophysics with emphasis on the geoidal field. We have demonstrated that wavelets can very easily pick up the sites of plate convergence and divergence, which would require many harmonics in a bandpass filtered spectral window. The earth's magnetic field is potentially another area where wavelets can shed much light, especially over continental interiors and margins.

The near future holds many potential usage's of wavelets with the upcoming gravity satellite missions, CHAMP and GRACE, which will produce time-dependent high resolution gravity fields, which must be interpreted on a weekly basis. Some new geoid model, the GRIM5-C1 (GRUBER *et al.*, 2000), is already available as a reference model for processing the CHAMP data. Isotropic spherical wavelets can considerably aid in the task of immensely compressing the gravity data by using a few useful local basis functions instead of the numerous spherical harmonic coefficients. The use of E_{max} and k_{max} maps can help us to carry out pattern recognition for earthquake prone regions from GRACE and CHAMP gravity missions. Instead of processing all of the voluminous information, we can use low-dimensional parameterization of E_{max} and k_{max} maps to pick up the strongest time-varying signals: for instance the monthly history of the maximum gravitational perturbation energy at the relevant scale of several hundred kilometers, in areas of high earthquake risk such as the plate margins. This type of strategy is in the spirit of the wavelet data compression procedure, and the spatial-temporal fluctuations of E_{max} and k_{max} fields can be compared with the modeling changes of the gravitational potential energy by earthquakes (TANIMOTO and OKAMOTO, 2000). Besides monitoring the global environment, wavelets will be of immense value in understanding the complicated spatial-temporal patterns generated in the large-scale numerical modeling of earthquake dynamical processes (e.g., BEN-ZION *et al.*, 1999; RUNDLE *et al.*, 2000). Inversion of wavelet results by means of onthogonal wavelets would entail the usage of multi-resolution analysis (FREENEN and WINDHUSER, 1996; CHIAO and KUO, 2001).

We emphasize that, as many geophysical processes, such as convection and mixing, are multiple-scale in nature because of the intrinsic nonlinearity, the analysis, inversion, and visualization of these phenomena can be greatly facilitated by the multiple-scale representation provided by wavelets. They allow for a systematic and efficient representation of a wide class of functions which can be easily constructed, such as the Gaussian function and its higher order derivatives. In sum, wavelets can open up many new possibilities in both forward and inverse modeling of geophysical phenomena.

Acknowledgments

We are grateful for numerous discussions with our Canadian colleague Dr. Stephen Y. Bergeron concerning many issues pertaining to wavelets and other matters, and also Cathy Hier for having stimulated us to look at scalograms in multidimensional convection. Other people we would like to acknowledge are Oleg Vasilyev, Claudia Piromallo, Fabien Dubuffet, Gordon Erlebacher, Shige Maruyama and Tetsu Seno for various discussions. This research has been supported by the DOE program in Complex Fluids, the geophysics program of the NSF, the NSF International program for Japan-U.S. and a NATO grant.

REFERENCES

ALEXANDRESCU, M., GIBERT, D., HULOT, G., LE MOUEL, J.-L., and SARACCO, G. (1995), *Detection of Geomagnetic Jerks Using Wavelet Analysis*, J. Geophys. Res. *100*, 12,557–12,572.

BAUDRY, N. and KROENKE, L. (1991), Intermediate Wavelength (400-600 km) *South Pacific Geoidal Undulations: Their Relationship to Linear Volcanic Chains*, Earth Planet. Sci. Lett. *102*, 430–443.

BEN-ZION, Y., DAHMEN, K., LYAKHOVSKY, V., ERTAS, D., and AGNON, A. (1999), *Self-driven Mode Switching of Earthquake Activity on a Fault System*, Earth and Planet. Sci. Lett. *172*, 11–21.

BERCOVICI, D. (1996), *Plate Generation in a Simple Model of Lithosphere-Mantle with Dynamic Self-lubrication*, Earth Planet. Sci. Lett. *144*, 41–51.

BERGERON, S. Y., VINCENT, A. P., YUEN, D. A., TRANCHANT, B. J. S., and TCHONG, C. (1999), *Viewing Seismic Velocity Anomalies with 3-D Continuous Gaussian Wavelets*, Geophys. Res. Lett. *26*, (15), 2311–2314.

BERGERON, S. Y., YUEN, D. A., and VINCENT, A. P. (2000a), *Looking at the Inside of the Earth with 3-D Wavelets: A New Pair of Glasses for Geoscientists*, Electr. Geosci. 5:3.

BERGERON, S. Y., YUEN, D. A., and VINCENT, A. P. (2000b), *Capabilities of 3-D Wavelet Transforms to Detect Plume-like Structures from Seismic Tomography*, Geophys. Res. Lett. *27*, (20), 3433–3436.

BURKE-HUBBARD, B., *The World according to Wavelets*, Second ed. (A. K. Peters, Wellesley, Ma, 1998).

CADEK, O. P., KYVALOVA, H., and YUEN, D. A. (1995), *Geodynamical Implications from the Correlation of Surface Geology and Seismic Tomographic Structure*, Earth Planet. Sci. Lett. *136*, 615–627.

CADEK, O. P. and FLEITOUT, L. (1999), *A Global Geoid Model with Imposed Plate Velocities and Partial Layering*, J. Geophys. Res. *104*, (B12), 29,055–29,075.

CAZENAVE, A., SOURIAU, A., and DOMIAH, K. (1989), *Global coupling of Earth Surface Topography with Hotspots, Geoid and Mantle Heterogeneities*, Nature 340, 54–57.

CAZENAVE, A., PARSONS, B., and CALCAGNO, P. (1995), *Geoid Lineations of 1000 km Wavelength over the Central Pacific*, Geophys. Res. Lett. *22*, (2), 97–1000.

CALMANT, S. and CAZENAVE, A. (1987), *Anomalous Elastic Thickness of the Oceanic Lithosphere in the South-central Pacific*, Nature *328*, 236–238.

CHASE, C. G. (1979), *Subduction, the Geoid, and Lower Mantle Convection*, Nature *282*, 464–468.

CHIAO, L.-Y. and KUO, B.-Y. (2001), *Multiscale Seismic Tomography*, Geophys. J. Int. *145*, 517–527.

CROUGH, S. T. and JURDY, D. M. (1980), *Subducted Lithosphere, Hotspots and the Geoid*, Earth Planet. Sci. Lett. *48*, 15–22.

CSEREPES, L. and YUEN, D. A. (2000), *On the Possibility of a Second Kind of Mantle Plume*, Earth Planet. Sci. Lett. *183*, 61–71.

EVERSON, R., SIROVICH, L., and SREENIVASAN, K. R. (1990), *Wavelet Analysis on the Turbulent Jet*, Phys. Lett. A *145*, Nos. 6 and 7, 314–322.

FARGE, M. (1992), *Wavelet Transforms and their Applications to Turbulence*, Annu. Rev. Fluid Mech. *24*, 395–457.

FORTE, A. M. and MITROVICA, J. X. (1996), *New Inferences of Mantle Viscosity from Joint Inversion of Long-wavelength Mantle Convection and Post-glacial Rebound Data*, Geophys. Res. Lett. *23*, 1147–1150.

FREENEN, W. and WINDHUSER, U. (1996), *Spherical Wavelet Transform and its Discretization*, Adv. Comput. Math. *5*, 51–94.

FREENEN, W. and WINDHUSER, U. (1997), *Combined Spherical Harmonic and Wavelet Expansion – A Future Concept in Earth's Gravitational Determination*, Appl. Comput. Harmon. Anal. *4*, 1–37.

GIBERT, D., HOLSCHNEIDER, M., and LE MOUEL, J.-L. (1998), *Wavelet Analysis of the Chandler Wobble*, J. Geophys. Res. *103*, (B11), 27,069–27,089.

GOUPILLAUD, P., GROSSMANN, A., and MORLET, J. (1984), *Cycle-Octave and Related Transforms in Seismic Signal Analysis*, Geoexploration *23*, 85–102.

GRAHAM, D. W., LUPTON, J. E., SPERA, F. J., and CHRISTIE, D. M. (2001), *Upper-mantle Dynamics Revealed by Helium Isotope Variations along the Southeast Indian Ridge*, Nature *409*, 701–703.

GROSSMAN, A. and MORLET, J. (1984), *Decomposition of Hardy Functions into Square Integrable Wavelets of Constant Shape*, SIAM J. Math. Anal. *15*, 723–736.

GRUBER, T., BODE, A., REIGBER, C., SCHWINTZER, P., BALMINO, G., BIANCALE, R., and LEMOINE, J.-M. (2000), GRIM5-C1: *Combination Solution of the Global Gravity Field to Degree and Order* 120, Geophys. Res. Lett. *27*, (24), 4005–4008.

HAGER, B. H. (1983), *Global Isostatic Geoid Anomalies for Plate and Boundary Layer Models of the Lithosphere*, Earth Planet. Sci. Lett. *63*, 97–109.

HAGER, B. H. (1984), *Subducted Slabs and the Geoid: Constraints on Mantle Rheology and Flow*, J. Geophys. Res. *89*, 6003–6016.

HOLSCHNEIDER, M., *Wavelets: An Analysis Tool* (Clarendon, Oxford, England, 1995).

HWANG, C., KAO, E.-C., and PARSONS, B. (1998), *Global Derivation of Marine Gravity Anomalies from Seasat, Geosat, ERS-1 and TOPEX/POSEIDON Altimeter Data*, Geophys. J. Int. *134*, 449–459.

KAISER, G., *A Friendly Guide to Wavelets* (Birkhauser, 1994).

KAULA, W. M., *Theory of Satellite Geodesy* (Blaisdell, London, 1966.)

KIDO, M. and CADEK, O. (1997), *Inferences of Viscosity from the Oceanic Geoid: Indication of a Low Viscosity Zone Below the 660-km Discontiuity*, Earth Planet. Sci. Lett. *151*, 125–138.

KIDO, M. and YUEN, D. A. (2000), *The Role Played by a Low Viscosity Zone under a 660 km Discontinuity in Regional Mantle Layering*, Earth Planet. Sci. Lett. *181*, 573–583.

KING, S. D. (1995), *Radial Models of Mantle Viscosity: Results from a Genetic Algorithm*, Geophys. J. Int. *122*, 725–734.

KUMAR, R. and FOUFOULA-GEORGIOU, E. (1997), *Wavelet Analysis for Geophysical Applications*, Rev. Geophysics *35*, 385–412.

LEMOINE, F. G., SMITH, D. E., KUNZ, L., SMITH, R., PAVLIS, E. C., PAVLIS, N. K., KLOSKO, S. M., CHINN, D. S., TORRENCE, M. H., WILLIAMSON, R. G., COX, C. M., RACHLIN, K. E., WANG, Y. M., KENYON, S. C., SALMAN, R., TRIMMER, R., RAPP, R. H., and NEREM, R. S., In *Gravity, Geoid and Marine Geodesy* Vol. 117, (ed. J. Segawa, H. Fujimoto, and S. Okubo) (International Association of Geodesy Symposia, 1997) pp. 461–469.

LE STUNFF, Y. and RICARD, Y. (1995), *Topography and Geoid due to Lithospheric Mass Anomalies*, Geophys. J. Int. *122*, 982–990.

MALLAT, S., *A Wavelet Tour of Signal Processing* (Academic Press, Boston, 1998).

MOREAU, F., GIBERT, D., HOLSCHNEIDER, M., and SARACCO, G. (1997), *Wavelet Analysis of Potential Fields*, Inverse Problems *13*, 165–178.

MOREAU, F., GIBERT, D., HOLSCHNEIDER, M., and SARAACO, G. (1999), *Identification of Sources of Potential Fields with the Continuous Wavelet Transform: Basic Theory*, J. Geophys. Res. *104*, (B3), 5003–5013.

MOSER, J., YUEN, D. A., LARSEN, T. B., and MATYSKA, C. (1997), *Dynamical Influences of Depth-dependent Properties on Mantle Upwellings and Temporal Variations of the Moment of Inertia*, Phys. Earth Plan. Int. *102*, 153–170.

PIROMALLO, C., VINCENT, A. P., YUEN, D. A., and MORELLI, A. (2001), *High-resolution tomography picture of the transition zone under Europe using cross-spectra involving two-dimensional wavelets*, Phys. Earth Planet. Inter. *125*, 125–139.

PUSTER, P., JORDAN, T. H., and HAGER, B. H. (1995), *Characterization of Mantle Convection Experiments Using Two Point Correlation Functions*, J. Geophys. Res. *100*, 6351–6365.

RAPP, R. H. and PAULIS, N. K. (1990), *The Development and Analysis of Geopotential Coefficient Models to Spheric Harmonic 360 degrees*, J. Geophys. Res. *95*, 21885–21911.

REGENAUER-LIEB, K. and YUEN, D. A. (2000), *Fast mechanisms for the Formation of New Plate Boundaries*, Tectonophysics *321*, (1), 53–67.

RICARD, Y., FLEITOUT, L., and FROIDEVAUX, C. (1984), *Geoid Heights and Lithospheric Stresses for a Dynamic Earth*, Ann. Geophys. *2*, 267–286.

RICHARDS, M. A. and HAGER, B. H. (1984), *Geoid Anomalies in a Dynamic Earth*, J. Geophys. Res. *89*, 5987–6002.

RUNDLE, J. B., KLEIN, W., TIAMPO, K., and GROSS, S. (2000), *Linear Pattern Dynamics in Nonlinear Threshold Systems*, Phys. Rev. E, *61*, (3), 2418–2143.

SABADINI, R., DOGLIONI, C., and YUEN, D. A. (1990), *Eustatic Sea Level Fluctuations Induced by Polar Wander*, Nature *345*, 708–710.

SANDWELL, D. T. and SMITH, W. H. F. (1997), *Marine Gravity from Geosat and ERS-1 Altimetry*, J. Geophys. Res. *102*, 10,039–10,054.

SIMONS, F. J., ZIELHUIS, A., and VAN DER HILST, R. D. (1999), *The Deep Structure of the Australian Continent Inferred from Surface Wave Tomography*, Lithos *48*/1–4, 17–43.

SIMONS, F. J., ZUBER, M. T., and KORENAGA, J. (2000), *Isostatic Response of the Australian Lithosphere: Estimation of Effective Elastic Thickness and Anisotropy Using Multitaper Spectral Analysis*, J. Geophys. Res. *105*, (B8), 19,163–19,184.

SIMONS, M. and HAGER, B. H. (1997), *Localization of the Gravity Field and the Signature of Glacial Rebound*, Nature *390*, 500–504.

SIMONS, M., SOLOMON, S. C., and HAGER, B. H. (1997), *Localization of Gravity and Topography: Constraints on the Tectonics and Mantle Dynamics of Venus*, Geophys. J. Int. *131*, 24–44.

SMITH, W. H. F. and SANDWELL, D. T. (1997), *Global Sea Floor Topography from Satellite Altimetry and Ship Depth Soundings*, Science *277*, 1956–1962.

STRANG, G. and NGUYEN, T. *Wavelets and Filter Banks* (Wellesley- Cambridge Press, 1996).

TANIMOTO, T. and OKAMOTO, T. (2000), *Change of Crustal Potential Energy by Earthquakes: An Indicator for Extensional and Compressional Tectonics*, Geophys. Res. Lett. *27*, (15), 2313–2316.

TEN, A., YUEN, D. A., LARSEN, T. B., and MALEVSKY, A. V. (1996), *The Evolution of Material Surfaces on Convection with Variable Viscosity as Monitored by a Characteristics-based Method*, Geophys. Res. Lett. *23*, 2001–2004.

TEN, A., YUEN, D. A., PODLADCHIKOV, Yu., LARSEN, T. B., PACHEPSKY, E., and MALVESKY, A. V. (1997), *Fractal Features in Mixing of Non-Newtonian and Newtonian Mantle Convection*, Earth Planet. Sci. Lett. *146*, 401–414.

TORRENCE, C. and COMPO, G. P. (1998), *A Practical Guide to Wavelet Analysis*, Bull. Am. Met. Soc. *79*, 61–78.

TURCOTTE, D. L. (1999), *Self-organized Criticality*, Rep. Prog. Phys. *62*, 1377–1429.

VASILYEV, O. V., YUEN, D. A., and PAOLUCCI, S. (1997), *Solving PDE's Using Wavelets*, Computers in Physics *11*, (5), 429–435.

VECSEY, L. and MATYSKA, C. (2001), *Wavelet Spectra and Chaos in Thermal Convection Modelling*, Geophys. Res. Lett. *28*, (2), 395–398.

WESSEL, P., BERCOVICI, D., and KROENKE, L. W. (1994), *The Possible Reflection of Mantle Discontinuities in Pacific Geoid and Bathymetry*, Geophys. Res. Lett. *21*, 1943–1946.

YUEN, D. A., HANSEN, U., ZHAO, W., VINCENT, A. P., and MALEVSKY A. V. (1993), *Hard Turbulent Thermal Convection and Thermal Evolution of the Mantle*, J. Geophys. Res. 98, (E3), 5355–5373.

YUEN, D. A., VINCENT, A. P., BERGERON, S. Y., DUBUFFET, F., TEN, A. A., STEINBACH, V. C. and STARIN, L., *Crossing of Scales and Nonlinearities in Geophysical Processes*, In *Problems in Geophysics for the New Millenium* (eds. Boschi, E. Ekström G. and Morelli A.) (Editrice Compositori, Bologna, Italy, 2000) pp. 403–462.

(Received February 20, 2001, revised June 11, 2001, accepted June 15, 2001)

To access this journal online:
http://www.birkhauser.ch

Pure appl. geophys. 159 (2002) 2311–2333
0033–4553/02/102311–23 $ 1.50 + 0.20/0

▌Pure and Applied Geophysics

Large Amplitude Folding in Finely Layered
Viscoelastic Rock Structures

HANS-BERND MÜHLHAUS,[1] LOUIS MORESI,[1]
BRUCE HOBBS,[1] and FRÉDÉRIC DUFOUR[1]

Abstract—We analyze folding phenomena in finely layered viscoelastic rock. Fine is meant in the sense that the thickness of each layer is considerably smaller than characteristic structural dimensions. For this purpose we derive constitutive relations and apply a computational simulation scheme (a finite-element based particle advection scheme; see MORESI *et al.*, 2001) suitable for problems involving very large deformations of layered viscous and viscoelastic rocks. An algorithm for the time integration of the governing equations as well as details of the finite-element implementation is also given. We then consider buckling instabilities in a finite, rectangular domain. Embedded within this domain, parallel to the longer dimension we consider a stiff, layered plate. The domain is compressed along the layer axis by prescribing velocities along the sides. First, for the viscous limit we consider the response to a series of harmonic perturbations of the director orientation. The Fourier spectra of the initial folding velocity are compared for different viscosity ratios.

Turning to the nonlinear regime we analyze viscoelastic folding histories up to 40% shortening. The effect of layering manifests itself in that appreciable buckling instabilities are obtained at much lower viscosity ratios (1:10) as is required for the buckling of isotropic plates (1:500). The wavelength induced by the initial harmonic perturbation of the director orientation seems to be persistent. In the section of the parameter space considered here elasticity seems to delay or inhibit the occurrence of a second, larger wavelength.

Finally, in a linear instability analysis we undertake a brief excursion into the potential role of couple stresses on the folding process. The linear instability analysis also provides insight into the expected modes of deformation at the onset of instability, and the different regimes of behavior one might expect to observe.

Key words: Folding, instability, viscoelasticity, layered material, cosserat continuum, finite elements, particle in cell.

1. Introduction

A common feature of mechanical systems producing infrequent, catastrophic releases of free energy such as earthquakes or rockbursts is a shared capacity to store energy over prolonged periods. This capacity is absent from models with purely dissipative, viscous rheologies which have conventionally been used to simulate

---
[1] CSIRO Division of Exploration and Mining, 39 Fairway, Nedlands WA 6009, Australia.
E-mails: hans@ned.dem.csiro.au; l.moresi@ned.dem.csiro.au; bruce.hobbs@per.dem.csiro.au; frederic@ned.dem.csiro.au

large-deformation geodynamic processes such as mantle convection, large amplitude folding, necking and so on. On the largest scale, catastrophic transitions in plate motion may occur when accumulated gravitational potential energy in the cool thermal boundary layer is released when the yields strength of the lithosphere is exceeded. On a smaller scale, the vigor and the geometric nature of the buckling event depend crucially on the capacity of the folding structure to store and release elastic energy.

Overview

In the following section we develop a mechanical model including a large deformation formulation for viscoelastic, multi-layered rock. The formulation is devised with the goal of describing materials with fine internal layering, which can be described by a single director orientation. This constitutive model is new; specifically designed for geological deformation problems involving very large deformations. Although there are more general descriptions possible, this formulation is, in fact, very broadly applicable to crustal rocks, where the preponderance of layering arises from deposition of one rock type onto another under the inescapable control of gravity. Indeed, one of the most enduring tenets of geology is the existence of an organized stratigraphy in the rock record. Another generalization we might make about the mechanical origins of geological formations is that deformation is almost certain to involve very high strain. For example, Figure 1 shows the large-amplitude folding of multi-layered rock on the scale of a few tens of centimeters. The macroscopic deformation is assumed volume preserving for convenience.

In the following two sections we turn to a computational method capable of following the evolution of elastic stresses, macroscopic interfaces and the internal layering direction introduced in the constitutive relationship of section 2. The method needs to be able to deal with very large strains associated with large-amplitude folding, while faithfully tracking material history and interfaces. Versatility and robustness are usually associated with the various formulations of the Finite Element Method (FEM). The need to track material history strongly suggests a Lagrangian formulation, which provides a reference frame, locked with the material itself. Unfortunately, large deformations are quite difficult to handle within the FEM because mesh distortion and remeshing are required to maintain optimal element configurations.

In section 3 we revisit the basic finite element formulation for viscous materials and demonstrate how the standard element vectors and matrices can be extended to include anisotropy and elasticity. In section 4 we describe an extension to standard finite element methods, which incorporates moving integration points to carry director orientation and other history variables.

The Particle-In-Cell (PIC) finite element method, as this technique is known, is a hybrid scheme which falls somewhere between the Finite Element Method (FEM)

Figure 1
Large-amplitude folding in a multilayered rock.

and a purely Lagrangian particle method such as the Discrete Element Method (DEM). The PIC scheme attempts to combine the versatility of the continuum FEM with the geometrical flexibility of DEM.

In Section 5, we explore scenarios from global to internal buckling and from elastic to purely viscous folding in nonlinear finite element studies. It turns out that the characteristic length scale of the emerging folding pattern tends to zero with increasing relaxation time. An explanation for this is attempted in section 6 within the framework of a linear stability analysis.

2. Mathematical Formulation

Layered materials are ubiquitous in geological formations. Accordingly, virtually every mathematically-minded structural geologist has, at some stage, contributed to this field (CHAPPLE, 1969; FLETCHER, 1974, 1982; JOHNSON and FLETCHER, 1994; SCHMALHOLZ and PODLADCHIKOV, 1999; HUNT et al., 1997; VASILYEV et al., 1998). Layering may be caused by purely mechanical, hydro-mechanical or chemo-mechanical means (e.g., WILLIAMS, 1972; ROBIN, 1979; ORTOLEVA, 1994). From a mechanical point of view, the salient feature of such materials is that there exists a distinguished orientation given by the normal vector field $n_i(x_k, t)$ of the layer planes, where (x_1, x_2, x_3) are Cartesian coordinates, and t is the time. Initially we assume linear viscous behavior and designate with η the normal viscosity and η_S the shear viscosity in the layer planes normal to n_i. The orientation of the normal vector, or director as it is sometimes called in the literature on oriented materials, changes with deformation. Using a standard result of continuum mechanics, the evolution of the director of the layers is described by

$$\dot{n}_i = W^n_{ij} n_j \quad \text{where} \quad W^n_{ij} = W_{ij} - (D_{ki}\lambda_{kj} - D_{kj}\lambda_{ki}) \quad \text{and} \quad \lambda_{ij} = n_i n_j \quad (1)$$

where $\mathbf{L} = \mathbf{D} + \mathbf{W}$ is the velocity gradient, $\mathbf{D}$ is the stretching and $\mathbf{W}$ is the spin. The superscripted n distinguishes the spin $\mathbf{W}^n$ of the director $\mathbf{n}$ (the unit normal vector of the deformed layer surfaces) from the spin $\mathbf{W}$ of an infinitesimal volume element dV of the continuum. The 2-D matrix representation of (1) as needed for our computational applications is represented in Appendix A.2 for easy reference. We define a corotational stress rate as:

$$\dot{\sigma}^n_{ij} = \dot{\sigma}_{ij} - W^n_{ik}\sigma_{kj} + \sigma_{ik}W^n_{kj} \ . \quad (2)$$

Again the superscripted n distinguishes the stress rate $\dot{\sigma}^n$ as observed by an material observer corotating with the director $\mathbf{n}$ from the material stress rate $\dot{\sigma}$ observed by a spatially fixed observer.

Specific Viscous and Viscoelastic Constitutive Relations

We consider layered, viscous and viscoelastic materials. The layering may be in the form of an alternating sequence of hard and soft materials or in the form of a superposition of layers of equal width of one and the same material, which are weakly bonded along the interfaces. We designate the normal shear modulus and the normal shear viscosity as μ and η, respectively; the shear modulus and the shear viscosity measured in simple, layer parallel shear we designate as μ_S and η_S.

In the following simple model for a layered viscous material we correct the isotropic part $2\eta D'_{ij}$ of the model by means of the Λ tensor (see Appendix A.1 for derivation) to consider the mechanical effect of the layering; thus

$$\sigma_{ij} = 2\eta D'_{ij} - 2(\eta - \eta_S)\Lambda_{ijlm}D'_{lm} - p\delta_{ij} \ , \quad (3)$$

where a prime designates the deviator of the respective quantity, and

$$\Lambda_{ijkl} = \left(\frac{1}{2}(n_i n_k \delta_{lj} + n_j n_k \delta_{il} + n_i n_l \delta_{kj} + n_j n_l \delta_{ik}) - 2n_i n_j n_k n_l\right) \ . \quad (4)$$

Similarly, a viscoelastic constitutive relationship for a layered medium may be written as:

$$D'_{ij} = \frac{1}{2\mu}\dot{\sigma}^{n\prime}_{ij} + \frac{1}{2\eta}\sigma'_{ij} + \left(\frac{1}{2\mu_S} - \frac{1}{2\mu}\right)\Lambda_{ijkl}\dot{\sigma}^{n\prime}_{kl} + \left(\frac{1}{2\eta_S} - \frac{1}{2\eta}\right)\Lambda_{ijkl}\sigma'_{kl} \ , \quad (5)$$

where $\dot{\sigma}^n_{ij}$ is the corotational stress rate introduced at the beginning of this section. We could have equally well used the Jaumann derivative of σ which is obtained by replacing $\mathbf{W}^n$ in (2) by the spin of an infinitesimal element of the continuum $\mathbf{W}$. The present choice seems more natural in the context of layered continua and also leads to algebraically simpler expressions in the linear stability analysis in section 6. A remark on the notation: We use index notation which is less ambiguous than symbolic notation when vectors, second and fourth-order tensors (such as Λ_{ijkl}) appear simultaneously in the equations.

3. Finite Element Formulation

The constitutive relationships derived in the previous section translate naturally into standard finite element matrix formulation for almost incompressible materials as follows:

$$\begin{pmatrix} \mathbf{K} & \mathbf{G} \\ \mathbf{G}^{\mathbf{T}} & \mathbf{0} \end{pmatrix} \begin{pmatrix} \mathbf{u} \\ \mathbf{p} \end{pmatrix} = \begin{pmatrix} \mathbf{F} \\ \mathbf{0} \end{pmatrix} . \tag{6}$$

$\mathbf{K}$ is the so-called global stiffness matrix which contains all the material property parameters, $\mathbf{G}$ is the divergence expressed in matrix form, $\mathbf{u}$ and $\mathbf{p}$ are the unknown velocities and pressure respectively, and $\mathbf{F}$ is a vector of driving terms comprising body forces and surface tractions.

The matrices $\mathbf{K}$ and $\mathbf{G}$ are global matrices composed in the usual way of elemental matrices; in the following we designate element matrices and vectors by a superscripted E. The components of an element stiffness matrix may be written as

$$\mathbf{K}^E = \int_{\Omega^E} \mathbf{B}^T(\mathbf{x})\mathbf{C}(\mathbf{x})\mathbf{B}(\mathbf{x})d\Omega . \tag{7}$$

The matrix $\mathbf{B}$ consists of the appropriate gradients of interpolation functions which transform nodal point velocity components to strain-rate pseudo-vectors at any point in the element domain.

The constitutive operator corresponding to (3) is composed of two parts $\mathbf{C} = \mathbf{C}_{\text{iso}} + \mathbf{C}_{\text{layer}}$ representing the isotropic part of the constitutive model and a correction term considering the influence of layering an. In two dimensions,

$$\mathbf{C}_{\text{iso}} = 2\eta \begin{bmatrix} 1 & & \\ & 1 & \\ & & \frac{1}{2} \end{bmatrix}; \quad \mathbf{C}_{\text{layer}} = -2\eta \left(\frac{\eta_S}{\eta} - 1 \right) \begin{bmatrix} -\Delta_0 & \Delta_0 & -\Delta_1 \\ \Delta_0 & -\Delta_0 & \Delta_1 \\ -\Delta_1 & \Delta_1 & -\frac{1}{2} + \Delta_0 \end{bmatrix} \tag{8}$$

in which $\Delta_0 = 2n_1^2 n_2^2$ and $\Delta_1 = (n_1 n_2^3 - n_1^3 n_2)$.

A viscoelastic equivalent of the viscous equations can be obtained by inserting the corotational rate (2) into (5) and subsequently replace the time derivative by the corresponding first-order difference quotient. After rearranging, the constitutive equations can be written in the form (3) with the viscosities replaced by effective viscosities and an additional term which we define below representing the stress history and the influence of the stress rotation (see equations (10) and (11)). The result reads:

$$\sigma_{ij}^{t+\Delta t} = 2\eta^{eff} D_{ij}^{\prime t+\Delta t} - 2(\eta^{eff} - \eta_S^{eff})\Lambda_{ijlm}D_{ij}^{\prime t+\Delta t} + s_{ij} - p\delta_{ij} \tag{9}$$

where

$$s_{kl}^t = \frac{\eta^{eff}}{\mu\Delta t}\left(\left[\frac{1}{2}(\delta_{ki}\delta_{lj} + \delta_{kj}\delta_{li}) + \left(\frac{\eta_S^{eff}\mu}{\eta^{eff}\mu_S} - 1\right)\Lambda_{klij}\right]\left(\sigma_{ij}^t + \Delta t\left(W_{ik}^{n^t+\beta\Delta t}\sigma_{kj}^t - \sigma_{ik}^t W_{kj}^{n^t+\beta\Delta t}\right)\right)\right). \tag{10}$$

Details of the derivation of (9) are represented in the Appendix 3. The effective viscosities read:

$$\eta^{eff} = \eta\frac{\Delta t}{\alpha + \Delta t} \quad \text{and} \quad \eta_S^{eff} = \eta_S\frac{\Delta t}{\frac{\eta_S}{\mu_S} + \Delta t} = \eta_S\frac{\Delta t}{\alpha_S + \Delta t}, \tag{11}$$

where the superscripted t refers to the previous time step and Δt is the time increment. The time increment is limited by a Courant condition ensuring that during one time step a particle does not travel further than a typical element dimension (this limitation ensures that changes to the stiffness matrix, $\mathbf{K}$, are small during a time step). In MORESI et al. 2001 the time increment for the advection of the material points usually differs from the time increment used for the integration of the stress history. This is necessary to ensure time scale of interest for elastic processes is determined by the physics of the problem and not by the mesh dependent Courant conditon. The parameters $\alpha = \eta/\mu$ and $\alpha_S = \eta_S/\mu_S$ are the relaxation times for pure and simple shear, respectively. The superscripted parameter $0 \le \beta \le 1$ in the corotational term in (10) was introduced to provide flexibility in the numerical treatment of the problem: during a predictor step we put $\beta = 0$ and during subsequent iterations we put $\beta = 1$.

Furthermore, we need to modify the force vector $\mathbf{F}^{t+\Delta t}$ for elasticity and also correct for the observer rotation (see section 2 equations (1, 2)). To achieve this we introduce the auxiliary vector $\mathbf{s}$ as defined by (10) and write:

$$\mathbf{F}^{E^{t+\Delta t}} = \mathbf{F}_{ext}^{E^{t+\Delta t}} - \int_{\Omega^E} \mathbf{B}^T \mathbf{s}^t \, d\Omega, \tag{12a}$$

where $\mathbf{F}_{ext}^E$ is the external load vector. After each time step the incremental solution may be improved iteratively. In this case, for t fixed, we replace $\mathbf{s}^t$ by $\mathbf{s}^{t+\Delta t}$, which during iteration is defined by $\mathbf{s}^{t+\Delta t} \leftarrow \sigma_{ij}^{t+\Delta t}$ as defined by (9). During iteration the director spin, particle positions etc. are replaced by their values at $t + \Delta t$ and are continuously updated until the increment of the velocities between two successive iterative steps are sufficiently small in the sense of a suitable norm. The above strategy allows one to modify existing codes for viscous materials without major interference with the rest of the code. There are many possibilities for refinement but for the purpose of this paper, for the examples presented in section (5), this simple formulation is perfectly applicable.

4. The Particle-in-Cell Finite Element Method

Some difficulties arise in devising a practical implementation of the finite element formulation described in section 3 for the large deformation modeling of layer folding. In particular, since the **C** matrix is a continuously evolving function of position through its dependence on director orientation, it is necessary that we are able to track an evolving vector function of the material during deformation.

4.1. Possible Numerical Schemes

In fluid dynamics, where strains are generally very large, but do not appear in the constitutive relationship of the material, it is common to transform the equations to an Eulerian mesh and deal with convective terms explicitly. Problems arise whenever advection becomes strongly dominant over diffusion since an erroneous numerical diffusion dominates. In our case, the advection of material boundaries and the stress tensor are particularly susceptible to this numerical diffusion problem. Mesh-based Lagrangian formulations alleviate this difficulty, but at the expense of remeshing and the eventual development of a less-than optimal mesh configuration. This increases complexity and can hinder highly efficient solution methods such as multigrid iteration. The Natural Element Method eliminates remeshing difficulties but is associated with considerable complexity of implementation, particularly in 3-D. A number of alternatives are available which dispense with a mesh entirely: smooth particle hydro dynamics and discrete element methods are common examples from the fluid and solid mechanics fields, respectively. These methods are extremely good at simulating the detailed behavior of highly deforming materials with complicated geometries (e.g., free surfaces, fracture development), and highly dynamic systems. They are, in general, formulated to calculate explicitly the interactions between individual particles which ultimately means that a great many time steps would be required to study creeping flow where the time scales associated with inertial effects are very many orders of magnitude smaller than typical flow times. We have therefore developed a hybrid approach – a particle-in-cell finite element method that uses a standard Eulerian finite element mesh (for fast, implicit solution) and a Lagrangian particle framework for carrying details of interfaces, the stress history, etc.

4.2. The Particle-in-Cell Approach

Our particle-in-cell finite element method is based closely on the standard finite element method, and is a direct development of the Material Point Method of SULSKY et al. (1995). The standard mesh is used to discretize the domain into elements, and the shape functions interpolate node points in the mesh in the usual fashion. The problem is formulated in a weak form to give an integral equation, and the shape function expansion produces a discrete (matrix) equation. For the discretized problem, these integrals occur over subdomains (elements) and are

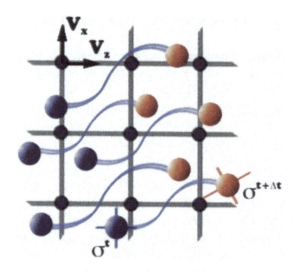

Figure 2
Schematic of Particle-in-Cell Method for representing large deformation in materials with interfaces and material history (including storing/transport of tensorial information such as stress). Mesh points remain fixed; particles move relative to the mesh and carry interface information via their relative positions and directional information directly on the particles.

calculated by summation over a finite number of sample points within each element. For example, in order to integrate equation (7), over the element domain Ω^E we replace the continuous integral by a summation

$$\mathbf{K}^E = \sum_p w_p \mathbf{B}^T(\mathbf{x}_p) \mathbf{C}_p(\mathbf{x}_p) \mathbf{B}(\mathbf{x}_p) \ . \tag{12b}$$

In standard finite elements, the positions of the sample points, $\mathbf{x}_p$, and the weighting, w_p are optimized in advance. In our scheme, the $\mathbf{x}_p$'s correspond precisely to the Lagrangian points embedded in the fluid, and w_p must be recalculated at the end of a time step for the new configuration of particles. Constraints on the values of w_p come from the need to integrate polynomials of a minimum degree related to the degree of the shape function interpolation, and the order of the underlying differential equation (e.g., HUGHES, 1987). These Lagrangian points carry the history variables including the director orientation and corotational stress rate, which are therefore directly available for the element integrals without the need to interpolate from nodal points to fixed integration points.

We therefore store an initial set of w_p's based on a measure of local volume and adjust the weights slightly to improve the integration scheme. MORESI *et al.* (2000) give a full discussion of the implementation of the particle-in-cell finite element scheme used here including full details of the integration scheme and its assumptions. They also discuss the specific modifications to the material point method required to handle a convecting fluid.

5. Numerical Simulations

We present an example of a simulation of folding of a layer of anisotropic viscoelastic material sandwiched between two isotropic viscous layers of equal viscosity. To accommodate the shortening of the system, one of the isotropic layers is compressible. In benchmarking of viscous folding, this sandwich of incompressible and compressible embedding material was found to give good agreement with analytic results assuming an infinite domain (MORESI et al., 2000). Throughout, we deal with a special case where the relaxation time for the normal and shear components of the rheology are always identical (i.e., $\alpha = \eta/\mu = \alpha_s = \eta_s/\mu_s$).

We first examine the viscous limit ($\alpha \to 0$, $\alpha_S \to 0$) for infinitesimal deformation of an embedded layer with a normal viscosity contrast of $\eta/\eta_M = 1$, 10 or 100 to the embedding medium. The boundary conditions in this case are slippery, undeforming boundaries on the vertical sides and the base. No density variations are assumed. For reference, in an isotropic sample with a viscosity contrast of 1 or 10 between the layer and the embedding material no appreciable folding occurs during shortening by 50%, as predicted by linear instability analysis (e.g., BIOT (1965a); see also section 6 in which we explore the potential influence of couple stresses within the framework of a linear instability analysis).

The initial orientation of the internal layering is approximately parallel to the macroscopic layering of the system with a small harmonic perturbation, introduced particle-by-particle:

$$\delta\theta_p = (\pi/100)\sin(qx_p) \ . \tag{13}$$

Figure 4 plots the dominant components of the Fourier transform of the vertical velocity along the mid-line of the embedded layer in each of three cases at a time $t = 0$. Case 1 has no normal viscosity contrast to the embedding medium ($\eta = \eta_M$), and a shear-to-normal viscosity contrast of 100 ($\eta = 100\eta_s$). The growth rate is strongly peaked for a perturbation wavenumber $q = \pi$. Case 2 has a normal viscosity contrast of 10 between the embedding and embedded materials ($\eta = 10\eta_M$) and $\eta = 10\eta_s$. There are two strong signals in the Fourier transform of the vertical velocity: one at the wavenumber of the director perturbation and another, stronger signal at $q = \pi/2$. The low wavenumber signal has a growth rate which is almost independent of the perturbation wavenumber. Growth at the wavenumber of the director perturbation falls off with increasing q, but reaches a plateau by around $q = 8\pi$. At a higher contrast between embedded and embedding materials, Case 3, $\eta = 100\eta_M$, $\eta = 10\eta_s$ shows a similar pattern to case 2, except that the growth rates are amplified.

The initial growth rates described in Figure 4 correspond to infinitesimal deformation of the layer. At finite shortening, the initial deformation pattern may be modified. For each of the three cases above, we plot the finite amplitude response to shortening in Figure 5.

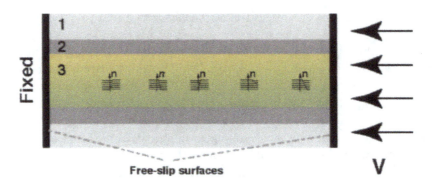

Figure 3

Initial geometry for the folding experiment. Layer 1 is compressible, viscous (η_M) background material, layer 2 is identical to layer 1 but incompressible (see text for an explanation), layer 3 is the test sample: viscoelastic (η, η_s, μ, μ_s) with a director orientation (**n**). The anisotropic layer contains small perturbations to the otherwise horizontal internal layering. $V = 10$ is constant during any given experiment and unchanged between different experiments.

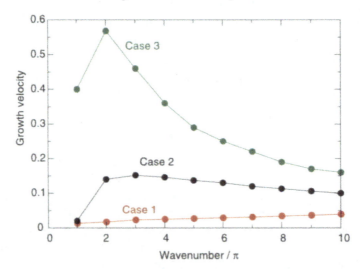

Figure 4

Rate of growth at wavenumbers introduced through perturbation to the director orientation expressed as Fourier coefficients of vertical velocity at the mid-line at time $t = 0$ for purely viscous cases ($\eta/\mu \rightarrow 0$). Isotropic embedding material has viscosity 1. Case 1, no contrast in normal viscosity ($\eta/\eta_s = 100$, $\eta/\eta_M = 1$). Case 2, $\eta/\eta_s = 10$, $\eta/\eta_M = 10$ Case 3, $\eta/\eta_s = 10$, $\eta/\eta_M = 100$.

For case 1 (Fig. 5a), the small growth rates observed at the outset produce only very small deflection of the layer interfaces after 40% shortening. Of interest, however, is the fact that a perturbation with wavenumber $q = 10\pi$ produced a corresponding layer deflection of larger magnitude than a perturbation at $q = 2\pi$, suggesting that the growth rate remains flat as a function of q even at finite amplitude deformation.

For case 2 (Fig. 5b), the growth rate is higher for smaller wavenumber in the infinitesimal deformation limit, and this persists with finite amplitude deformation. Growth at wavenumber $q = 2\pi$ is significantly more developed after 40% shortening than growth at $q = 10\pi$.

For case 3 (Fig. 5c), we observe the same overall trend as case 2: high wavenumber perturbations do not grow as fast as low wavenumber. However, we also observe that low wavenumber modes are excited in the finite deformation limit irrespective of the perturbation wavenumber. The perturbation causes a secondary variation in the interface deflection.

Shortening by 40% is examined in Figure 6 for a viscoelastic layer. This case corresponds to case 3 of the purely viscous simulations, but with finite relaxation time: $\eta = 100\eta_M$, $\eta = 10\eta_s$, $\mu = 100$–100000. The introduction of elasticity strongly reduces the tendency to generate long-wavelength buckling modes in the layer. For example, when $\mu = 1000$, the perturbation wavenumber is dominant even for $q = 10\pi$, and the amplitude of the deformation is considerably larger for small wavenumber than the corresponding deformation for the viscous case.

The linear instability analysis predicts a strong tendency to generate very high wavenumber (short wavelength) folds. The numerical simulations uphold this prediction, amplifying the deformation at the finest available wavelength: the one provided by the finite element mesh. The anisotropic layer itself shortens almost uniformly, but the internal layering direction develops an extremely strong periodicity in the shortening direction. Couple stresses (neglected in our numerical analysis) would stabilize the solution at a finite wavenumber and, thus, at least ameliorate the mesh-sensitivity of the result (MÜHLHAUS, 1993). In the viscous simulations, the natural development of low-wavenumber buckling of the anisotropic layer from initially fine-scale perturbations suggests that the role of couple stresses is considerably less important. We conclude the main body of this paper with a brief excursion into the potential role of couple stresses within the framework of a linear instability analysis.

6. Couple Stresses and Linear Instability Analysis

Here we restrict ourselves to an incrementally elastic constitutive relation for the couple stress, partially for algebraic convenience but also because the numerical difficulties mentioned above seem to occur only in connection with significant elastic deformations. Couple stresses may be significant in situations where the gradient of n_i changes strongly over a short distance (limiting case: disinclination).

In such cases we have to take the variations of the normal stresses across the layer cross sections into consideration (e.g., MÜHLHAUS, 1993). The couple stress theories (see e.g., MINDLIN and THIERSTEN, 1962; MÜHLHAUS, 1993; MÜHLHAUS and

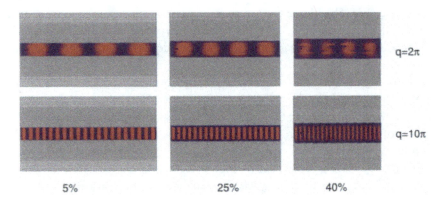

Figure 5a

Evolution of folding in anisotropic viscous layer. Isotropic embedding material has viscosity 1, layer has shear viscosity 0.01, normal viscosity 1. Results are shown for perturbation to the director orientation with wavenumber $q = 2\pi$ and $q = 10\pi$.

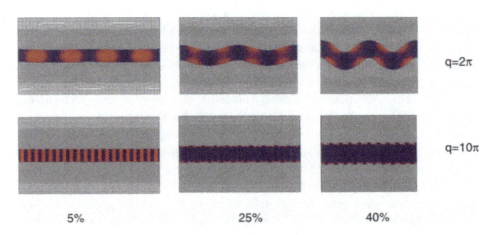

Figure 5b

Evolution of folding in anisotropic viscous layer. Isotropic embedding material has viscosity 1, layer has shear viscosity 1, normal viscosity 10. Results are shown for perturbation to the director orientation with wavenumber $q = 2\pi$ and $q = 10\pi$.

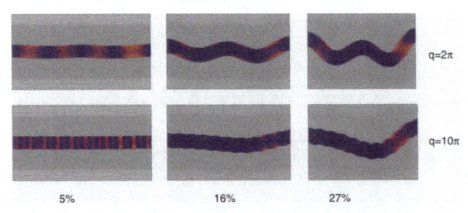

AIFANTIS, 1991a,b) provide a convenient framework for the consideration of stress fluctuations on the layer-scale without having to abandon the homogeneity properties of the anisotropic standard continuum approach. In the present case the couple stress enhancement leads naturally to the superposition of viscoelastic bending stiffness on our standard continuum model. In connection with layered materials the internal length scale introduced by the couple stresses is proportional to the layer thickness (ranging from microns to kilometers in geological applications) and to the differences between the viscosities and shear moduli governing pure and simple shear respectively (see e.g., MÜHLHAUS, 1993). In layered materials the explanation why the stress tensor is nonsymmetric in couple stress materials is straightforward: In a continuum description the stresses represent average values over multiples of the layer thickness. In bending the shear stress obtained by averaging normal to the layering is different in general from the shear stress parallel to the layering. The latter is even zero for instance in the case of a stack of perfectly smooth cards (a standard continuum model would break down in this case). Within the framework of a couple stress theory one considers the variation of the normal stress across the layer thickness (in much the same way as in the standard engineering beam and plate theories), introduces statically equivalent couple stresses balancing the difference between the shear stresses.

As usual in plate theories the couple stresses (moment per unit area) are conjugate in energy to a suitable measure for the rate of curvature. We assume that in the reference configuration the layer normal is parallel to the Cartesian x_3 axis. Without going into details we mention the following definition of the curvature rate, which is suitable in the context of layered materials

$$\kappa = \text{symm}((\text{grad}(\mathbf{n} \times \varphi)(1 - \lambda)) \quad \text{where } \varphi = \mathbf{n} \times \dot{\mathbf{n}} , \tag{14}$$

which upon linearization, and nothing else is required for our linear instability analysis, reduces to the familiar expression

$$\kappa_{11} = v_{2,11} . \tag{15}$$

All other components of κ vanish in the linear case. In (14) symm(.) means the symmetric part of the argument and $\lambda = \mathbf{n}\mathbf{n}^T$ (see equation (1)).

We assume the simplest possible constitutive relationship for the couple stress by putting $\dot{m}_{11}^n = \dot{m}_{11} = \Theta \mu h^2 \kappa_{11}$. The corotational rate is equal to the material rate because $m_{11} = 0$ in the ground state; Θ is a dimensionless scalar typically smaller than unity and h is the width of a periodic cell of the layered material. In the linear instability analysis (see Appendix 4 for details) we assume one-dimensional modes of

◀

Figure 5c

Evolution of folding in anisotropic viscous layer. Isotropic embedding material has viscosity 1, layer has shear viscosity 10, normal viscosity 100. Results are shown for perturbation to the director orientation with wavenumber $q = 2\pi$ and $q = 10\pi$.

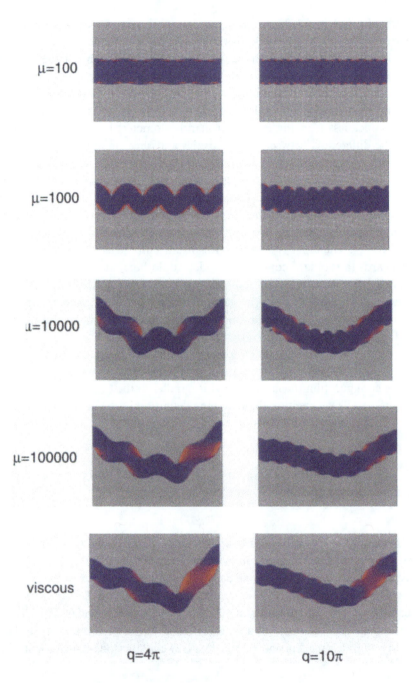

Figure 6

Layered viscoelastic case. Isotropic embedding material has viscosity 1, layer has shear viscosity 10, normal viscosity 100. Results are shown for a range, μ and for perturbation to the director orientation with wavenumber $q = 4\pi$ and $q = 10\pi$. The ratio of elastic moduli is constant throughout: $\mu/\mu_s = 10$.

the form

$$\delta u_2 = U \exp(\omega t) \cos \frac{2\pi}{L} x_1 = U \exp(\omega t) \cos q x_1, \quad \delta u_1 = 0 \tag{16}$$

reflecting the fact that typical folding scenarios are usually one-dimensional (Fig. 2), at the onset at least, depending only on the coordinate parallel to the layer surfaces, x_1 (e.g., BIOT, 1964, 1965a,b, 1967; HILL and HUTCHINSON, 1975). The one-dimensionality itself does not represent an assumption *per se*; it does however in the context of our specific problem: Here and in the computational examples we consider a multilayered structure embedded in an isotropic, incompressible viscous medium of infinite extent. Again, within the context of this specific problem, the assumption of one-dimensionality expressed in (16) precludes gross changes in layer thickness during folding as is observed in many natural folds.

Next in our stability analysis we insert (16) into the constitutive relations and the result into the incremental equilibrium conditions (see Appendix 4). Because of the one-dimensionality of our problem equilibrium needs to be considered in the x_2 direction only. The resistance of the embedding medium against the folding (see Fig. 7) of the structure is considered BIOT-style (1965a) by the traction $2Q = -4\eta_M \omega q \exp(\omega t) U \cos(q x_1)$. Again, details of this derivation are included in the Appendix.

Insertion of (A.18) into (A.23) and nondimensionalising leads to the following characteristic equation for the dimensionless growth coefficient of the folding instability:

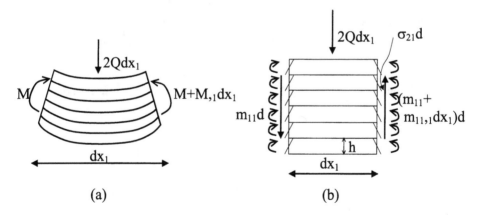

(a)   (b)

Figure 7

Moment equilibrium diagram for a slice of the layered structure. The macroscopic bending moment M (a); couple stress m_{11} (b); the traction $2Q$ represent the resistance of the embedding medium against the deformation of the layered structure.

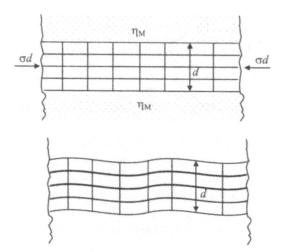

Figure 8
Multilayered structure embedded in an infinite, isotropic medium of viscosity η_M; t is the width of the layered structure.

$$\hat{\omega}^2 - \hat{\omega}\left((\hat{\sigma}-1)\hat{q} - 1 - \frac{\Theta\mu}{\mu_S}\left(\frac{4\eta_M}{\eta_S}\right)^2\left(\frac{h}{d}\right)^2\hat{q}^3\right) - \hat{\sigma}\hat{q} + \frac{\Theta\mu}{\mu_S}\left(\frac{4\eta_M}{\eta_S}\right)^2\left(\frac{h}{d}\right)^2\hat{q}^3 = 0 \ . \tag{17}$$

where d is the thickness of the folding plate (Fig. 7) and

$$\hat{\omega} = \omega\alpha_S = \omega\frac{\eta_S}{\mu_S}, \quad \hat{q} = \frac{\eta_S}{4\eta_M}tq, \quad \hat{\sigma} = \frac{\sigma}{\mu_S} \ . \tag{18}$$

For $h/d \to 0$ we recover the standard continuum case (equation (A.19)). However, unlike in the case of the standard continuum (Fig. 9) the positive branch of (17) tends to zero as $q \to \infty$ i.e., the ill-posed nature of the folding problem is removed.

A maximum of $\omega(q)$ exists as long as $h > 0$. For illustration we consider the extreme case $\eta_S = 0$. From (17) we obtain:

$$\omega = \frac{1}{4\eta_M}(\sigma dq - \Theta\mu h^2 dq^3) \ . \tag{19}$$

Unstable modes are obtained for $q \le (\sigma/(\Theta\mu h^2))^{1/2}$ and the maximum growth rate $\omega_{max}(q_{max})$ is obtained as:

$$\omega_{max} = \frac{\sigma}{6\eta_M}\sqrt{\frac{\sigma}{3\Theta\mu}\frac{d}{h}} \quad \text{and} \quad q_{max} = \frac{1}{h}\sqrt{\frac{\sigma}{3\Theta\mu}} \ . \tag{20}$$

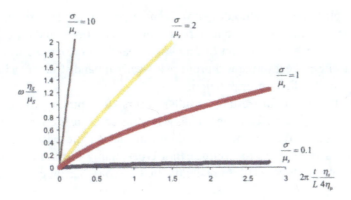

Figure 9

Dispersion function for viscoelastic, layered material with bending resistance, η_M = viscosity of embedding material, η_S and μ_S are the shear viscosity and the weak shear modulus, t and h are the width of the structure and the individual layer, respectively, L = wavelength, ω = growth coefficient. The growth coefficient ω is bounded if $\sigma/\mu_S < 1$ (see Appendix 3).

The wavelength predicted by (20) for the extreme case $\eta_S = 0$ ranges between 15 to 20 times the thickness h (which is well within the ranges of wavelength observed in real rocks) if the stress ratio σ/μ ranges between 1/10 and 1/20. The maximum shifts to the longer wavelength side with increasing magnitude of the coefficient of $\hat{q}^3$ in (20). If $h = 0$ the maximum degenerates to a boundary maximum at $q \to \infty$. In this case the perturbations with the shortest wavelength grow fastest, are dominant, and we have to expect a strong dependency of finite element solutions on the mesh size. A similar situation occurs in connection with strain localisation and strain softening (DE BORST et al., 1993). However, in many cases the maximum occurs at wavenumbers, which are beyond the applicability of the concept of equivalent (or homogenized) continua and prior to the maximum, the situation is usually satisfactorily described as a standard continuum.

7. Conclusions and Future Research

We have presented a simple formulation for the consideration of viscoelasticity in deforming layered systems. The combination of the basic model with a large deformation, particle-in-cell finite element method allows the simulation of a diverse range of crustal deformation problems.

Our demonstration examples include a realistic treatment of folding which includes the mechanical influence of fine-scale laminations and viscoelasticity. The model is relatively simple in its present form but still gives a useful insight into the physical processes involved in certain types of folding.

While we have noted the importance of considering couple stresses for the elastic-dominated simulations, the story will not be complete unless we also consider that the large couple stresses produced by layer bending could also lead to failure of the rock. The consideration of layered materials with couple stresses and a yield criterion in large deformation is the subject of ongoing research, however, we would predict that yielding within the layering will produce a localized band structure with a strong discontinuity in the director orientation.

One of the most interesting results occurs for purely viscous, layered simulations where low-wavenumber folding is induced even for very low viscosity contrasts between embedded and embedding media. In the past, the very large viscosity contrasts required to produce Biot-type folding in purely viscous media have led people to discount the possibility that viscous buckling occurs at all in geology.

Viscoelastic layered buckling tends to emphasize the finest scale imposed by explicitly assumed perturbations as well as discretization artifacts. The latter implies mesh-dependency not at all uncommon in numerical simulation but unwelcome all the same. The possible cure would be to extend the numerical representation to include terms which, in nature, become dominant at large wavenumber (in this case consideration of the bending stiffness of the individual layers).

The particle-in-cell formulation for including couple stresses is conceptually no different from the one presented here though more complicated in detail.

Linear instability analysis gives a good insight into the expected modes of deformation at the onset of instability, and the different regimes of behavior one might expect to observe.

Appendix A

A.1. Derivation of the Anisotropy Tensor Λ

In preparation for the derivation of constitutive relations for our layered material we define some auxiliary relationships. The corotational simple shearings (spin: $\mathbf{W}$) in the layer orthogonal to $\mathbf{n}$ read:

$$d_i = D_{ik}n_k - D_{kl}n_k n_l n_i \ . \tag{A.1}$$

We define

$$\mathbf{s} = \frac{\mathbf{d}}{|\mathbf{d}|} \ . \tag{A.2}$$

Note that $\mathbf{s}^T \mathbf{n} = 0$.

$$\begin{aligned}
D_{ns}(n_i s_j + s_i n_j) &= D_{jk}n_i n_k + D_{ik}n_j n_k - 2D_{kl}n_k n_l n_i n_j \\
&= (n_i n_k \delta_{lj} + n_j n_k \delta_{il} - 2n_i n_j n_k n_l)D_{lk} \ .
\end{aligned} \tag{A.3}$$

Symmetry $(ij) \rightarrow (lk)$

$$D_{ns}(n_i s_j + s_i n_j) = (n_i n_k \delta_{lj} + n_j n_k \delta_{il} - 2n_i n_j n_k n_l)D_{lk}$$

$$= \left(\frac{1}{2}(n_i n_k \delta_{lj} + n_j n_k \delta_{il} + n_i n_l \delta_{kj} + n_j n_l \delta_{ik}) - 2n_i n_j n_k n_l\right)D_{lk}$$

$$= \Lambda_{ijkl}D_{lk} \tag{A.4}$$

A.2. Matrix Representations of Constitutive Relations

We assume plane deformations in the (x_1, x_2) plane. The matrix representations of (1) reads:

$$\begin{bmatrix} \dot{n}_1 \\ \dot{n}_2 \end{bmatrix} = \begin{bmatrix} 0 & -(v_{1,1} - v_{2,2})n_1 n_2 - v_{2,1}n_2^2 + v_{1,2}n_1^2 \\ (v_{1,1} - v_{2,2})n_1 n_2 + v_{2,1}n_2^2 - v_{1,2}n_1^2 & 0 \end{bmatrix} \begin{bmatrix} n_1 \\ n_2 \end{bmatrix}. \tag{A.5}$$

$$[\Lambda] = \begin{bmatrix} 2n_1^2 n_2^2 & -2n_1^2 n_2^2 & (n_1 n_2^3 - n_2 n_1^3) \\ -2n_1^2 n_2^2 & 2n_1^2 n_2^2 & (n_2 n_1^3 - n_1 n_2^3) \\ (n_1 n_2^3 - n_2 n_1^3) & (n_2 n_1^3 - n_1 n_2^3) & \frac{1}{2} - 2n_1^2 n_2^2 \end{bmatrix}. \tag{A.6}$$

In the ground state we have

$$[\Lambda] = \begin{bmatrix} 0 & 0 & 0 \\ 0 & 0 & 0 \\ 0 & 0 & \frac{1}{2} \end{bmatrix}; \tag{A.7}$$

thus:

$$[\delta\Lambda] = \begin{bmatrix} 0 & 0 & \delta n_1 \\ 0 & 0 & -\delta n_1 \\ \delta n_1 & -\delta n_1 & 0 \end{bmatrix} = \begin{bmatrix} 0 & 0 & -u_{2,1} \\ 0 & 0 & u_{2,1} \\ -u_{2,1} & u_{2,1} & 0 \end{bmatrix} \tag{A.8}$$

A.3. Inversion of Constitutive Operators

The tensor relating the stretching to the stress in (3) may be written symbolically as

$$\mathbf{C} = \mathbf{1} + (a - 1)\Lambda , \tag{A.9}$$

where a is an arbitrary real scalar. From the definition of Λ (see A.4) it follows that $\Lambda^2 = \Lambda$. Furthermore

$$\mathbf{C}^{-1} = \mathbf{1} + \left(\frac{1}{a} - 1\right)\Lambda . \tag{A.10}$$

A.4. Linear Instability Analysis

The calculations in this section are based on three simplifying assumptions:
1) The characteristic time scale of the instability is much shorter then the time scale of the deformation of the ground state (the general case was treated e.g., by HOBBS et al., 2001). In the present, simplified case we have:

$$\delta D_{ij} \approx \tfrac{1}{2}(u_{i,tj} + u_{j,ti}) \ .$$

2) The viscosity η_S and the shear modulus μ_S are much smaller than the normal viscosity η and the normal shear modulus μ. Thus:

$$\frac{1}{\eta_S} - \frac{1}{\eta} \approx \frac{1}{\eta_S} \quad \text{and} \quad \frac{1}{\mu_S} - \frac{1}{\mu} \approx \frac{1}{\mu_S}$$

3) In the linear instability analysis we assume one-dimensional modes of the form

$$\delta u_2 = U \exp(\omega t) \cos \frac{2\pi}{L} x_1 = U \exp(\omega t) \cos q x_1, \quad \delta u_1 = 0 \ .$$

Neglecting couple stresses in the first instance;
First we write down the perturbed form of the constitutive relations (5). Neglecting couple stresses in the first instance, we find

$$\delta D'_{ij} = \frac{1}{2\mu} \delta \dot{\sigma}^{n'}_{ij} + \frac{1}{2\eta} \delta \sigma'_{ij} + \left(\frac{1}{2\mu_S} - \frac{1}{2\mu} \right) (\Lambda_{ijkl} \delta \dot{\sigma}^{n'}_{kl} + \delta \Lambda_{ijkl} \dot{\sigma}^{n'}_{kl})$$

$$+ \left(\frac{1}{2\eta_S} - \frac{1}{2\eta} \right) (\delta \Lambda_{ijkl} \sigma'_{kl} + \Lambda_{ijkl} \delta \sigma'_{kl}) \tag{A.11}$$

Using the results of appendix two, (A.11) becomes

$$\delta D'_{11} = \frac{1}{2\mu} \delta \ddot{\sigma}'_{11} + \frac{1}{2\eta} \delta \sigma'_{11}$$

$$\delta D'_{22} = \frac{1}{2\mu} \delta \ddot{\sigma}'_{22} + \frac{1}{2\eta} \delta \sigma'_{22} \tag{A.12}$$

$$\delta D_{12} = \frac{1}{2\mu_S} (\delta \dot{\sigma}_{12} + \sigma v_{2,1}) + \frac{1}{2\eta_S} \delta \sigma_{12} + \left(\frac{1}{2\eta_S} - \frac{1}{2\eta} \right) u_{2,1} \sigma \ .$$

The incremental form of the equilibrium conditions read:

$$\delta \sigma_{11,1} + \delta \sigma_{12,2} = 0$$

$$\delta \sigma_{21,1} + \delta \sigma_{22,2} = 0 \ , \tag{A.13}$$

where $\delta\sigma_{12} = \delta\sigma_{21}$ if couple stresses are neglected. We integrate the second line of the equations (A.11) over the width of the layered structure. Thus:

$$\int_{-d/2}^{d/2} \delta\sigma_{21,1}\, dx_2 \mp \delta\sigma_{22}\Big|_{d/2} - \delta\sigma_{22}\Big|_{-d/2} = 0 \ . \tag{A.14}$$

For the normal traction we assume BIOT's (1965) result for an isotropic viscous half space (see Fig. 3 in the main part of the paper):

$$\delta\sigma_{22}\Big|_{d/2} - \delta\sigma_{22}\Big|_{-d/2} = 2Q = -4\eta_M q\omega U \cos qx_1 \ . \tag{A.15}$$

Next we consider 1-D instability modes of the form:

$$\delta u_2 = U \exp(\omega t) \cos\frac{2\pi}{L} x_1 = U \exp(\omega t) \cos qx_1, \quad \delta u_1 = 0 \ . \tag{A.16}$$

Insertion into (A.12) yields the modal form of the constitutive relations:

$$\delta\sigma_{11} - \delta\sigma_{22} = 4\hat{\eta}\omega u_{1,1}$$
$$\delta\sigma_{12} = \delta\sigma_{21} = \hat{\eta}_S \omega u_{2,1} - \sigma u_{2,1} \tag{A.17}$$

where

$$\hat{\eta} = \frac{\eta}{1 + \omega\frac{\eta}{\mu}} \quad \text{and} \quad \hat{\eta}_S = \frac{\eta_S}{1 + \omega\frac{\eta_S}{\mu_S}} \ . \tag{A.18}$$

Substitution of (A.17) into (A.15) and insertion of the result into (A.14) gives:

$$((4\eta_M + \hat{\eta}_S dq)\omega - \sigma dq) = 0 \ . \tag{A.19}$$

From this, we obtain the dispersion relation as

$$\omega = \frac{\sigma dq}{4\eta_M + \hat{\eta}_S dq} \ . \tag{A.20}$$

We use the notations $\alpha = \eta/\mu$ and $\alpha_S = \eta_S/\mu_S$ for the normal and shear relaxation times, respectively. If $\omega \ll 1/\alpha_S$ then the dispersion relationship reduces to the one for a viscous layer in a viscous matrix, i.e.:

$$\omega = \frac{\sigma dq}{4\eta_M + \eta_S dq} \ . \tag{A.21}$$

Conversely, if $\omega \gg 1/\alpha_S$, we obtain

$$\omega = \frac{\mu_S}{4\eta_M}\, dq\left(\frac{\sigma}{\mu_S} - 1\right) \ . \tag{A.22}$$

Finally, in the presence of couple stresses we have $\delta\sigma_{21} = \delta\sigma_{12} - \delta m_{11,1}$. With $\delta m_{11} = \Theta\mu h^2 u_{2,11}$ and inserting into (A.14) yields the modified momentum balance equation:

$$\left(\left(4\eta_M + \hat{\eta}_S dq + \Theta\frac{\mu}{\omega}h^2 dq^3\right)\omega - \sigma dq\right) = 0 . \tag{A.23}$$

REFERENCES

BIOT, M. A. (1964), *Theory of Internal Buckling of a Confined Multilayered Structure*, Geol. Soc. Am. Bull. *75*, 563–568.

BIOT, M. A., *The Mechanics of Incremental Deformations* (John Wiley, New York 1965a).

BIOT, M. A. (1965b), *Theory of Similar Folding of the First and Second Kind*, Geol. Soc. Am. Bull. *76*, 251–258.

BIOT, M. A. (1967), *Rheological Stability with Couple Stresses and its Application to Geological Folding*, Proc. Royal Soc. London *A 2298*, 402–2423.

DE BORST, R., SLUYS, L. J., MÜHLHAUS, H.-B., and PAMIN, J. (1993), *Fundamental Issues Finite Element Analyses of Localization of Deformation*, Engin. Comput. *10*, 99–121.

CHAPPLE, W. N. (1969), *Fold Shape and Rheology: The Folding of an Isolated Viscous-Plastic Layer*, Tectonophysics *7*, 97–116.

FLETCHER, R. C. (1974), *Wavelength Selection in the Folding of a Single Layer with Power Law Rheology*, Am. J. Sci. *274*, 1029–1043.

FLETCHER, R. C. (1982), *Coupling of Diffusional Mass Transport and Deformation in Tight Rock*, Tectonophysics *83*, 275–291.

GREEN, A. E. and ZERNA, W., *Theoretical elasticity*. 2nd Edition (Oxford at the Clarendon Press 1968).

HILL, R. and HUTCHINSON, J. W. (1975), *Bifurcation Phenomena in Plane Tension Test*, J. Mech. Phys. Sol. *23*, 239–264.

HOBBS, B. E., MÜHLHAUS, H.-B, ORD, A., and MORESI, L. (2001), *The Influence of Chemical migration upon Fold Evolution in Multilayered Materials*, Vol. 11, Yearbook of Self-Organisation (eds., H. J. Krug and J. H. Kruhl), Duncker & Humblot, Berlin; ISBN 3-428-10506-0, 229–253.

HOBBS, B. E., MEANS, W. D., and WILLIAMS, P. F., *An Outline of Structural Geology* (Wiley, 1976).

HUNT, G. W., MÜHLHAUS, H.-B. and WHITING, A. I. M. (1997), *Folding Processes and Solitary Waves in Structural Geology*, Phil. Trans. of R. Soc. London *355*, 2197–2213.

JOHNSON, A. M. and FLETCHER, R. C. *Folding of Viscous Layers (Columbia University Press, New York 1994)*.

MINDLIN, R. D. and THIERSTEN, H. F. (1962). *Effects of Couple Stresses in Linear Elasticity*, Arch. Rat. Mech. Analysis. *11*, 415–448.

MORESI, L., MÜHLHAUS, H.-B., and DUFOUR, F. (2001), *Particle- in-Cell Solutions for Creeping Viscous Flows with Internal Interfaces*. In Proceedings of the 5th International Workshop on *Bifurcation and Localization in Geomechanics (IWBL'99)*, Perth, W. A., Australia (Eds. H.-B. Mühlhaus, A. Dyskin and E. Pasternak) (Balkema, Lisse) ISBN 90265 18234, 345–355.

MÜHLHAUS, H.-B., and AIFANTIS, E. C. (1991a), *A Variational Principle for Gradient Plasticity*, Int. J. Sol. and Struct. *28*, 845–857.

MÜHLHAUS, H.-B., and AIFANTIS, E. C. (1991b), *The Influence of Microstructure-induced Gradients on the Localization of Deformation in Viscoplastic Materials*, Acta Mechanica *89*, 217–231.

MÜHLHAUS, H.-B., *Continuum models for layered and blocky rock*. In *Comprehensive Rock Engineering*. Invited Chapter for Vol. II: *Analysis and Design Methods* (Pergamon Press, 1993) pp. 209–230.

ORTOLEVA. P. J., *Geochemical Self-Organization* (Oxford University Press, 1994).

ROBIN, P.-Y. F. (1979), *The Theory of Metamorphic Segregation and Related Processes*, Geochim. Cosmochim. *43*, 1587–1600.

SCHMALHOLZ, S. M. and PODLADCHIKOV, Y. (1999), *Buckling versus Folding: Importance of Viscoelasticity*, Geophys. Res. Lett. *26*(17), 2641–2644.

SULSKY, D., ZHOU, S.-J., and SCHREYER, H. L. (1995), *Application of a Particle-in-cell Method to Solid Mechanics*, Comput. Phys. Commun. *87*, 236–252.

VASILYEV, O. V., PODLADCHIKOV, Y. Y., and YUEN, D. A. (1998), *Modelling of Compaction Driven Flow in Poro-viscoelastic Medium Using Adaptive Wavelet Collocation Method*, Geophys. Res. Lett. *17*, 32–39.

WILLIAMS, P. F. (1972), *Development of Metamorphic Layering and Cleavage in Low-grade Metamorphic Rocks at Barmagui, Australia*, Am. J. Sci. *272*, 1–47.

(Received February 20, 2001, revised June 11, 2001, accepted June 25, 2001)

To access this journal online:
http://www.birkhauser.ch

Pure appl. geophys. 159 (2002) 2335–2356
0033–4553/02/102335–22 $ 1.50 + 0.20/0

© Birkhäuser Verlag, Basel, 2002

❘ Pure and Applied Geophysics

Mantle Convection Modeling with Viscoelastic/Brittle Lithosphere: Numerical Methodology and Plate Tectonic Modeling

Louis Moresi,[1]* Frédéric Dufour,[1] and Hans-Bernd Mühlhaus[1]

Abstract — The earth's tectonic plates are strong, viscoelastic shells which make up the outermost part of a thermally convecting, predominantly viscous layer. Brittle failure of the lithosphere occurs when stresses are high. In order to build a realistic simulation of the planet's evolution, the complete visco-elastic/brittle convection system needs to be considered. A particle-in-cell finite element method is demonstrated which can simulate very large deformation viscoelasticity with a strain-dependent yield stress. This is applied to a plate-deformation problem. Numerical accuracy is demonstrated relative to analytic benchmarks, and the characteristics of the method are discussed.

Key words: Mantle convection, viscoelasticity, brittle failure, finite element, Lagrangian, geodynamics.

Introduction

Solid state convection in the earth's mantle drives the surface motion of a cool thermal boundary layer comprising a number of distinct lithospheric plates. Motions in the mantle are described by the equations of fluid dynamics for substantial deformation. The rheology needed to describe deformation in the lithosphere is highly nonlinear, and near the surface where temperatures are less than approximately 600°C it becomes necessary to consider the role of elasticity (WATTS *et al.*, 1980). The strong correlation between seismicity and plate boundaries (e.g., BARAZANGI and DORMAN, 1969) makes it seem likely that plate motions are associated with localization of deformation occurring when stresses reach the yield strength of the lithosphere.

This picture of the earth's interior is widely accepted by geophysicists (Fig. 1). It clearly indicates that the fundamental process is thermal convection; plate tectonics is the manner in which the system organizes. Therefore, a consistent model of plate

[1] CSIRO Exploration and Mining, PO Box 437, Nedlands, WA 6009, Australia.
E-mail: l.moresi@ned.dem.csiro.au, http://www.ned.dem.csiro.au/research/solidmech/
F. Dufour, E-mail: frederic@ned.dem.csiro.au; H-B. Mühlhaus, E-mail: hans@ned.dem.csiro.au
*Present Address: School of Mathematical Sciences, Monash University, PO Box 28M Clayton, Victoria, 3800 Australia

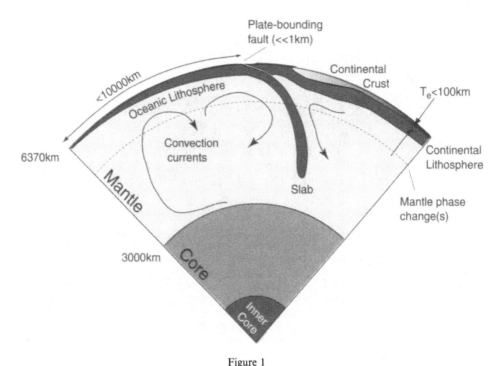

Figure 1

A simplified cross section of the earth with major layerings shown to scale except for the upper boundary layer which is exaggerated in thickness by a factor of roughly two.

behavior should contain a description of the convection system of which the plate is a part. The principle difficulty for modeling is that plate tectonics is itself only a kinematic description of the observations: a fully consistent dynamic description of the motion of the plates is still sought. Another issue is the very large range of time- and length-scales which must be considered in a complete model. For example, plates span up to 10,000 km horizontally, the elastic thickness is less than 100 km, and the relevant scale of plate bounding faults may be a few hundreds of metres.

There have been major steps towards the simulation of plate tectonics in recent years by solving brittle/viscous fluid flow equations (e.g., TACKLEY, 1998, 2000; MORESI and SOLOMATOV, 1998; OGAWA, 2001). The importance of elasticity has not been quantified by such modeling. In the past, strongly viscoelastic convection simulations with a lithosphere component have been limited to models with explicit layering in which a nonconvecting viscoelastic layer is coupled to a viscous convecting domain (PODLADCHIKOV *et al.*, 1993). Viscoelastic mantle convection simulations have been limited to considering constant viscosity (HARDER, 1991). Models of subduction zones which incorporate viscoelasticity, faulting, and free-surface behavior have generally been limited to modest evolution times, after which further deformation produces severe remeshing problems (e.g., MELOSH, 1978, GURNIS *et al.*, 1996).

Having identified the need for efficient, large-scale convection simulations with elastic effects in an evolving cool lithosphere, we present a method for simulating viscoelastic-brittle materials in extreme deformation. The method is first tested on a number of very simple benchmark cases in which analytic solutions are known, or where the accuracy can otherwise be quantified in large deformation. We then demonstrate the application of the method to mantle convection with a viscoelastic, brittle lithosphere.

Mathematical Model

We begin our analysis in a general way with the classical momentum conservation equation:

$$\nabla . \boldsymbol{\sigma} = \mathbf{f} \tag{1}$$

where $\boldsymbol{\sigma}$ is the stress tensor and $\mathbf{f}$ a force term. As we are interested only in very slow deformations of highly viscous materials, (infinite Prandlt number) we have neglected all inertial terms in (1). It is convenient to split the stress into a deviatoric part, $\boldsymbol{\tau}$, and an isotropic pressure, p,

$$\boldsymbol{\sigma} = \boldsymbol{\tau} - p\mathbf{I} \ , \tag{2}$$

where $\mathbf{I}$ is the identity tensor.

Viscoelasticity

We will employ a Maxwell viscoelastic model which has been used in previous studies of lithospheric deformation where viscous and elastic effects are important such as post-glacial rebound (PELTIER, 1974). SCHMALHOLZ *et al.* (2001) provide an excellent discussion of the formulation of numerical schemes for large-deformation viscoelastic modeling of geological folding, which highlights many of the issues which our method is designed to overcome.

This model assumes that the strain rate tensor, $\mathbf{D}$, defined as:

$$D_{ij} = \frac{1}{2} \left(\frac{\partial V_i}{\partial x_j} + \frac{\partial V_j}{\partial x_i} \right) \tag{3}$$

is the sum of an elastic strain rate tensor $\mathbf{D}_e$ and a viscous strain rate tensor $\mathbf{D}_v$. The velocity vector, $\mathbf{V}$, is the fundamental unknown of our problem and all these entities are expressed in the fixed reference frame x_i. Now we decompose each strain tensor

$$\mathbf{D}_e = \tfrac{1}{3}\mathrm{tr}(\mathbf{D}_e)\mathbf{I} + \hat{\mathbf{D}}_e \quad \text{and} \quad \mathbf{D}_v = \tfrac{1}{3}\mathrm{tr}(\mathbf{D}_v)\mathbf{I} + \hat{\mathbf{D}}_v \ , \tag{4}$$

where $\hat{\mathbf{D}}$ is the deviatoric part of $\mathbf{D}$ and $\mathrm{tr}(\mathbf{D})$ represents the trace of the tensor.

Individually we express each deformation tensor as a function of the deviatoric stress tensor τ and pressure p:

$$\frac{\overset{\triangledown}{\tau}}{2\mu} + \frac{\tau}{2\eta} = \hat{\mathbf{D}}_e + \hat{\mathbf{D}}_v = \hat{\mathbf{D}} \ , \tag{5}$$

where $\overset{\triangledown}{\tau}$ is the Jaumann corotational stress rate for an element of the continuum, μ is the shear modulus and η is shear viscosity.

$$\overset{\triangledown}{\tau} = \frac{D\tau}{Dt} + \tau\mathbf{W} - \mathbf{W}\tau \ , \tag{6}$$

where $\mathbf{W}$ is the material spin tensor,

$$W_{ij} = \frac{1}{2}\left(\frac{\partial V_i}{\partial x_j} - \frac{\partial V_j}{\partial x_i}\right) \ . \tag{7}$$

The $\mathbf{W}$ terms account for material spin during advection which reorients the elastic stored-stress tensor.

The Jaumann derivative is an objective (observer independent) rate, although others exist which account further for the effects of a deforming coordinate system. In particular, the Olroyd derivative contains terms to account for material stretching during advection. HARDER (1991) discusses the influence of the different objective derivatives on convection solutions and finds that the stretching terms strongly effect rapidly shearing boundary layers near corners. In our proposed application where viscoelasticity is important in the stiff, cool thermal boundary layer but not elsewhere, the simpler Jaumann derivative is probably adequate. However, we plan to test this assumption in further studies.

The isotropic part provides a scalar equation for the pressure:

$$\frac{1}{K_e}\frac{Dp}{Dt} + \frac{p}{\xi} = -\mathrm{tr}(\mathbf{D}) \ , \tag{8}$$

where K_e is the bulk modulus, ξ is the bulk viscosity and D/Dt is a material time derivative. We note that the form of equation (8) is unsuited to conventional fluids as the material has no long-term resistance to compression. This behavior is, however, relevant to the simulation of the coupled porous-flow, matrix deformation problem where it is common to ascribe an apparent bulk viscosity to the matrix material in order to model compaction effects on large scales (e.g., MCKENZIE, 1984).

Brittle Failure

Rocks in the cool lithosphere have a finite strength which may be exceeded by tectonic stresses. For a full description of the lithosphere, therefore, it is necessary to include a description of the brittle nature of the near-surface material. Here we use the term "brittle" quite loosely to distinguish fault-dominated deformation which may result in seismic activity, from ductile creep which occurs at higher temperature and pressure. In all recent studies of mantle convection where the brittle lithospheric rheology has been considered, the brittle behavior has been parameterized using a nonlinear effective viscosity which is introduced whenever the stress would otherwise exceed the yield value τ_{yield}. This approach ignores details of individual faults, and treats only the influence of fault systems on the large-scale convective flow. It produces a lithospheric strength profile in accord with Byerlee's law (e.g., BYERLEE, 1978).

To determine the effective viscosity we extend (5) by introducing a Prandtl-Reuss flow rule for the plastic part of the stretching, $\mathbf{D}_p$:

$$\frac{\overset{\triangledown}{\tau}}{2\mu} + \frac{\tau}{2\eta} + \lambda\frac{\tau}{2|\tau|} = \hat{\mathbf{D}}_e + \hat{\mathbf{D}}_v + \hat{\mathbf{D}}_p = \hat{\mathbf{D}} \ , \tag{9}$$

where λ is a parameter to be determined such that the stress remains on the yield surface, and $|\tau| \equiv (\tau_{ij}\tau_{ij}/2)^{(1/2)}$. The plastic flow rule introduces a nonlinearity into the constitutive law which, in general, requires iteration to determine the equilibrium state.

Numerical Implementation

As we are interested in solutions where very large deformations may occur – including thermally driven fluid convection, we would like to work with a fluid-like system of equations. Hence we obtain a stress/strain-rate relation from (5) by expressing the Jaumann stress-rate in a difference form:

$$\overset{\triangledown}{\tau} \approx \frac{\tau^{t+\Delta t^e} - \tau^t}{\Delta t^e} - \mathbf{W}^t\tau^t + \tau^t\mathbf{W}^t \tag{10}$$

where the superscripts $t, t + \Delta t^e$ indicate values at the current and future timestep, respectively. Equations (5) and (8) become respectively

$$\tau^{t+\Delta t^e} = 2\frac{\eta\Delta t^e}{\alpha + \Delta t^e}\hat{\mathbf{D}}^{t+\Delta t^e} + \frac{\alpha}{\alpha + \Delta t^e}\tau^t + \frac{\alpha\Delta t^e}{\Delta t^e + \alpha}(\mathbf{W}^t\tau^t - \tau^t\mathbf{W}^t) \tag{11}$$

and

$$p^{t+\Delta t^e} = -\frac{\xi\Delta t^e}{\beta + \Delta t^e}D_{kk}^{t+\Delta t^e} + \frac{\beta}{\beta + \Delta t^e}p^t \ , \tag{12}$$

where $\alpha = \eta/\mu$ is the shear relaxation time and $\beta = \xi/K_e$ is the bulk relaxation time. We can simplify the above equations by defining an effective viscosity η_{eff} and an effective compressibility ξ_{eff}:

$$\eta_{eff} = \eta\frac{\Delta t^e}{\Delta t^e + \alpha} \quad \text{and} \quad \xi_{eff} = \xi\frac{\Delta t^e}{\Delta t^e + \beta} \tag{13}$$

Then the deviatoric stress is given by

$$\tau^{t+\Delta t^e} = \eta_{\text{eff}} \left(2\hat{\mathbf{D}}^{t+\Delta t^e} + \frac{\tau^t}{\mu \Delta t^e} + \frac{\mathbf{W}^t \tau^t}{\mu} - \frac{\tau^t \mathbf{W}^t}{\mu} \right) \tag{14}$$

and the pressure by

$$p^{t+\Delta t^e} = \xi_{\text{eff}} \left(D_{kk}^{t+\Delta t^e} - \frac{p^t}{\Delta t^e K_e} \right) . \tag{15}$$

To model an incompressible material K_e and ξ are made very large such that $D_{kk} \approx 0$.

Our system of equations is thus composed of a quasi-viscous part with modified material parameters and a right-hand-side term depending on values from the previous timestep. This approach minimizes the modification to the viscous flow code. Instead of using physical parameters for viscosity and bulk modulus, we use effective material properties (13) to take into account elasticity. The discretization of the stress rates in this manner produces an additional "force" term in the right-hand side of (1) commonly called the internal force which consist of stresses from the previous timestep or from initial conditions (which we refer to as stored stresses).

$$\mathbf{f}^{e,t} = \frac{\xi_{\text{eff}}}{K_e \Delta t^e} \nabla p^{t-\Delta t^e} - \frac{\eta_{\text{eff}}}{\mu \Delta t^e} \nabla \cdot \tau^{t-\Delta t^e} . \tag{16}$$

Implementation of Yielding

Starting from equation (9), we again express the Jaumann stress rate in difference form (in the Lagrangian particle reference frame) to give:

$$\tau^{t+\Delta t^e} \left[\frac{1}{2\mu \Delta t^e} + \frac{1}{2\eta} + \frac{\lambda}{2|\tau|} + \right] = \hat{\mathbf{D}}^{t+\Delta t^e} + \frac{1}{2\mu \Delta t^e} \tau^t + \frac{1}{2\mu} (\mathbf{W}^t \tau^t - \tau^t \mathbf{W}^t) . \tag{17}$$

No modification to the isotropic part of the problem is required when the von Mises yield criterion is used. At yield we use the fact that $|\tau| = \tau_{\text{yield}}$ to write

$$\tau^{t+\Delta t^e} = \eta' \left[2\hat{\mathbf{D}}^{t+\Delta t^e} + \frac{1}{\mu \Delta t^e} \tau^t + \frac{1}{\mu} (\mathbf{W}^t \tau^t - \tau^t \mathbf{W}^t) \right] \tag{18}$$

using an effective viscosity, η' given by

$$\eta' = \frac{\eta \tau_{\text{yield}} \mu \Delta t^e}{\eta \tau_{\text{yield}} + \tau_{\text{yield}} \mu \Delta t^e + \lambda \eta \mu \Delta t^e} . \tag{19}$$

We determine λ by equating the value of $|\tau^{t+\Delta t^e}|$ with the yield stress in (18). Alternatively, in this particular case, we can obtain η' directly as

$$\eta' = \tau_{\text{yield}} / |\hat{\mathbf{D}}_{\text{eff}}| , \tag{20}$$

where

$$\hat{\mathbf{D}}_{\mathrm{eff}} = 2\hat{\mathbf{D}}^{t+\Delta t^e} + \frac{1}{\mu\Delta t^e}\tau^t + \frac{1}{\mu}(\mathbf{W}^t\tau^t - \tau^t\mathbf{W}^t) \tag{21}$$

and $|\mathbf{D}| = (2D_{ij}D_{ij})^{1/2}$.

The value of λ or η' is iterated to allow stress to redistribute from points which become unloaded. The iteration is repeated until the velocity solution is unchanged to within the error tolerance required for the solution as a whole.

Computational Method

Having devised a suitable mathematical representation of the class of problems we wish to model, we need to choose a numerical algorithm which can obtain an accurate solution for a wide range of conditions. Our formulation, while simple, still contains many difficult problems which require particular attention.

Numerical Scheme

In fluid dynamics, where strains are generally very large however not important in the constitutive relationship of the material, it is common to transform the equations to an Eulerian mesh and deal with convective terms explicitly. Problems arise whenever advection becomes strongly dominant over diffusion since an erroneous numerical diffusion dominates. In our case, the advection of material boundaries and the stress tensor are particularly susceptible to this numerical diffusion problem. Mesh-based Lagrangian formulations alleviate this difficulty, although at the expense of remeshing and the eventual development of a less-than optimal mesh configuration which increases complexity and can hinder highly efficient solution methods such as multigrid iteration.

A number of mesh-free alternatives are available: smooth particle hydrodynamics (e.g., see MONAGHAN, 1992 and references therein) and discrete element methods (CUNDALL and STRACK, 1979) are common examples from the fluid and solid mechanics fields, respectively. These methods are extremely good at simulating the detailed behavior of highly deforming materials with complicated geometries (e.g., free surfaces, fracture development), and highly dynamic systems. They are generally formulated to calculate explicitly interactions on a particle-particle scale, which is usually impossible for creeping flow which has no inherent timescale for stress transfer.

We have developed a hybrid approach – a particle in cell finite element method which uses a standard Eulerian finite element mesh (for a fast, implicit solution) and a Lagrangian particle framework for carrying details of interfaces, the stress history, etc.

The Particle in Cell Approach

Our method is based closely on the standard finite element method, and is a direct development of the Material Point Method of SULSKY *et al.* (1995). Our particular formulation could best be described as a finite element representation of the equations of fluid dynamics with moving integration points.

A mesh is used to discretize the domain into elements, and the shape functions interpolate node points in the mesh in the standard fashion. Material points embedded in the fluid are advected using the nodal point velocity field interpolated by the shape functions. A typical updating scheme for the location, $\mathbf{x}_p$ of particle p is

$$\mathbf{x}_p^{t+\Delta t^p} = \mathbf{x}_p^t + \Delta t^p \sum_{\text{nodes}} \mathbf{v}_n N_n(\mathbf{x}_p) \;, \tag{22}$$

where $\mathbf{v}$ is the nodal velocity and N are the shape functions associated with the nodes, n, of the element in which the particle currently resides. Δt^p is the timestep used in advecting the particles. In practice, a higher order scheme such as second- or fourth-order Runge-Kutta produces a more accurate result. Particle updates can be done in a predictor-corrector fashion, although to date we have found no benefit in doing this.

The problem is formulated in a weak form to devise an integral equation, and the shape function expansion produces a discrete (matrix) equation. Equation (1) in weak form, using the notation of (2) becomes

$$\int_{\Omega} N_{(i,j)} \tau_{ij} \, d\Omega - \int_{\Omega} N_{,i} p \, d\Omega = \int_{\Omega} N_i f_i \, d\Omega \tag{23}$$

where Ω is the problem domain, and the trial functions, N, are the shape functions defined by the mesh; we have assumed that no nonzero traction boundary conditions are present. For the discretized problem these integrals occur over subdomains (elements) and are calculated by summation over a finite number of sample points within each element. For example, in order to integrate a quantity, ϕ over the element domain Ω^e we replace the continuous integral by a summation

$$\int_{\Omega^e} \phi \, d\Omega \leftarrow \sum_p w_p \phi(\mathbf{x}_p) \tag{24}$$

In standard finite elements the positions of the sample points, $\mathbf{x}_p$, andthe weighting, w_p are optimized in advance. In our scheme the $\mathbf{x}_p$'s correspond precisely to the Lagrangian points embedded in the fluid, and w_p must be recalculated at the end of a timestep for the new configuration of particles.

Constraints on the values of w_p originate from the need to integrate polynomials of a minimum degree related to the degree of the shape function interpolation, and the order of the underlying differential equation (e.g., HUGHES, 1987). These Lagrangian points carry the history variables which are therefore directly available for the element integrals without the need to interpolate from nodal points to fixed integration points. In our case, the distribution of particles is usually not ideal, and a unique solution for w_p cannot be found, or we may find we have negative weights which are not suitable for integrating physical history variables. We therefore store an initial set of w_p's based on a measure of local volume and adjust the weights slightly to improve the integration scheme.

There are additional complications involved in allowing integration points to move through the mesh. These include: the need to divide a particle in two when local strain becomes comparable to the size of the element in which it resides; and the need to develop an inverse mapping scheme to compute particle coordinates in the master element domain. For more details and benchmarks, see MORESI *et al.* (2001).

Elastic Timestep

Note that the timestep used in advecting particles, Δt^p, may differ from the timestep used to calculate elastic stress rates, Δt^e. This is a reflection of the fact that the elastic timestep is chosen from a physical perspective whereas the advection timestep comes from the numerical representation. In general we choose Δt^p to ensure that particles do not travel further than the typical dimensions of their local elements – a mesh based measure which it would be undesirable to see reflected in the elastic timestep. We may wish to impose an arbitrarily small timestep to resolve particle advection within a very fine mesh, or to account for a different physical process such as thermal diffusion or chemical reactions. To make Δt^e correspondingly small would produce a very low effective viscosity, increase the role of the explicit elastic terms making up the internal forces, and potentially destabilize the solution. For problems in which elasticity does not dominate the physical response of the system, this approach is not appropriate. In order to keep the two timesteps independent, however, an averaging scheme is necessary to ensure that the stress rate is computed over the appropriate interval. In the update of internal stress we now write:

$$\tau^t \leftarrow \phi \left(2\eta_{\text{eff}} \mathbf{D}^t + \frac{\eta_{\text{eff}}}{\mu \Delta t^e} \tau^{t-\Delta t^p} \right) + (1 - \phi)\tau^{t-\Delta t^p} , \tag{25}$$

where

$$\phi = \frac{\Delta t^p}{\Delta t^e} . \tag{26}$$

This amounts to a running average of the stress tensor at a material point over a time Δt^e, and accounts correctly for rotation.

BENCHMARKS

1-D Compression

Choosing a simple, first-order differencing scheme for the stress rate poses risks if we want to obtain a robust formulation. In particular, as elasticity comes to dominate a system (e.g., examining a shorter and shorter timescale) one might expect the first-order scheme to experience difficulties. Two questions arise. What is a suitable value of Δt^e to model a problem with a given relaxation time, and can we reliably approach sufficiently small values of Δt^e withoutloss of stability?

A simple system which can address these questions is the compression at constant velocity of a viscoelastic (relaxation time, $\alpha = 1$), compressibleunit square block. The velocity boundary condition, $v = 0.1$, was applied until the sample had undergone 90% shortening, then it was switchedoff and the stresses were allowed to relax without further deformation. We solved

$$\frac{1}{K_e}\frac{Dp}{Dt} + \frac{p}{\xi} = \frac{V}{h(t)} \quad , \tag{27}$$

where the height of the sample, $h(t)$ is computed from the applied compression velocity, V. The solution for p under loading $(V \neq 0)$ is

$$p(t) = \lambda(t) \exp\left(-\frac{K_e}{\xi}t\right) \quad , \tag{28}$$

where λ is determined by numerical integration using

$$\lambda = \int_0^t \frac{K_e V}{h(t)} \exp\left(\frac{K_e}{\xi}t\right) dt \quad . \tag{29}$$

During relaxation in response to past loading $(V = 0)$

$$p = p(t_0) \exp\left(-\frac{K_e}{\xi}t - t_0\right) \quad . \tag{30}$$

The pressure was benchmarked (Fig. 2) against the analytical solution for a given material ($\alpha = 1.0$), a given advection timestep ($\Delta t^p = 0.001$) and different values of the elastic observation time(Δt^e). It is not surprising that taking values of Δt^e comparable to or longer than the relaxation time produces inaccurate results. With $\Delta t^e = \alpha/10$ or smaller, very accurate results are obtained. There is no loss of accuracy associated with the fact that we take a running average of the stress-rate in order to decouple the elastic and particle-advection timesteps. The only limitation from elasticity on the choice of particle-advection timestep is to limit the maximum value such that $\Delta t^p \leq \Delta t^e$. We also observed no difference in the accuracy of the numerical solution for $\Delta t^e = 0.01$ for a range of $0.001 \leq \Delta t^p < 0.01$.

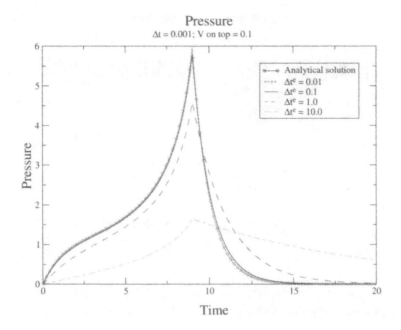

Figure 2

Numerical solution for pressure in a sample compressed in one direction for different elastic timesteps (Δt^e) with fixed relaxation time ($\alpha = 1$) and particle advection timestep ($\Delta t^p = 0.001$). Compression starts at $t = 0$ and continues until $t = 9.0$ (90% shortening) when the sample is allowed to relax without further deformation in either direction. For comparison, the analytic solution is shown by circles.

Cantilever Beam

We next benchmark a case with material interfaces moving across mesh boundaries, and where loading and relaxation are accompanied by rotation(evaluating the importance of corotational terms in the stress transport computation). A simple example is the gravitational loading of a thick, heavy beam of viscoelastic material. The beam dimensions were 0.1 wide by 0.75 thick, with viscosity of 10^7, elastic shear modulus of 10^6, density 2500 and perfectly rigid support at one end. Loading was conducted in a very low viscosity ($\eta = 0.01$) compressible, background medium. The beam was loaded by applying a gravitational body force at time $t = 0$, which was subsequently switched off when the deflection of the beam reached 0.35 at the unsupported end. We varied the intensity of the gravitational acceleration from $g = 5$ to $g = 500$ to produce responses with different characteristic times(owing to the viscous resistance of the embedding medium).

Figure 3 shows superimposed snapshots of the loading (a) and unloading (b) of the beam for $g = 500$. Under such strong loading, the beam reached the maximum allowed deflection very rapidly (relative to the viscous relaxation time) and the load was then released. Once unloaded, unrelaxed elastic stress in the beam caused it to

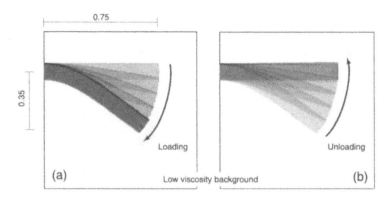

Figure 3

A heavy viscoelastic beam of length 0.75 is subjected to a gravitational load at time $t = 0$ which is released when the end deflection reaches 0.35. The superimposed configurations correspond to approximately equal time intervals.

straighten out. The loading time was sufficiently short that negligible viscous deformation occurred, allowing the beam to return to within 0.001 of its original location.

The deflection of the mid-point at the unsupported end of the beam is shown in Figure 4 for the range of loading intensities. Under light loading ($g = 5, 9$) the initial deflection of the beam is rapid, dominated by elastic deformation of the beam and the viscous resistance of the embedding medium. In the absence of viscous terms, the elastic stresses in the beam would come to an equilibrium with gravitational loading. However, After $t = 0.0005$, there was a small but observable rate of end deflection still occurring through gradual viscous relaxation of the elastic stresses. When the gravitational load was released, this relatively long period of viscous relaxation resulted in a permanent deflection of 0.1.In the case $g = 5$, we observed the initial rapid deflection due to elastic deformation of the beam, and the gradual deflection after $t = 0.0005$. However, the rate of viscous deformation was so low that the end deflection of 0.35 was not reached within the time limit imposed for the experiment.

The fast-loading cases, $g = 20, g = 500$, suffered very little viscous relaxation before unloading occurred. The stress state at the time of unloading was therefore similar for the two cases. Although the loading rates were very different, the unloading displacement curves remained almost parallel until most the elastic stresses had been released.

To demonstrate the importance of computing the stress update correctly, we ran the simulations omitting the corotational terms in the stress rate (dashed lines in Fig. 4). The results were qualitatively similar, however the cumulative result of not allowing for stress rotation was greater viscous deformation. This resulted in larger permanent deformation at the end of the simulation, particularly noticeable in the

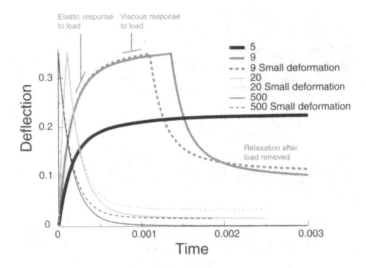

Figure 4

Deflection as a function of time for the heavy viscoelastic beam for numerous different load intensities. Dashed lines correspond to simulations where the corotational terms were omitted from the computation of the stress rate.

$g = 500$ case where there is virtually no final deformation if the corotational stress rate is computed correctly.

Extension of Test Sample

The yielding algorithm is benchmarked by measuring the second invariant of the stress and displacement at points within a viscoelastic beam which was extended or compressed at a fixed rate, $v = 5$, by an imposed velocity boundary condition at one end. Figure 5 indicates the geometry of the numerical experiment: the mesh was initially 3 units long by 1 unit high. The sample was 0.5 units thick, occupying the central half of the mesh, and was surrounded by a low viscosity, compressible

Figure 5

Geometry for simulation of the extension of a viscoelastic bar with yield stress.

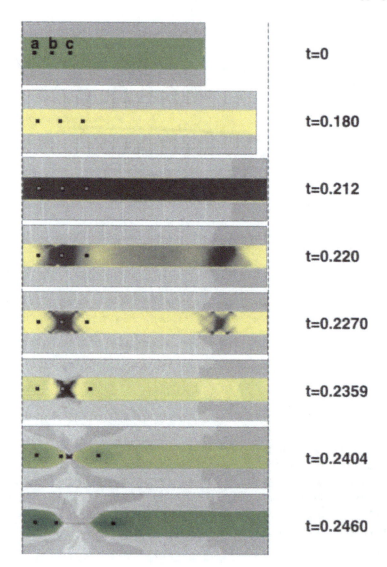

Figure 6

Simulation of the extension of a viscoelastic bar with yield stress. Black shading indicates regions deforming at yield. Embedded marker points which follow the material deformation are indicated by a,b,c.

material. Three sampling points (a,b,c) for recording the stress invariant and displacement were chosen within the sample initially placed along the mid-line at $x = 0.2, 0.5, 0.8$.

The material parameters of the sample ($\eta = 10^8$, $\mu = 10^6$) were chosen such that the relaxation time was long ($\alpha = \eta/\mu = 100$) compared to the duration of the experiment (0.25) so that the material behaved nearly as an elastic solid.

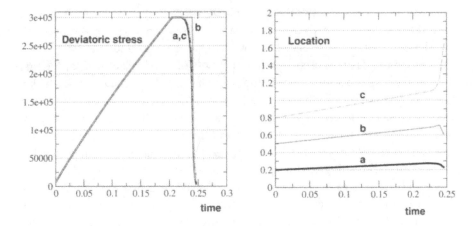

Figure 7

Stress and displacement at sample points a,b,c as a function of time for the extension experiment of
Figure 6.

Figure 6 shows the progress of the experiment. Initially, deformation was
uniform, resulting in gradual stretching of the sample ($t \leq 0.180$). The entire sample
reached the yield point at the same time ($t = 0.212$) and initially deformed uniformly
with all points yielding. However, the deformation soon localized to a number of
shear bands ($t = 0.220$), then to two places along the sample ($t = 0.227$), and finally
to a single location ($t = 0.2359$)which focussed all subsequent deformation until the
sample failed entirely ($t \geq 0.2404$). The frames are not uniformly spaced in time since
the post-failure behavior occurred on a considerably shorter timescale than the
gradual loading. Note, for example, that the necking and separation of the two parts
of the sample occurred with barely any movement of the end boundary.

Even with a material which has no strain softening, there is a tendency for
deformation to localize in a particle-in-cell representation of the sample. This occurs
because the sample boundary is never perfectly flat (as in real life) due to numerical
fluctuations in the particle locations, and to a mild interference (Moiré) effect
between the array of particles and the underlying grid. These effects produce small
fluctuations in the stress field which can result in early failure at certain points. Once
nucleated, shear bands can propagate from these points – ultimately resulting in
necking and complete separation of the two halves of the sample.

In Figure 7a, we plot the stress at each of the sample points in the material as a
function of time for a fixed end velocity. The evolution of stress within the beam was
close to linear – apart from the influence of the changing of the beam thickness during
deformation. The yield stress of the material was 3×10^5. The stress increased within
the sample at the same rate for all the sample points until the yield stress was reached.
At this stage, the stress was not able to increase any further, and the material
deformed uniformly at the yield stress. Once localization had occurred, however,

points outside the necking area begin to unload, and the stress dropped dramatically. The rate at which stress drops from yield back to zero is governed by the viscous part of the rheology, and the presence of a low viscosity background material.

The unloading is more clearly seen in the plots of the displacement of the sample points through time in Figure 7b. Before yielding, the displacement of each sample point increases monotonically. Once yielding occurs, and the deformation localizes, the sample points on the left of the break (a,b), under the action of stored elastic stresses, rapidly retreat towards their original locations. The sample point on the right of the break (c) moves rapidly to the right as the elastic deformation relaxes.

It is worth discussing at this point a consequence of the fact that the yield criterion only applies to the deviatoric stress. During the separation of the layer, the pressure becomes enormous at the constriction, which obviously could not occur in a real material. To model this situation in a more realistic manner we would need to complement the yield criterion on the deviatoric stress with a suitable tension cutoff condition.

Application to Plate Dynamics

Physical Model Description

We treat the earth on a large scale as an incompressible, viscoelastic Maxwellfluid with infinite Prandtl number in which motions are driven by internal temperature variations. The force term from equation (1) is a gravitational body force due to density changes. We assume that these arise, for any given material, through temperature effects:

$$\nabla \cdot \tau - \nabla p = g\rho_0(1 - \alpha_T T)\hat{\mathbf{z}} , \tag{31}$$

where g is the acceleration due to gravity, ρ_0 is material density at a reference temperature, α_T is the coefficient of thermal expansivity, and T is temperature. $\hat{\mathbf{z}}$ is a unit vector in the vertical direction. We have also assumed that the variation in density only needs to be considered in the driving term (the Boussinesq approximation: BOUSSINESQ, 1903).

The equation of motion is then

$$\nabla(\eta_{\text{eff}}\mathbf{D}^{t+\Delta t^e}) - \nabla p = g\rho_0(1 - \alpha_T T)\hat{\mathbf{z}} - \nabla\left(\eta_{\text{eff}}\left[\frac{\tau^t}{\mu\Delta t^e} + \frac{\mathbf{W}^t\tau^t}{\mu} - \frac{\tau^t\mathbf{W}^t}{\mu}\right]\right) . \tag{32}$$

The velocity field $\mathbf{u}$ and pressure at $t + \Delta t^e$ can be solved for a given temperature distribution and the stress history from the previous step.

Motion is driven by the heat escaping from the interior. The energy equation governs the evolution of the temperature in response to diffusion of heat through the fluid. For a given element of fluid,

$$\frac{DT}{Dt} = -\kappa\nabla^2 T \ , \tag{33}$$

where κ is the thermal diffusivity of the material.

Rheology

The viscosity of the mantle at long timescale is known to be a complicated function of temperature, pressure, stress, grain-size,composition (particularly water content) etc. KARATO and WU (1993) give the following expression for mantle deformation

$$\eta = \frac{1}{2A}\left(\frac{\mu}{\tau}\right)^{n-1}\left(\frac{h}{b^*}\right)^m \exp\frac{E^* + PV^*}{RT} \ , \tag{34}$$

where A is a constant, μ is shear modulus, b^* is the Burgers vector, T is temperature, τ is the second invariant of the deviatoric stress tensor, E^* is an activation energy, V^* and activation volume, R is the gas constant, h is the grain size, n is a stress exponent, and m a grain-size exponent. Despite this complexity, the dominant effect on the viscosity from the point of view of the large-scale dynamics of the system is the effect of temperature (e.g., SOLOMATOV, 1995). To minimize the dimensionality of the parameter space we need to study, it is common to use the single-parameter Frank-Kamenetskii (McKENZIE, 1977; MORESI and SOLOMATOV, 1995) approximation to the viscosity law to obtain the following simplification,

$$\eta = A' \exp(-CT) \ . \tag{35}$$

In making these assumptions, we have implicitly required that the shear modulus, μ does not vary with temperature. This is an assumption justified by seismological observations which ascribe no more than a few percent variation in seismic velocity to thermal perturbations (e.g., WOODHOUSE and DZIEWONSKI, 1984). In comparison to its influence on viscosity, the influence of temperature on shear modulus can be neglected. Consequently, we assume that the relaxation time of the mantle and lithosphere is entirely controlled by the viscosity. We expect to see high relaxation times only in the slowly-deforming lithosphere, and negligible relaxation times in the fast moving interior. It is this fact which makes it whatever possible to solve this problem: without relaxation of stresses in the rapidly evolving interior, we would expect unresolvably small-scale structures dominated by elastic stresses to control convection in the mantle.

Mantle Convection Simulation

The benchmarking demonstrated how complex behavior is present even in systems with the most elementary boundary conditions. We cannot hope to present a comprehensive examination of the influence of viscoelastic effects in the lithosphere when driven by mantle convection, and where yielding may take place at high stresses. Instead we present three convection models with different relaxation times as an indication of the styles of interaction we expect to see.

The reference simulation is for viscous convection with highly temperature dependent viscosity given by

$$\eta = 10^5 \exp(-11.5129T') \tag{36}$$

where T' is dimensionless temperature varying from 0 at the surface to 1 at the base. This produces a viscosity contrast of 10^5 across the entire layer. The Rayleigh number based on the interior viscosity is 3×10^6. The yield stress is given by

$$\tau_y = 10^4 + 0.4p' \;, \tag{37}$$

where p' is the dimensionless hydrostatic pressure. There is also a strain-softening effect: the yield stress is reduced by a factor of two linearly with strain accumulated at the yield point up to a strain of 0.5. Above this strain, the yield stress remains constant. If a material point is heated above a dimensionless temperature of $T' = 0.5$, the accumulated strain is reset.

Simulations of this kind are capable of generating considerable realism in certain respects. For example, LENARDIC *et al.* (2000), examine the interaction between subducting slabs and mobile belts using this approach with a viscous formulation. The first-order behavior of such systems is similar to that of the earth's tectonic plates: surface velocity are comparable to interior ones, large regions of the surface deform at very low strain rates compared to narrow boundary regions, and so on. However, in more detail the simulations fail to predict the dynamic behavior of subducting slabs. For example, slabs rarely roll-back (retrograde subduction) in the virtual mantle, whereas this is a near ubiquitous behaviour in the real Earth (ELSÄSSER, 1971). Can elastic effects, which tend to unbend the slab, play a role in producing roll-back?

In the absence of elasticity the convection simulation evolved to a steady state condition with a single downwelling, a nearby surface divergence and a characteristic "rolling-forward" of the downwelling. The upper thermal boundary layer was inverted onto the lower boundary (Fig. 8a). When we introduced elastic stresses the downwelling was immediately seen to be considerably straighter (Fig. 8b) in the upper part of the mantle due to the strong elastic stresses (Fig. 8c). The evolution was also backward-rolling, i.e., with the upper thermal boundary layer landing the right way up on the lower boundary after 'subduction'.

This status was temporary, however, as the system failed to reach a thermal steady state. Strong oscillations in the surface heat transport were observed (Fig. 9a compared to 9b). The system soon evolved from rolling-back to a state with a more

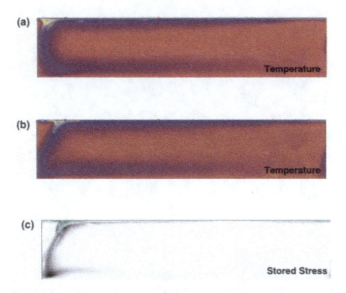

Figure 8

Example: Snapshot from two convection simulations showing (a,b) the thermal field (and regions of yielding), and (c) the magnitude of the stored stresses. The simulation in (a) has no elasticity, the simulation in (b,c) has a maximum relaxation time at the surface of 0.002 compared to a surface velocity of 100. Regions with high strain rate due to yielding are shown in yellow.

symmetrical, stationary downwelling, and subsequently to a highly time-dependent state with rolling-forward of the downwelling.

When the relaxation time was increased from 0.002 to 0.005, the system evolution became unexpectedly like the viscous case. The downwelling continued to roll forwards, with slight fluctuations emerging from short spells where the downwelling began to unbend. These fluctuations in the geometry of the downwelling were minor, however, with a mild influence on surface velocity, and, consequently, observed heat flow (Fig. 9c).

The influence of elasticity is to produce a pronounced tendency for the downwellings to roll backwards. In purely viscous models the tendency is to roll forwards. However, the effect of increasing the elastic stresses in the end is to modify the manner in which the lithosphere yields. This highlights the unpredictability of complex nonlinear systems and serves as a warning that application of these simulations to modeling of plate tectonics requires considerable care and a thorough attention to data which constrain the evolution of specific plate boundaries.

Obviously, this model is substantially simplified compared to the earth – the lack of a third dimension, curvature, and continents are clearly deficiencies which must be addressed. Fortunately, however, the introduction of these complicating details does not require any new algorithm development, only considerably greater computational resources.

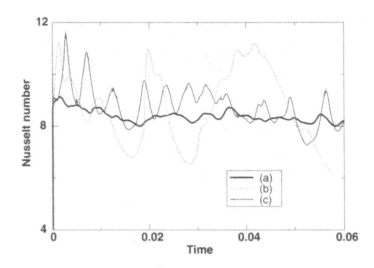

Figure 9

Plot of Nusselt number (a measure of hear transport efficiency) against time for three simulations with (a) purely viscous rheology, (b) relaxation time 0.002, (c) relaxation time 0.005. Relaxation times are based on the viscosity at the surface.

Discussion

We have demonstrated a model consisting of a set of rheological relations and a computational scheme for viscoelastic-plastic materials. The algorithm is designed to introduce elastic effects into convection simulations where temperature-dependent viscosity and yielding dominate the mechanical behavior. The viscosity of the mantle and the mantle lithosphere is very strongly dependent on temperature(several orders of magnitude variation over 1000°C) whereas the shear modulus is not strongly affected (there is only a modest change in seismic wavespeed due to temperature). Therefore, elastic effects become unimportant outside the cold thermal boundary layer where viscosity is extremely large. The influence of elastic stresses is likely to be felt at the subduction zones where the lithosphere is bent into the interior of the earth. In these regions stresses are typically close to the yield stress – a fact which allows the plates to move in the first place.

Our methodology is limited to a coarse continuum description of the subduction zone system at a resolution of a few km. This may give us valuable information into the nature of plate tectonics, the thermal conditions in and around subducting lithosphere, and the stress state of the system. However, the resolution is too coarse to provide any information about the detailed mechanics of the failure of lithospheric fault zones and the conditions for a major failure to occur. For this we require a coupling of the large-scale code with an engineering-scale code (e.g., DEM or small-deformation Lagrangian FEM) using the large-scale to provide boundary conditions for the small scale. The issue of scale-bridging is important in many areas of numerical simulation. Essentially the same difficulties arise in material science where the atomic scale is best treated by molecular dynamics codes however the large scale must be treated as a continuum (e.g., BERNHOLC, 1999).

REFERENCES

BARAZANGI, M. and DORMAN, J. (1969), *World seismicity maps compiled from ESSA, Coast and Geodetic Survey, Epicenter Data 1961–1967*, Bull. Seismol. Soc. Am. *59*, 369–380.

BERNHOLC, J. (1999), *Computational Materials Science: The Era of Applied Quantum Mechanics*, Physics Today *52*, (9), 30–35.

BOUSSINESQ, J. *Théory analytique de la chaleur mise en harmonie avec la thermodynamique et avec la théory de la Lumière*. Vol II, (Gauthier-Villars, Paris (1903)).

BYERLEE, J. D. (1978), *Friction of Rocks*, Pure Appl. Geophys. *116*, 615–629.

CUNDALL, P. A. and STRACK, O. D. L. (1979), *A Discrete Numerical Model for Granular Assemblies*, Geotechnique *29* (1), 47–64.

ELSASSER, W. M. (1971), *Sea-floor Spreading as Thermal Convection*, J. Geophys. Res. *76*, 1101–1112.

GURNIS, M., ELOY, C., and ZHONG, S. (1996), *Free-surface formulation of mantle convection—II. Implications for subduction zone observables*, Geophys. J. Int. *127*, 719–727.

HARDER, H. (1991) *Numerical Simulation of Thermal Convection with Maxwellian Viscoelastity*, J. Non-Newt. Fl. Mech. *39*, 67–88.

HUGHES, T. J. R. The Finite Element Method, (Prentice-Hall, New Jersey, 1987)

KARATO, S.-I. and WU, P. (1993), *Rheology of the Upper Mantle: A Synthesis*, Science *260*, 771–778.

LENARDIC, A., MORESI, L., and MÜHLHAUS, H.-B. (2000), *The Role of Mobile Belts for the Longevity of Deep Cratonic Lithosphere: The Crumple Zone Model*, Geophys. Res. Lett. *27*, 1235–1239.

MCKENZIE, D. P. (1977), *Surface Deformation, Gravity Anomalies, and Convection*, Geophys. J. R. Astr. Soc. *48*, 211–238.

MCKENZIE, D. (1984), *The Generation and Compaction of Partially Molten Rock*, J. Petrology *25*, 713–765.

MELOSH, H. J. (1978), *Dynamic Support of the Outer Rise*, Geophys. Res. Lett. *5*, 321–324.

MONAGHAN, J. J. (1992), *Smoothed Particle Hydrodynamics*, Annu. Rev. Astron. Astrophys. *30*, 543–574.

MORESI, L., MÜHLHAUS, H.-B., and DUFOUR, F., *Viscoelastic Formulation for Modelling of Plate Tectonics*, In *Bifurcation and Localization in Soils and Rocks 99. (Mühlhaus, Dyskin, A. and Pasternak, E., ed, (Balkema, Rotterdam, 2001)*.

MORESI, L. and SOLOMATOV, V. S. (1995), *Numerical Investigations of* 2-D *Convection in a Fluid with Extremely Large Viscosity Variations*, Phys. Fluids *7*, 2154–2162.

MORESI, L. and SOLOMATOV, V. S. (1998), *Mantle Convection with a Brittle Lithosphere: Thoughts on the Global Tectonics Styles of the Earth and Venus*, Geophys. J. Int. *133*, 669–682.

OGAWA, M. (2001), *The Plate-like Regime of a Numerically Modelled Thermal Convection in a Fluid with Temperature-, Pressure-, and Stress-history-dependent Viscosity*, J. Geophys. Res., submitted.

PELTIER, W. R. (1974), *The Impluse Response of a Maxwell Earth*, Rev. Geophys. Space Phys. *12*, 649–669.

PODLACHIKOV, YU., LENARDIC, A., YUEN, D. A., and QUARENI, F. (1993), *Dynamical Consequences of Stress Focussing for Different Rheologies: Earth and Venus Perspectives*, EOS Trans. AGU, *74* (43) Suppl., 566.

SCHMALHOLTZ, S. M., PODLADCHIKOV, Y. Y., and SCHMID, D. W. (2001), *A Spectral/Finite Difference Method for Simulating Large Deformations of Heterogeneous, Viscoelastic Materials*, Geophys. J. Int. *145*, 199–208.

SOLOMATOV, V. S. (1995), *Scaling of Temperature- and Stress-dependent Viscosity Convection*, Phys. Fluids, *7*, 266–274.

SULSKY, D., ZHOU, S.-J., and SCHREYER, H. L. (1995), *Application of a Particle-in-cell Method to Solid Mechanics*, Comput. Phys. Commun. *87*, 236–252.

TACKLEY, P. J. (1998), *Self-consistent Generation of Tectonic Plates in Three-dimensional Mantle Convection*, Earth Planet. Sci. Lett. *157*, 9–22.

TACKLEY, P. J. (2000), *The quest for self-consistent generation of plate tectonics in mantle convection models*, AGU Monograph on *The History and Dynamics of Global Plate Motions*, (ed. M. Richards), in press.

WATTS, A. B., BODINE, J. H., and RIBE, N. M. (1980), *Observations of Flexure and the Geological Evolution of the Pacific Basin*. Nature, 283, 532–537.

WOODHOUSE, J. H. and DZIEWONSKI, A. M. (1984), *Three dimensional Mapping of Earth Structure by Inversion Of Seismic Waveforms*, J. Geophys. Res. *89*, 5953–5986.

(Received February 20, 2001, revised June 11, 2001, accepted June 15, 2001)

 To access this journal online:
http://www.birkhauser.ch

Pure appl. geophys. 159 (2002) 2357–2381
0033–4553/02/102357–25 $ 1.50 + 0.20/0

| Pure and Applied Geophysics

GEM Plate Boundary Simulations for the Plate Boundary Observatory: A Program for Understanding the Physics of Earthquakes on Complex Fault Networks via Observations, Theory and Numerical Simulation

JOHN B. RUNDLE,[1] PAUL B. RUNDLE,[2] WILLIAM KLEIN,[3]
JORGE DE SA MARTINS,[4] KRISTY F. TIAMPO,[4]
ANDREA DONNELLAN,[5] and LOUISE H. KELLOGG[6]

Abstract — The last five years have seen unprecedented growth in the amount and quality of geodetic data collected to characterize crustal deformation in earthquake-prone areas such as California and Japan. The installation of the Southern California Integrated Geodetic Network (SCIGN) and the Bay Area Regional Deformation (BARD) network are two examples. As part of the recently proposed Earthscope NSF/GEO/EAR/MRE initiative, the Plate Boundary Observatory (PBO) plans to place more than a thousand GPS, strainmeters, and deformation sensors along the active plate boundary of the western coast of the United States, Mexico and Canada (http://www.earthscope.org/pbo.com.html). The scientific goals of PBO include understanding how tectonic plates interact, together with an emphasis on understanding the physics of earthquakes. However, the problem of understanding the physics of earthquakes on complex fault networks through observations alone is complicated by our inability to study the problem in a manner familiar to laboratory scientists, by means of controlled, fully reproducible experiments. We have therefore been motivated to construct a numerical simulation technology that will allow us to study earthquake physics via numerical experiments. To be considered successful, the simulations must not only produce observables that are maximally similar to those seen by the PBO and other observing programs, but in addition the simulations must provide dynamical predictions that can be falsified by means of observations on the real fault networks. In general, the dynamical behavior of earthquakes on complex fault networks is a result of the interplay between the geometric structure of the fault network and the physics of the frictional sliding process. In constructing numerical simulations of a complex fault network, we will need to solve a variety of problems, including the development of analysis techniques (also called data mining), data assimilation, space-time pattern definition and analysis, and visualization needs. Using

[1] Colorado Center for Chaos and Complexity, CIRES, and Department of Physics, CB 216, U.S.A. E-mail: rundle@cires.colorado.edu
[2] Department of Physics, 301 E 12th St., Harvey Mudd College, Claremont, CA 91711, U.S.A. E-mail: paulrun@cires.colorado.edu
[3] Department of Physics, Boston University, Boston, MA 02215, and Center for Nonlinear Science, Los Alamos National Laboratory, Los Alamos, NM 87545, U.S.A. E-mail: klein@cnls.lanl.gov
[4] Colorado Center for Chaos and Complexity, and CIRES, CB 216, University of Colorado, Boulder, CO 80309, U.S.A. E-mails: jorge@cires.colorado.edu; kristy@cires.colorado.edu
[5] Exploration Systems Autonomy Division, Jet Propulsion Laboratory, Pasadena, CA 91109-8099, U.S.A. E-mail: donnellan@jpl.nasa.gov
[6] Department of Geology, University of California, Davis, CA 95616, U.S.A. E-mail: kellogg@geology.ucdavis.edu

simulations of the network of the major strike-slip faults in southern California, we present a preliminary description of our methods and results, and comment upon the relative roles of fault network geometry and frictional sliding in determining the important dynamical modes of the system.

Key words: Earthquakes, numerical simulations, earthscope.

Introduction

Earthquakes in urban centers are capable of causing enormous damage. The recent January 16, 1995 Kobe, Japan earthquake was only a magnitude 6.9 event and yet produced damages estimated to range from $95 billion to $147 billion or more (http://www.eqe.com/publications/kobe/economic.htm). Despite an active earthquake prediction program in Japan, this event was a complete surprise. Similar scenarios are possible in Los Angeles, San Francisco, Seattle, and other urban centers around the Pacific plate boundary. It has been estimated (http://www.eqe.com/revamp/main.htm) that a repeat of the 1906 San Francisco earthquake on the northern San Andreas fault, or a repeat of the 1857 Fort Tejon earthquake near Los Angeles, may well cause in excess of $1 trillion in damages, together with many thousands of casualties and deaths. In an increasingly economically interdependent nation and world, such destruction can no longer be tolerated.

The Earthscope Plate Boundary Observatory (PBO)

The Plate Boundary Observatory of the NSF Earthscope initiative (http://www.eqe.com/revamp/main.htm) is a project that plans to deploy more than 1000 GPS, strainmeter, and ancillary sensors in several clusters of locations along the western boundary of the United States, Canada, and Mexico in an effort to understand processes associated with plate boundaries, including the physics of earthquakes. Current plans for the PBO entail comprehensive, systematic observations along the plate boundary. However, the PBO proposal, which outlines both the goals and the proposed design of the network, also states that numerical simulations are necessary to plan instrument deployments that can effectively answer the major scientific questions, to develop testable hypotheses, and to understand the basic physics of earthquakes. To date, the effort put towards simulations has been limited to the scale of individual faults or regions, and we are not aware of any significant planning for simulations of the entire San Andreas fault system as a coherent entity. For that reason, we describe here the development and preliminary use of just such a *Plate Boundary Simulator* (PBS).

A major thrust of the PBO is to gain a vastly improved knowledge of the dynamics and physical processes along the San Andreas fault system, with a view towards significant progress in earthquake forecasting and prediction. An observational system at this scale and expense demands careful thought in the design of the instrumental arrays, the conduct of the observations, and the analysis of the resulting data. Simulations of earthquake processes will result in development of multiple models and scientific questions, while providing guidelines for design of instrument deployments that can distinguish between competing hypotheses.

Numerical simulation approaches (e.g., RUNDLE, 1988; RUNDLE and KLEIN, 1995; LAPUSTA et al., 2000; WARD, 2000) also provide a "laboratory" in which the physics of earthquakes can be investigated using a series of controlled and fully reproducible numerical experiments. Physical processes of complex fault networks on a wide range of spatio-temporal scales such as friction, fluid flow and the branching and interaction of fractures could then be integrated into a common framework. Using these simulations, diverse data sets taken at different scales can be interrelated and reconciled, with predictions extrapolated to motivate and guide further observational studies. As a specific example, the optimal placement of sensors along the fault system can be determined from simulations, given the scientific questions being asked. Intuitively, it is clear that some locations for GPS instruments will yield considerably more information than others about earthquake nucleation zone properties, earthquake physics and fault zone dynamics, as well as structural properties such as asthenosphere viscosities, lithosphere thicknesses, and crustal permeabilities, for given fixed time and spatial scales. Realistic simulations will also allow hypotheses to be framed in ways not currently possible, as will become apparent below. While we recognize that, in practice, sensor network design may depend to a large extend on logistical considerations including local stability and legal property rights issues, we feel that this provides an even more important role for simulation to ensure that the proposed networks address the scientific questions being asked.

Nature of Earthquake Physics — Space and Time Scales

Complex, nonlinear earthquake fault networks exhibit a wealth of emergent, dynamical phenomena over a large range of spatial and temporal scales, including space-time clustering of events, self-organization and scaling (e.g., SCHOLZ, 1990). Examples of the latter include the Gutenberg-Richter magnitude-frequency relation, and the Omori law for aftershocks (and foreshocks). The physical processes associated with earthquakes also occur on a wide range of spatial and temporal scales.

A few of the obvious spatial scales for physical fault geometries include:

The microscopic scale ($\sim 10^{-6}$ m to 10^{-1} m) associated with static and dynamic friction (the primary nonlinearities associated with the earthquake process).

The fault-zone scale ($\sim 10^{-1}$ m to 10^{2} m) that features complex structures containing multiple fractures and crushed rock.

The fault-network scale (10^2 m to 10^4 m), in which faults are seen to be neither straight nor simply connected, but in which bends, offsetting jogs and subparallel strands are common and known to have important mechanical consequences during fault slip.

The regional fault-network scale (10^4 m to 10^5 m), where seismicity on an individual fault cannot be understood in isolation from the seismicity on the entire regional network of surrounding faults, and where concepts such as "correlation length" and "critical state" borrowed from statistical physics have led to new approaches to understanding regional seismicity.

The tectonic plate-boundary scale (10^5 m to 10^7 m), at which Planetary Scale boundaries between plates can be approximated as thin shear zones and the motion is uniform at long time scales.

Computational simulations must address the same range of spatial scales as for the San Andreas fault studies of the PBO, roughly between the regional fault-network scale and the tectonic plate-boundary scale, and link the physics operating on these scales with the physics observed in the laboratory and at other scales. Simulations must also lead to understanding the origins and implications of the space-time correlations and dynamical patterns in these fundamentally multi-scale phenomena.

Plate Boundary Data

The best current examples of the kind of data that can be collected on the San Andreas plate boundary is found in southern California. In that region, there exists an instrumental record of activity dating to $\sim$ 1930; a historically recorded (primarily in newspapers and other literature) record of activity dating to $\sim$ 1800; and a paleoseismic record of activity dating to $\sim$ 600 AD (SIEH *et al.*, 1989; SCHOLZ, 1990). An example that includes all epicenters of the major historical events larger than magnitude 4.5 is shown in Figure 1 (http://www.scecdc.scec.org/clickmap/). More recently, GPS data from the SCIGN array has recorded deformation from events such as the October 16, 1999 Hector Mine, CA event (Figure 2). Surface deformation from this event was also recorded by InSAR images (http://milhouse.jpl.nasa.gov/hector/). These data, which characterize both the seismicity, as well as deformation changes in the crust of southern California, are examples of the kind of data that will be gathered by the PBO. Therefore, the construction of useful models and simulations to explain these data must be capable of representing the detailed, three-dimensional geometry of the fault network that produced these observations; it must use modern ideas about friction that arise both from laboratory studies, field constraints, and theoretical ideas; it must utilize current ideas about fault system loading due to plate tectonic forces; and it must of course reproduce all types of data (e.g., seismicity, surface

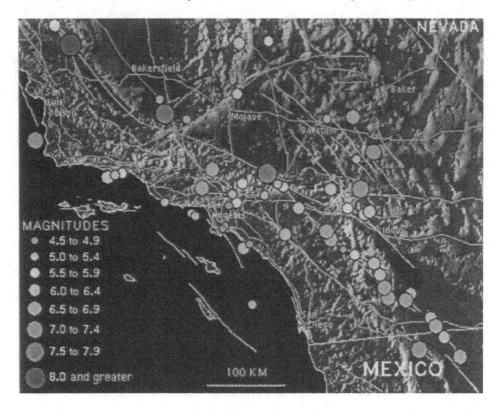

Figure 1

Historic seismicity superposed on the major active faults in the Southern California region. Circles are centered on epicenters of events, and size of circle is keyed to magnitude of event. (Source: http://www.scecdc.scec.org/clickmap.html)

deformation, etc.) that will be measured by the PBO and other instruments, to appropriately and arbitrarily detailed degrees.

The types of data discussed here include only very limited information about seismic waveforms, the only use of which in Figure 1 is in locating earthquakes and computing their magnitudes. Since the data collection activities of the PBO, and the scientific questions being asked are fundamentally at periods considerably longer than typical seismic wave periods, we focus our simulation efforts at time scales that are significantly longer than source process times (tens of seconds). Moreover, since the initial stress state in the earth, the frictional properties of faults, and the effects of small-scale structure will basically always be unknown at the level of detail required, we are necessarily led to consider a theoretical framework in which the dynamics of sliding is treated as a stochastic process, the implications of which we elaborate below. These considerations lead to the construction of numerical simulations that are geometrically realistic, based on laboratory and field data, and yet remain computationally tractable.

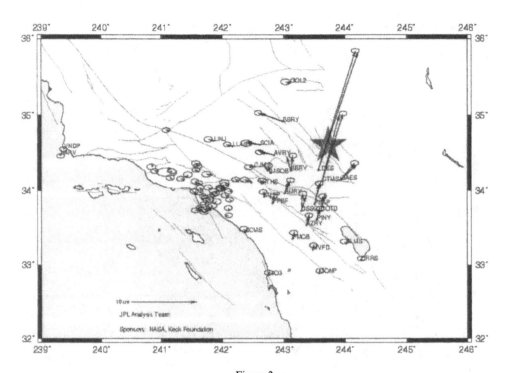

Figure 2
Deformation vectors associated with the Hector Mine earthquake of October 16, 1999 in the Mojave Desert of California. Epicenter of the earthquake is at the center of the star. Twin sets of vectors indicate the data (vector tip located at center of ellipse) together with best-fitting model (for details, refer to: http://milhouse.jpl.nasa.gov/hector/). Figure produced by the SCIGN analysis group at the Jet Propulsion Laboratory.

Simulation Methods and Problems

Physical processes that are known to be important in the earthquake generation process (RUNDLE, 1988; GROSS and KISSLINGER, 1994; RUNDLE *et al.*, 2000a; STEIN, 1999; TULLIS, 1996; KARNER and MARONE, 2000; PERSSON, 1998; SCHOLZ, 1990; HARRIS, 1998) include: 1) Transfer of stress from one fault to another by means of elastic, viscoelastic, and other interactions; 2) frictional and/or other material instabilities acting on the fault surface that prevent stress from becoming larger than a spatially (and possibly temporally) varying threshold value $\sigma^F(\mathbf{x}, t)$, resulting in sudden, overdamped sliding to a new state of displacement; 3) persistently increasing stress accumulation due to action of the tectonic plates driven by convection forces in the earth's mantle; and 4) the detailed three-dimensional geometry of the fault surfaces themselves, leading to branching, scale-invariant geometric structures, and new dynamical modes of behavior arising from excitation and inhibition from the dynamical influence of one fault upon another.

   These physical processes exist at all scales, and the parameters that describe their physical effects must be considered scale-dependent quantities whose ranges of values must be obtained from observations (RUNDLE et al., 2000b, 2001). However, numerical simulations, and to a much greater degree laboratory experiments, can only be carried out over a finite range of scales. Of necessity, there will then be some minimum scale at which the physics must be considered as unmodeled subgrid scale processes. So in addition to the four physical processes above must be added 5) subgrid scale processes, which must be treated as either random or correlated noise.

   To be useful in addressing the scientific questions discussed above, the Plate Boundary Simulator must include the physical processes described above, particularly the requirement of realistic geometric complexity. As a consequence, these requirements, together with the finite nature of present computational resources, mean that many physical processes can either not be represented, can be represented only at large scales, or can be represented in only simple ways. Part of the task of research in constructing and using a Plate Boundary Simulator (PBS) is then to evaluate which physical processes must be present in the simulations and to what degree, depending on the problem to be solved. For example, other physical processes that are known to be characteristic of earthquakes that have not been included in the list above include elastic waves; chemical alteration, grinding and wearing of rocks in the fault zone; creation of fresh fault surface through fracturing processes, and destruction of older active fault surfaces by various means. In particular, elastic waves, which transmit the stress from the active slip site throughout the earth's crust, are subject to a multiplicity of reflection, refraction and attenuation processes due to the material heterogeneity in the rock, and thus demand enormous computational resources to accurately compute the resulting motion.

   Fortunately however, elastic wave motion turns out not to be a critical component of a PBS for several reasons. First, the PBO data stream from GPS, strainmeters, and InSAR will be sensitive only to deformation changes at time intervals of several tens of minutes and longer, intervals that are considerably longer than any earthquake source process times. Second, the primary goal of a PBS can be stated as understanding the multi-scale properties of fault interaction, stress transfer, space-time earthquake patterns, and slip partitioning (http://www.earthscope.org/pbo.com.html). To understand these effects, one must understand the dynamical space-time correlations that build up throughout the fault system over many earthquake cycles, time intervals (days to thousands of years) that are again much longer than earthquake rupture times (RUNDLE et al., 2000a, 2000b; TIAMPO et al., 2000). Third, the presence of small-scale short-time noise and random influences affecting the physics of sliding on fault surfaces implies that sliding must be treated not as a deterministic process, but instead as a stochastic process. Fourth, recent work on geometrically simple, infinitely long faults (LAPUSTA et al., 2000) subject to rate-and-state friction has clearly shown that the inclusion of elastic waves in their simulations does not qualitatively change the results when compared to earlier, quasistatic simulations, a result that agrees with conclusions drawn by other groups as well (WARD, 2000; PRESTON, 2000). As a consequence, as long as we regard frictional parameters determined by assimilating field observations into the simulations as *effective* or *renormalized* values, we can expect to achieve meaningful results.

Simple Models

The physics of simple planar fault models provides the underlying physical basis for understanding the dynamics of the more realistic models embodied by the Virtual_California simulation and the PBS. Beginning with the Burridge-Knopoff (BK) slider block model (BURRIDGE and KNOPOFF, 1967), simulation studies of planar earthquake faults have been carried out in which each spatially coarse-grained site on the fault is represented by a block sliding on a frictional surface. Physically, the blocks are intended to represent the sticking points, or asperities, on the fault surface. All blocks are connected to a loader plate by means of a spring having spring constant K_L. In addition, each block is also connected to q other blocks by means of coupling springs, each spring having constant K_C. For these models, the shear stress builds up until the static friction stress reaches the failure threshold σ^F. Sliding of each block commences at the threshold, and continues until the block stress is reduced to the kinetic friction stress σ^R. During the sliding process, stress is transmitted from the sliding block(s) to other blocks by means of the coupling springs. Since the BK model represents a planar fault, all inter-block interactions are excitatory, meaning that $K_C > 0$. Sliding of a given block then increases the stress on other blocks to which it is coupled, and therefore an avalanche of failing blocks may be initiated by a single unstable block. The avalanche of failing blocks represents an earthquake in the model. While the original BK specified massive blocks with inertial, recent models are more commonly of the stochastic cellular automaton type (RUNDLE and JACKSON, 1977; RUNDLE and BROWN, 1991), in which the sliding block travels a distance $(\sigma - \sigma^R)/K + W$, where $K = K_L + qK_C$, and W is a random number representing random overshoot or undershoot during the slip process (RUNDLE *et al.*, 1995).

We are particularly interested in the physics of self-organizing processes in *mean field* threshold systems (RUNDLE *et al.*, 1995; KLEIN *et al.*, 1997; FERGUSON *et al.*, 1999; KLEIN *et al.*, 2000), which in the case of the slider block model is described by the condition $qK_C \rightarrow \infty$. In the mean field regime, the range of interaction becomes large, leading to a damping of fluctuations, and the appearance of a mean field *spinodal*, the classical limit of stability of a spatially extended system (PENROSE and LEIBOWITZ, 1979; GUNTON and DROZ, 1983). Examined in this limit, driven threshold systems appear to be locally ergodic, and display equilibrium behavior when driven at a uniform rate. Following the initial discovery (RUNDLE *et al.*, 1995) that driven mean field slider block systems display equilibrium properties, other studies have confirmed local ergodicity and the existence of Boltzmann fluctuations in both these and other systems (KLEIN et al., 1997, 2000; MOREIN and TURCOTTE, 1997; FERGUSON et al., 1999; MAIN et al., 2000; EGOLF, 2000). Thus the origin of the physics of scaling, critical phenomena and nucleation appears to lie in the ergodic properties of these mean field systems. Further studies have shown that mean field and near-mean field systems are also associated with the appearance of an *energy landscape* (KLEIN et al., 2000), similar to other equilibrium systems (HOPFIELD, 1982).

Spinodal Equations

Consider the $d = 2$ mean field slider block system described above, in which each block is connected to q other blocks with springs having constants K_C. Each block is also connected to a loader plate translating at an externally imposed load velocity $V(t)$ by a spring with constant K_L. The coarse-grained dynamical equation is (KLEIN et al., 2000):

$$\frac{\partial \sigma(\mathbf{x}, t)}{\partial t} = K_L V - f[\sigma(\mathbf{x}, t), V(t)] + \eta(\mathbf{x}, t) . \tag{1}$$

As described in KLEIN et al. (1997), FERGUSON et al. (1999), and KLEIN et al. (2000), $\sigma(\mathbf{x}, t)$ is the coarse-grained stress within a volume of q blocks centered at $\mathbf{x}$, and time is coarse-grained over a temporal window centered at t whose duration is roughly half the time interval between slips for an average block. It can be seen that (1) expresses the balance between the rate $K_L V$ at which stress is accumulated at $\mathbf{x}$ by the loader plate motion, and the rate $f[\sigma(\mathbf{x}, t), V]$ at which stress is dissipated by the sliding blocks, and $\eta(\mathbf{x}, t)$ is a stochastic noise term. Denoting the spatial average over the sliding surface by $\langle \rangle$, we define $\langle \sigma(\mathbf{x}, t) \rangle = \sigma(t)$. Note that the sliding velocity on the surface is $u(\mathbf{x}, t)$, with spatial average $u(t) = \langle u(\mathbf{x}, t) \rangle = \partial \langle s(\mathbf{x}, t) \rangle / \partial t = ds(t)/dt$, where $s(\mathbf{x}, t)$ is the coarse-grained slip on the surface at $(\mathbf{x}, t)$, and in general, $u \neq V$. $f[\sigma(\mathbf{x}, t), V]$ is a general functional of $\{\sigma(\mathbf{x}, t), V\}$ that depends also on σ^F, σ^R, as well as upon a thermalizing noise amplitude β. In view of the equations described in KLEIN et al. (2000), we expect a rather weak dependence of $f[\sigma(\mathbf{x}, t), V(t)]$ upon $V(t)$. Indeed, laboratory experiments at both the atomic scale (GNECCO et al., 2000) and the laboratory scale (DIETERICH, 1979; RUINA, 1983; TULLIS, 1996; BLANPIED et al., 1987, 1998; SLEEP, 1997; TULLIS, 1988) indicate a dependence of stress $\sigma(t) \sim \log V(t)$.

Mean Field and Thermodynamics

For the most part, we are interested here in modeling laboratory experiments carried out on elastically stiff, strongly correlated samples that are often viewed as *lumped parameters*. Thus we take $qK_C \to \infty$, and in addition employ an equation describing the elasticity of the loader springs:

$$\frac{\partial \sigma(t)}{\partial t} = K_L(V(t) - u(t)) . \tag{2}$$

In the mean field limit, the noise term is suppressed in equation (1), which then becomes:

$$\frac{\partial \sigma(t)}{\partial t} = K_L V - f[\sigma(t), V(t)] \quad . \tag{3}$$

Combining (2) and (3), we obtain:

$$K_L u = f[\sigma, V] \quad . \tag{4}$$

As described above, this system, representing a driven mean field frictional contact zone, demonstrates ergodicity and Boltzmann fluctuations (RUNDLE *et al.*, 1995; KLEIN *et al.*, 1997, 2000; FERGUSON *et al.*, 1998, 1999; MOREIN *et al.*, 1997; MAIN *et al.*, 2000; EGOLF, 2000). Equation (4) can therefore be viewed as a thermodynamic *equation of state* for the contact zone that relates the three state variables: u, V, and σ.

Figure 3 shows a specific example of $f[\sigma, V]$, assuming steady-state conditions $V = $ constant. As explained in detail in KLEIN *et al.* (1997), the function $f[\sigma, V]$ depends on four parameters, the coupling spring constant between blocks K_C, the loader spring constant K_L, the loader plate velocity V, and the noise amplitude, which is measured (inversely) by β. In Figure 3, the value $\beta = 0.1$ corresponds to a relatively low temperature, giving rise to the typical Van der Waals-type spinodal structure having two extrema, one at a value of stress $\sigma \sim 30$, the other at higher stress $\sigma \sim 45$. Here $\sigma^F = 50$, $\sigma^R = 0$, and $V = 0.48$. The solution to equation (3) can be obtained graphically from Figure 3 as the intersection of the line labeled $K_L V$ with the curve $f[\sigma, V]$. It can be seen that there are three possible solutions: 1) A low stress, stable phase, 2) an intermediate stress, unstable phase, and 3) a high stress metastable phase. At higher values of $K_L V$, the high stress phase becomes stable and the low stress phase becomes metastable. Observing the high stress phase as the stable phase is difficult, but it can nevertheless be seen under the proper conditions (GOLDSTEIN *et al.*, 2001).

Leaky Threshold Equations

The leaky threshold equation for a single sliding block is obtained by expanding $f[\sigma(t), V(t)]$ in $\sigma(t)$, setting $V = $ constant, and parameterizing the threshold instability. Thus we have:

$$u(t) = \frac{f[\sigma(t), V]}{K_L} \approx \left(\frac{\sigma(t) - \sigma^R}{K_L}\right)[\alpha + \delta(t - t_F)] \quad , \tag{5}$$

where t_F is any of the times defined by the condition $\sigma(t_F) = \sigma^F$. When $\alpha = 0$, one has the simple threshold equation. The presence of nonzero α means, in the case of friction, that there exists stable, aseismic slip preceding the instability at $t = t_F$. Such slip has been observed in laboratory friction experiments, modifying the sawtooth form of the data for stress plotted against slip, to produce a concave-down curvature (TULLIS, 1996; KARNER and MARONE, 2000).

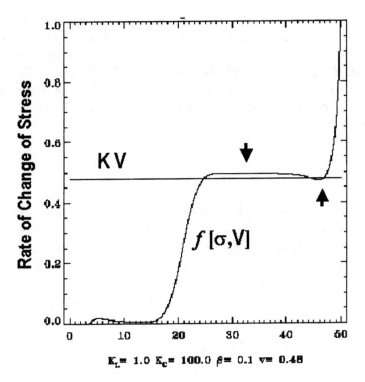

K_L= 1.0 K_C= 100.0 β= 0.1 v= 0.48

Figure 3
Plot of an example of the rate of stress dissipation function $f[\sigma, V]$ described in equation (3), for a mean field slider block model. Details of the method for computing this function can be found in KLEIN et al. (1997; 2000) and in FERGUSON et al. (1999). The arrows indicate the extrema of $f[\sigma, V]$, which are called spinodals. The horizontal line represents the rate of stress accumulation $K_L V$ due to the loader plate.

Now consider two slider blocks, connected to each other by a coupling spring K_C, and each connected to the loader plate by means of a loader spring K_L. Denoting the slip and stress of block 1 by s_1, σ_1, the equations for block 1 are then:

$$u_1 = \frac{ds_1}{dt} = \frac{\sigma - \sigma^R}{(K_L - K_C)} \alpha \tag{6}$$

$$\sigma_1 = K_L(V_L t - s_1) + K_C(s_2 - s_1) . \tag{7}$$

An analogous set of equations holds for block 2. If the difference in stress at time $t = 0$ is given by $\delta\sigma(0) = \sigma_2(0) - \sigma_1(0)$, then at a time t later:

$$\delta\sigma(t) = \delta\sigma(0)e^{-\gamma t}, \quad \text{where} \quad \gamma = \alpha\left(\frac{K_L + 2K_C}{K_L + K_C}\right) . \tag{8}$$

Equation (8) shows that under action of the leaky dynamics in (5), variations in stress in a slider block system are progressively reduced if $\alpha > 0$, and progressively increased if $\alpha < 0$. The case $\alpha > 0$ is therefore a process of stress smoothing. On the other hand, it may be that either 1) $\alpha < 0$, or 2) the geometric arrangement of the faults is such that the elastic coupling parameters corresponding to K_C are negative or inhibiting, a condition that can occur in more general elastic or frictional systems (KLEIN *et al.*, 2000; PENROSE and LEIBOWITZ, 1979; DIETERICH, 1979). In either of these latter cases, variations in stress can grow exponentially in time, a process of *stress roughening*. For general three-dimensional fault network models, both stress smoothing and stress roughening may in fact occur.

From (5), it can be seen that α is a thermodynamic derivative:

$$\alpha \equiv \left. \frac{\partial f}{\partial \sigma} \right|_V . \tag{9}$$

The slope α of the f-σ curve determines whether stress-smoothing or stress-roughening occurs under the leaky dynamics. For general three-dimensional fault network models, both stress smoothing and stress roughening should occur, as well as the possibility of smoothing-to-roughening transitions. Data from laboratory experiments on sliding granite (TULLIS, 1996; KARNER and MARONE, 2000) indicates that $\alpha \sim 0.1/T_R$, where T_R is the recurrence interval for unstable slip.

Virtual California Simulation and PBS

The lack of sensitivity of simulation results to the presence or absence of wave-mediated stress transfer (LAPUSTA *et al.*, 2000) has recently been shown to stem from the long-range elastic interactions (mean field models) rather than some accidental influence (PRESTON, 2000). In other words, for certain classes of models, typically mean field models (RUNDLE *et al.*, 1995; KLEIN *et al.*, 1997), the order of failures of sites is not important in determining the dynamical results. This effect arises because stress transfer to sites from a failed site is proportional to the inverse of the area over which the stress is transferred (PRESTON, 2000). However, changes in the order of failures of sites introduces an error of the order of the inverse area squared, a very small quantity. While the rigorous proof of this result is mathematically complicated, this statement contains the essential idea. For all these reasons, we conclude that, at the level at which simulations are needed to be useful for the PBO, the inclusion of elastic waves in the simulations will not have a substantive impact on the conclusions regarding space-time correlations and patterns that we obtain from the simulations, provided we view values of parameters in the simulations derived by assimilating field observations as *effective values*. In the following, we therefore use the standard quasistatic stress interaction (Green's function) tensors $T_{ij}^{kl}(\mathbf{x} - \mathbf{x}', t - t')$ appropriate to an elastic half space (OKADA, 1992), although it is in principle possible to use similar tensors appropriate to a layered viscoelastic medium. Propagation of stress interactions is carried out via a boundary element technique, but for the future we intend to recast the problem with finite element methods.

The current generation of earthquake simulation technology has been discussed in the literature (RUNDLE, 1988; RUNDLE et al., 2000a, 2001; WARD, 2000). One defines a fault model geometry (Figure 4) in an elastic half space, computes the elastic stress Greens functions $T_{ij}^{kl}(\mathbf{x} - \mathbf{x'}, t - t')$ (i.e., stress transfer coefficients), and assigns frictional properties to each fault as discussed below and as in Figure 5, which plots the difference between static and kinetic friction coefficients. The system is driven via the deficit in slip relative to the long-term offset on the fault, a type of model that has been called a "backslip model." Previous work has shown that the elastic interactions in the simulations produce mean-field dynamics (RUNDLE et al., 1995; KLEIN et al., 1997), which means that the most important dynamical properties of the system are associated with the longest wavelength fluctuations in stress. Since our focus is on understanding the mean-field stochastic space-time dynamics of the strongly correlated network of faults, we use a discrete dynamical evolution equation (cellular automaton) with additive random noise during sliding.

Our present efforts have focused on the major horizontally slipping strike-slip faults in southern California that produce by far the most frequent, largest magnitude events. For geometric and other characteristics, we used the tabulation of strike slip faults and fault properties as published in the literature (DENG and SYKES, 1997). Faults in southern California, together with the major historic earthquakes, are shown in Figure 1. Figure 4 shows our model fault network for southern California, in which we have used only the principal strike-slip faults to construct 215 distinct rectangular fault segments in an elastic half space. Note that at this point in the method, there is no physical or geological meaning to the internal segment boundaries, which simply divide a given fault into rectangular pieces approximately ~10 km long, extending from the surface to 20 km depth. The external boundaries of a fault ("fault ends") are determined by the mapped expressions of the faults.

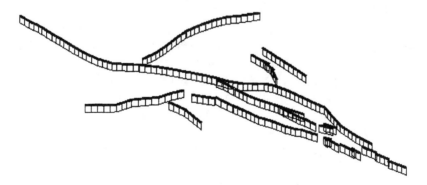

Figure 4

Model fault network for the major strike slip faults in southern California (compare to faults in Fig. 2). In this model, there are 215 fault segments, each 20 km deep, and each about 10 km in along-strike horizontal extent.

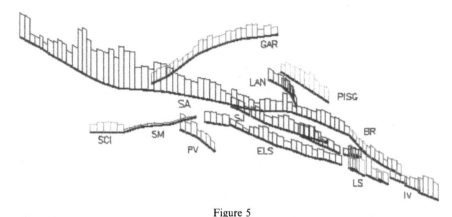

Figure 5
Plot of differences in coefficients of friction $\mu_S - \mu_K$ associated with each fault segment in the Southern California model.

A variety of friction laws has been described in the literature, including Coulomb failure (PERSSON, 1998), slip-dependent or velocity-dependent friction (SCHOLZ, 1990; PERSSON, 1998), and rate-and-state (DIETERICH, 1979). In our most current work (RUNDLE *et al.*, 2001), we have used a generalization of the leaky threshold equations discussed above. As noted, the solutions to these equations closely model the recent laboratory friction experiments by TULLIS (1996), and by KARNER and MARONE (2000), in which the stiffness of the loading machine is low enough to allow for unstable stick-slip when a failure threshold $\sigma^F(V)$ is reached, where $\sigma^F(V)$ is a weak (logarithmic) function of the load point velocity V. Sudden slip then occurs in which the stress decreases to the level of a residual stress (or kinetic friction) $\sigma^R(V)$, again a weak function of V. Stable precursory slip is observed to occur whose velocity increases with stress level, reaching a magnitude of a few percent of the driving load point velocity just prior to failure at $\sigma = \sigma^F(V)$.

Earthquake data obtained from the historical record as well as geological field studies represent the primary physical signatures of how the earthquake cycle is affected by the frictional properties that exist on the faults. The timing, magnitude and complexity of these historical events are a direct reflection of the values of the frictional parameters: α, σ^F, σ^R. Since the dynamics (1) depends on the characteristic length scale L for each segment, all of these frictional parameters should be regarded as scale-dependent functions of L: $\alpha = \alpha(\Delta\sigma, L)$, $\sigma^F = \sigma^F(L, V)$, $\sigma^R = \sigma^R(L, V)$. For simulations in which one or more distinct scales L are chosen for each fault segment (length and width, for example), one must choose α, σ^F, σ^R in such a way that the historical record of activity on the fault network is matched as closely as possible. This is the *data assimilation* problem that has been discussed in other contexts (GIERING and KAMINSKI, 1997; http://www.techtransfer.anl.gov/software/adi-for.html), and for which we have developed a simple, but physically motivated method that we describe in the following.

For historical earthquakes, there can be considerable uncertainty about where the event was located (SCHOLZ, 1990). Modern studies (SIEH et al., 1989; GRANT and SIEH, 1994; MASSONET and FEIGL, 1998) of earthquakes indicate that slip or seismic moment M_0, defined in equation (8) below, is often distributed regionally over a number of faults and sub-faults. Therefore our technique assigns a weighted average of the scalar seismic moments for given historic or prehistoric events during an observational period to *all* of the faults in the system. To be physically plausible, the weighting scheme should assign most of the moment M_0 to the nearby faults that slipped, and decay rapidly with distance. Since the seismic moment is the torque associated with one of the moment tensor stress-traction double couples, it is most reasonable to use the (inverse cube power of distance r) law that describes the decay of stress with distance ESHELBY (1957). Comparisons with data (SIEH et al., 1989; GRANT and SIEH, 1994) indicate that this method yields average recurrence intervals similar to those found in nature.

Step 1: Assignment of Moment Rates

We use all historical events in southern California since 1812 http://www.sce-cdc.scec.org/clickmap/). For each of the 215 fault segments in the model, the contribution of moment release rate from the j-th historical earthquake $dM_0(t_j)/dt$ to the rate on the i-th fault segment, dm_i/dt, is:

$$\frac{dm_i}{dt} = \Gamma \left[\frac{\sum_j \frac{dM_0(t_j)}{dt} r_{ij}^{-3}}{\sum_j r_{ij}^{-3}} \right] , \qquad (10)$$

where $r_{ij} = |\mathbf{x}_i - \mathbf{x}_j|$ is the distance between the event at the published location $\mathbf{x}_j$ and time t_j, and the fault segment at $\mathbf{x}_i$. The factor Γ accounts for the limited period of historical data available compared to the length of the earthquake cycle, and is determined by matching the total regional moment rate, $\sum_i dm_i/dt$, to the observed current regional moment rate. We find $\Gamma \approx 0.44$. Application of (10) when $\mathbf{x}_i \approx \mathbf{x}_j$ is understood to be in the limiting sense. Equation (10) arises if one regards r_{ij}^{-3} as a probability density function, and assumes that each earthquake is a point source. We correct for the largest events, which are long compared to the depth, by representing the large event as a summation of smaller events distributed along the fault.

Step 2: Determination of Friction Coefficients

The seismic moment is:

$$M_0(t_j) = \mu \langle s(t_j) \rangle A , \qquad (11)$$

where μ is shear modulus, $\langle s(t_j) \rangle$ is average slip at time t_j, and A is fault area. For a compact fault, the average slip in terms of stress drop $\Delta\sigma$ is (ESHELBY, 1957; KANAMORI and ANDERSON, 1975):

$$\langle s \rangle = \frac{f \Delta\sigma \sqrt{A}}{\mu} , \tag{12}$$

where f is a dimensionless fault segment shape factor having a value typically near 1. Standard assumptions of $f \sim 1$, $\Delta\sigma \sim 5 \times 10^6$ Pa, $\mu \sim 3 \times 10^{10}$ Pa yield reasonable slip values. The average slip is converted to a difference between static and kinetic friction, $(\mu_S - \mu_K)_i$ for the i-th fault segment via the relation:

$$(\mu_S - \mu_K)_i \approx \frac{m_i}{f A^{3/2} \chi_i}, \tag{13}$$

which is obtained by combining (11), (12), and $\Delta\sigma \approx \sigma^F - \sigma^R$. To compute $(\mu_S - \mu_K)_i$, a typical value of χ_i for each segment is computed from the average gravitationally-induced compressive stress. Since the stochastic nature of the dynamics depends only on the differences $(\mu_S - \mu_K)_i$ (RUNDLE, 1988; RUNDLE et al., 2000), we set $\mu_K = 0.001$ (all i). The result of this procedure is shown in Figure 5.

Step 3. Aseismic Slip Factor α

Earthquake faults are characterized by varying amounts of aseismic slip that arises from the "stress leakage" factor α. The most famous example of aseismic slip is the region of the San Andreas fault in central California, in which no seismic slip has ever been observed. The average fraction of slip on each fault that is nonseismic is equal to $\alpha_i/2$, and has been tabulated in some cases for southern California faults in DENG and SYKES (1997). We use these tabulated values for α_i. However, along the San Andreas fault proper, simulations indicate that the complex elastic fault interactions, together with the published values of α_i, do not lead to earthquakes that are large enough to satisfy the paleoseismic observations (SIEH et al., 1989; GRANT and SIEH, 1994). For these locations, we have therefore increased the values of the α_i on the faults until the stress smoothing effect produced satisfactory sequences of large events at roughly the observed recurrence intervals. This process is often referred to as "tuning" the simulations.

Example of Results from Current Plate Boundary Simulation Technology

Under the dynamics of the model, histories of synthetic earthquake events can be generated that are remarkably similar to actual events occurring on the actual southern California fault network. An example from the simulations of a great earthquake on the San Andreas fault is shown in Figure 6 (model year 7246), similar to the real Fort Tejon event of 1857. The surface deformation that could be observed in association with this event using standard land-based and space-based (GPS) techniques is shown in Figure 7. In addition, Interferometric Synthetic Aperture Radar (InSAR) data can also be used to reveal fine scale details of the surface deformation. In Figure 8 we show the InSAR surface deformation from the same event at simulation year 7246 as described in the preceding two figures. Similar GPS and InSAR images can be computed, for example, for the time periods leading up to, and just following, the major events. In this way, expected precursory space-time changes in deformation state can be computed that can be searched for with PBO deployments.

Figures 9 and 10 are illustrations of the space-time behavior of the Coulomb Failure Function $CFF(\mathbf{x}_i, t) = \{\sigma(\mathbf{x}_i, t) - \mu_S(\mathbf{x}_i)\chi(\mathbf{x}_i, t)\}$ for all of the 215 fault segments, placed end-to-end along the horizontal axis, in the southern California network model (see Figs. 4 and 5). A horizontal line represents an earthquake, which occurs on a segment when $CFF(\mathbf{x}_i, t) = 0$. In Figure 9, values for α_i have been assigned using the data in DENG and SYKES, (1997), whereas in Figure 12, all $\alpha_i = 0$.

From Figures 9 and 10, it can be seen that changing the values of α_i has a profound effect on the network dynamics. For larger α_i (Fig. 9) and excitatory interactions, the stress field is increasingly smoothed and the earthquakes tend to be larger ("decreasing complexity"). For smaller α_i or even inhibiting interactions, the stress field tends to roughen (Fig. 10) and the corresponding events are smaller ("increasing complexity"). With smaller α_i (Fig. 10), the various fault segments tend to behave more independently than for larger α_i (Fig. 9). One can speak of a "roughness length" for the stress field similar in many respects to a correlation length (RUNDLE et al., 1998). We predict that physical manifestations of friction laws on faults are revealed by the space-time patterns in the network dynamics. Dynamical switching of activity due to fault interactions should also be observed. An example is shown in Figure 9, in which the south-central region of the San Andreas (left arrow) tends to switch off activity on the eastern Garlock fault (right arrow). Dynamical switching of activity may have already been revealed through observations of real fault networks (HARRIS, 1998; STEIN, 1999).

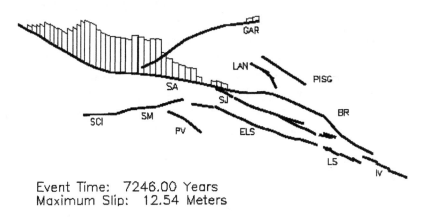

Event Time: 7246.00 Years
Maximum Slip: 12.54 Meters

Figure 6
Plot of slip as a function of position for a simulated great earthquake at year 7246, similar to the great 1857 Fort Tejon earthquake.

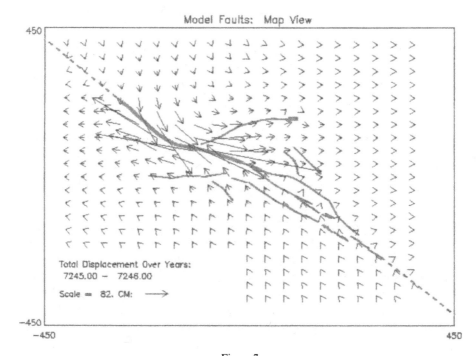

Figure 7
A view of horizontal surface deformation that would be observed via GPS-type observations of the simulated event shown in Figure 7.

Discussion

The results we have obtained thus far in southern California have encouraged us to pursue the construction of an integrated model for all the major strike slip faults in California. In particular, the evidence in (Figs. 9 and 10; DENG and SYKES, 1997) showing that coupling of earthquake activity can exist over hundreds of kilometers in southern California strongly suggests that great earthquakes such as the 1857 Fort Tejon and the 1906 San Francisco events on the southern and northern San Andreas faults, respectively, might mutually interact and produce significant space-time correlations that build up over many earthquake cycles. Earthquakes on the northern San Andreas fault might therefore be expected to influence the timing and the magnitudes of events on the southern San Andreas fault and vice versa. It is therefore of great importance to begin the construction of a model that integrates the entire fault network in all of California. As before, however, it would be wise to focus efforts in the beginning on the most active faults, which are all nearly pure strike-slip faults. We view the more infrequent activity on the major thrust faults in the state as a kind of perturbation on the strike-slip faults system, which represent by far the main sources of seismic activity in California. We have begun work on such a model,

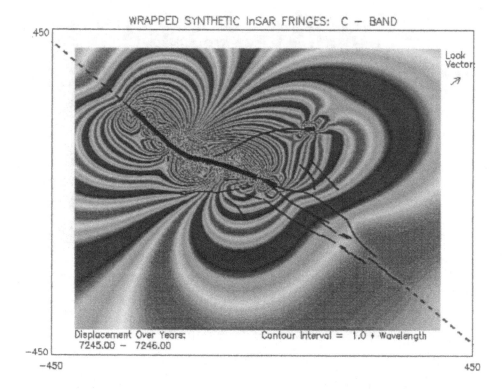

Figure 8
Interference fringes that would be observed using Synthetic Aperture Radar (InSAR) C-band observations of the simulated event shown in Figure 7. The horizontal projection of the look vector to the satellite is shown at the upper right.

shown in Figure 11, that incorporates the southern California faults in a self-consistent manner (compare Figs. 4 and 11). The model shown has 450 segments, as compared to the 215 segments making up the southern California model. Results from this work should be forthcoming shortly.

A second goal for future work is to understand the basic physics of geometrically realistic hierarchies of fault models that share the correct large scale features, as checked by the ability to reproduce the scaling, clustering, and correlated phenomena observed in real data, and which progressively incorporate microscopic elements on differing scales. This strategy allows for the evaluation of the impact of each of these elements by itself and in cooperation with others, developing fundamental new insights and leading to an understanding of the physics of fault segmentation processes. These elements can then be parameterized and tuned to match each particular data set in our catalogs through the data assimilation process (RUNDLE et al., 2001; http://www.tech-transfer.anl.gov/software/adifor.html; GIERING and KAMINISKI, 1997).

Building on these ideas, a primary task for the Plate Boundary Simulator is the identification and characterization of an *ideal*, minimum number of distinct fault

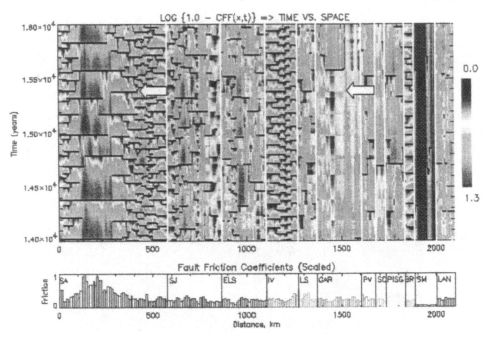

Figure 9

Plot of Coulomb Failure Functions plotted as a function of time vs. spatial location for 2000 simulation years for a set of values $\{\alpha_i\}$ obtained by approximately matching the ratio of aseismic slip to seismic slip on various fault segments from tabulated data in DENG and SYKES (1997). All 215 fault segments in the model are concatenated end-to-end along the horizontal axis. An earthquake on these plots is represented by a solid horizontal line. We actually plot Log$\{1 - \text{CFF}(\mathbf{x}, t)\}$ so that subtle differences can be seen. Due to limitations of black and white plotting, both low and high (near 0) values of CFF$(\mathbf{x}, t)$ are represented by darker shades, while intermediate values of CFF$(\mathbf{x}, t)$ are represented by lighter shades.

segments necessary to reproduce the historical seismicity patterns (RUNDLE *et al.*, 2000a, b, 2001). We expect to be substantially aided in this work by the eigenpattern analysis methods we developed in RUNDLE *et al.* (2000a, 2001). The identification of these fault segments is a major and labor-intensive procedure, and with our simulations we will be in a particularly advantageous position to deal with it. Each segment can be treated as a patched planar fault with long-range, geometrically realistic elastic interactions. Our recent work (PRESTON, 2000; RUNDLE *et al.*, 1995) with friction laws having the characteristic of precursory, stable aseismic slip due to nonzero α has moved us to a domain that can be characterized as incorporating important features of both the cellular automaton (CA) and "rate-and-state" (DIETERICH, 1979) approaches. The inclusion in the frictional physics of stable inter-event precursory slip generates deviations from pure scaling behavior and the progressive appearance of a characteristic time scale between rupture events (MARTINS *et al.*, 2001; MAIN *et al.*, 2000). The parameter α can be viewed as a tuning variable, that allows the form of the Gutenberg-Richter magnitude-frequency relation to be progressively adjusted.

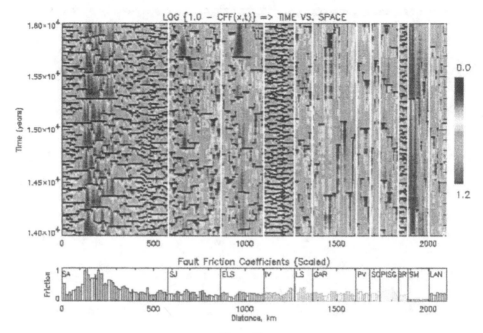

Figure 10
Same as Figure 3 but with all $\alpha_i = 0$.

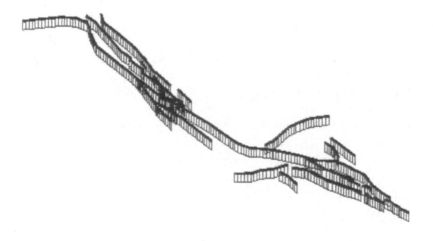

Figure 11
Model fault network for the major strike slip faults for all of California (compare to faults in Fig. 2 and Fig. 5). In this model, there are 450 fault segments, each 20 km deep, and each about 10 km in along-strike horizontal extent.

A third goal is to apply numerical simulations of deformation associated with earthquakes on a finer scale to specific regions, with the goal of understanding stress transfer and crustal rheology. Studies of this kind will provide important calibration

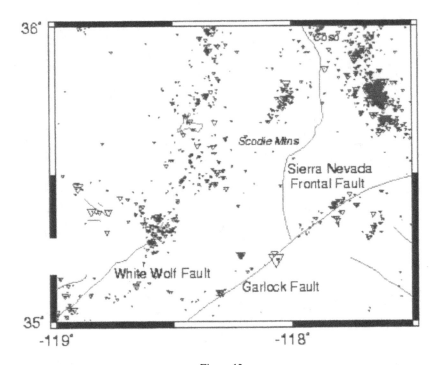

Figure 12
Recent seismicity observed in the southern Sierra Nevada mountain range between the White Wolf fault
and the Sierra Nevada frontal fault system near the Coso Geothermal Field.

and validation studies for the simulations. An example of this approach is our
current plan to develop simulations of the local fault network associated with
earthquakes on the 1952 M ∼ 7.2 Kern County of southern California and the
extension of the White Wolf fault to the northeast, towards the location of the 1946
Walker Pass earthquake. Currently, these faults are not included in our southern
California model, but this region is tectonically interesting because it spans the
transition between the big-bend section of the San Andreas fault system and
the eastern California Shear Zone. A band of recent seismicity extends across the
southern Sierra Nevada range between the White Wolf fault and the Sierra Nevada
frontal fault near the Coso geothermal field (Fig. 12; see BAWDEN *et al.*, 1999, for
details). Relocation of these events and analysis of the focal mechanisms indicates
that this seismic lineament may be an incipient blind strike-slip fault. Left-lateral
focal planes are rotated counterclockwise from the strike of the seismic lineament. A
similar phenomenon is observed in laboratory experiments (BARTLETT *et al.*, 1981;
SCHREURS, 1994; GLASSCOE *et al.*, 2000) in which Reidel shear fractures form during
deformation of material that lacks a single continuous fault surface. These fractures
form at an oblique angle to the direction of maximum shear, consistent with the
orientation of focal mechanisms in the seismic lineament in the Scodie Mountains. If

the seismic lineament is exhibiting Reidel shear, it would represent an unusual example of a newly forming seismogenic structure. Understanding the interaction among the White Wolf fault, the Garlock fault, and the Sierra Nevada frontal fault will provide insight into the origin of seismicity in this region, the development of new faults, and the partitioning of strain among mature and new seismogenic structures.

The simulations will produce a suite of deformation fields that vary with properties including fault geometry and possibly crustal rheology. The deformation fields represent a prediction of each model which can be compared with observations from the PBO and other data sources. Since there are currently few geodetic observations in this region, we will instead use the predictive outputs of the models to design optimal hypothetical geodetic networks that can distinguish among various models. We also plan to carry out similar calibration and validation exercises using detailed simulations for other regions along the San Andreas fault system.

Acknowledgements

Research by PBR was supported by the Southern California Earthquake Center (Contribution 540) under NSF grant EAR-8920136. Research by JBR was funded by USDOE/OBES grant DE-FG03-95ER14499 (theory), and by NASA grant NAG5-5168 (simulations). Research by WK was supported by USDOE/OBES grant DE-FG02-95ER14498. KFT and SM were funded by NGT5-30025 (NASA), and JdsM was supported as a Visiting Fellow by CIRES/NOAA, University of Colorado at Boulder. We would also like to acknowledge generous support by the Maui High Performance Computing Center, project number UNIVY-0314-U00.

REFERENCES

BARTLETT, W. L., FRIEDMAN, M., and LOGAN, J. M. (1981), *Experimental Folding and Faulting of Rocks Under Confining Pressure. 9. Wrench Faults in Limestone Layers*, Tectonophysics 79, 255–277.

BAWDEN, G. W., MICHAEL, A. J., and KELLOGG, L. H. (1999), *Birth of a Fault: Connecting the Kern County and Walker Pass, California, Earthquakes*, Geology 27, 601–604.

BLANPIED, M. H., TULLIS, T. E., and WEEKS, J. D. (1987), *Frictional Behavior of Granite at Low and High Sliding Velocities*, Geophys. Res. Lett. 14, 554–557.

BLANPIED, M. H., TULLIS, T. E., and WEEKS, J. D. (1998), *Effects of Slip, Slip Rate, and Shear Heating on the Friction of Granite*, J. Geophys. Res. 103, 489–511.

BURRIDGE, R. and KNOPOFF, L. (1967), *Model and Theoretical Seismicity*, Bull. Seism. Soc. Am. 57, 341–371.

DENG, J. S. and SYKES, L. R. (1997), *Evolution of the Stress Field in Southern California and Triggering of Moderate Size Earthquakes: A 200 Year Perspective*, J. Geophys. Res. 102, 9859–9886.

DIETERICH, J. H. (1979), *Modeling of Rock Friction 1: Experimental Results and Constitutive Equations*, J. Geophys. Res. 84, 2161–2168.

EGOLF, D. A. (2000), *Equilibrium Regained: From Nonequilibrium Chaos to Statistical Mechanics*, Science 287, 101–104.

ESHELBY, J. D. (1957), *The determination of the Elastic Field of an Ellipsoidal Inclusion and Related Problems*, Proc. Roy. Soc. Ser. A *241*, 376–396.

FERGUSON, C. D., KLEIN, W., and RUNDLE, J. B. (1999), *Spinodals, Scaling and Ergodicity in a Threshold Model with Long Range Stress Transfer*, Phys. Rev. E *60*, 1359–1373.

GIERING, R. and KAMINSKI, T. (1998), *Recipes for adjoint code construction*, ACM Trans. Math. Software *24*, 437–474.

GLASSCOE, M. T., DONNELLAN, A., PARKER, J., BLYTHE, A. E., and KELLOGG, L. H. (2000), *Two dimensional finite element modeling of strain partitioning in northern metropolitan Los Angeles*, EOS Trans. Am. Geophys. Un. (abstract), *81*, F326.

GNECCO, E., BENNEWITZ, R., GYALOG, T., LOPPACHER, C., BAMMERLIN, M., MEYER, E., and GUNTHERODT, H.-J. (2000), *Velocity Dependence of Atomic Friction*, Phys. Rev. Lett. *84*, 1172–1175.

GOLDSTEIN, J., KLEIN, W., GOULD, H., and RUNDLE, J. B. (2001), manuscript in preparation.

GRANT, L. B. and SIEH, K. E. (1994), *Paleoseismic Evidence for Clustered Earthquakes on the San Andreas Fault in the Carrizo Plain, California*, J. Geophys. Res. *99*, 6819–6841.

GROSS, S. J. and KISSLINGER, C. (1994), *Test of Models of Aftershock Rate Decay*, Bull. Seism. Soc. Am. *84*, 1571–1579.

GUNTON, J. D. and DROZ, M. (1983), *Introduction to the Theory of Metastable and Unstable States*, Lecture Notes in Physics *183*, Springer-Verlag, Berlin.

HARRIS, R. A. (1998), *Introduction to Special Section: Stress Triggers, Stress Shadows, and Implications for Seismic Hazard*, J. Geophys. Res. *103*, 24,347–24,358.

HOPFIELD, J. J. (1982), *Neural Networks and Physical Systems with Emergent Collective Computational Abilities*, Proc. Nat. Acad. Sci. USA *79*, 2554–2558.

KANAMORI, H. and ANDERSON, D. L. (1975), *Theoretical Basis of Some Empirical Relations in Seismology*, Bull. Seism. Soc. Am. *65*, 1073–1095.

KARNER, S. L. and MARONE, C. *Effects of loading rate and normal stress on stress drop and stick-slip recurrence interval*, pp. 187–198. In *Geocomplexity and the Physics of Earthquakes* (eds. Rundle, J. B., Turcotte, D. L., and Klein, W.) Geophysical Monograph *120* (American Geophysical Union, Washington, DC, 2000).

KLEIN, W., RUNDLE, J. B., and FERGUSON, C. D. (1997), *Scaling and Nucleation in Models of Earthquake Faults*, Phys. Rev. Lett. *78*, 3793–3796.

KLEIN, W., ANGHEL, M., FERGUSON, C. D., RUNDLE, J. B., and DE SA MARTINS, J. S. *Statistical analysis of a model for earthquake faults with long-range stress transfer*, pp. 43–72. In *Geocomplexity and the Physics of Earthquakes* (eds. J.B. Rundle, D.L. Turcotte and W. Klein, Geophysical Monograph *120* (American Geophysical Union, Washington, DC, 2000).

LAPUSTA, N., RICE, J. R., BEN-ZION, Y., and ZHENG, G. (2000), *Elastodynamic Analysis for Slow Tectonic Loading with Spontaneous Rupture Episodes on Faults with Rate- and State-dependent Friction*, J. Geophys. Res. *105*, 23,765–23,791.

MAIN, I. G., O'BRIEN, G. O., and HENDERSON, G. R. (2000), *Statistical Physics of Earthquakes: Comparison of Distribution Exponents for Source Area and Potential Energy and the Dynamics Emergence of Log-periodic Energy Quanta*, J. Geophys. Res. *105*, 6105–6126.

MARTINS, J. S., DE SA, RUNDLE, J. B., ANGHEL, M., and KLEIN, W. (2000), *Precursory Dynamics in Threshold Systems*, Phys. Rev. Lett., submitted.

MASSONET, D. and FEIGL, K.L. (1998), *Radar Interferometry and Its Application to Changes in the Earth's Surface*, Rev. Geophys. *36*, 441–500.

MOREIN, G. and TURCOTTE, D.L. (1997), *On the Statistical Mechanics of Distributed Seismicity*, Geophys. J. Int. *131*, 552–558.

OKADA, Y. (1992), *Internal Deformation due to Shear and Tensile Faults in a Half-space*, Bull. Seismol. Soc. Am. *82*, 1018–1040.

PENROSE, O. and LEIBOWITZ, J. L. *Towards a Rigorous Molecular Theory of Metastability*, Chap.5, In *Fluctuation Phenomena* (eds. Montroll, E.W. and J.L. Leibowitz, (North-Holland, Amsterdam 1979).

PERSSON, B. N. J. *Sliding Friction*, PHYSICAL PRINCIPLES AND APPLICATIONS (Springer-Verlag, Berlin 1998).

PRESTON, E. *Near Mean Field Earthquake Fault Models*, Ph.D. dissertation (University of Colorado, 2000).

RUINA, A. L. (1983), *Slip Instability and State Variable Friction Laws*, J. Geophys. Res. *88*, 10,359–10,370.

RUNDLE, J. B. and JACKSON, D. D. (1977), *Numerical Simulation of Earthquake Sequences*, Bull. Seismol. Soc. Am. *67*, 1363–1377.

RUNDLE, J. B. (1988), *A Physical Model for Earthquakes: 1. Fluctuations and Interactions*, J. Geophys. Res. *93*, 6237–6254.

RUNDLE, J. B. and BROWN, S. R. (1991), *Origin of Rate Dependence in Frictional Sliding*, J. Stat. Phys. *65*, 403–412.

RUNDLE, J. B. and KLEIN, W. (1995), *New Ideas About the Physics of Earthquakes*, Reviews of Geophysics and Space Physics Supplement, and Quadrennial Report to the IUGG and AGU 1991-1994 (invited), July, 283–286.

RUNDLE, J. B., KLEIN, W., GROSS, S. J., and TURCOTTE, D. L. (1995), *Boltzmann Fluctuations in Numerical Simulations of Nonequilibrium Threshold Systems*, Phys. Rev. Lett. *75*, 1658–1661.

RUNDLE, J. B., PRESTON, E., McGINNIS, S., and KLEIN, W. (1998), *Why Earthquakes Stop: Growth and Arrest in Stochastic Fields*, Phys. Rev. Lett. *80*, 5698–5701.

RUNDLE, J. B., KLEIN, W., TIAMPO, K. F., and GROSS, S. J. (2000a), *Linear Pattern Dynamics in Nonlinear Threshold Systems*, Phys. Rev. E *61*, 2418–2431.

RUNDLE, J. B., KLEIN, W., TIAMPO, K. F., and GROSS, S. J. *Dynamics of seismicity patterns in systems of earthquake faults*, pp. 127–146. In *Geocomplexity and the Physics of Earthquakes* (eds. Rundle, J. B., Turcotte, D. L., and Klein, W., Geophysical Monograph *120* (American Geophysical Union, Washington, DC, 2000b).

RUNDLE, P. B., RUNDLE, J. B., TIAMPO, K. F., DE SA MARTINS, J. S., McGINNIS, S., and KLEIN, W. (2001), *Nonlinear Network Dynamics on Earthquake Fault Systems*, Phys. Rev. Lett., *submitted.*

SCHREURS, G. (1994), *Experiments on Strike Slip Faulting and Block Rotation*, Geology *22*, 567–570.

SCHOLZ, C. H. *The Mechanics of Earthquakes and Faulting*, (Cambridge University Press, Cambridge, UK, 1990).

SIEH, K. E., STUIVER, M., and BRILLINGER, D. (1989), *A More Precise Chronology of Earthquakes Produced by the San Andreas Fault in Southern California*, J. Geophys. Res. *94*, 603–623.

SLEEP, N. H. (1997), *Application of a Unified Rate and State Friction Theory to the Mechanics of Fault Zones with Strain Localization*, J. Geophys. Res. *102*, 2875–2895.

STEIN, R. S. (1999), *The Role of Stress Transfer in Earthquake Occurrence*, Nature *402*, 605–609.

TIAMPO, K. F., RUNDLE, J. B., McGINNIS, S., GROSS, S. J., and KLEIN, W. *Observation of Systematic Variations in Non-local Seismicity Patterns from Southern California*, pp. 211–218. In *Geocomplexity and the Physics of Earthquakes* (eds. Rundle, J.B., Turcotte, D.L. and Klein, W. Geophysical Monograph *120* (American Geophysical Union, Washington, DC, 2000).

TULLIS, T. E. (1996), *Rock Physics and its Implications for Earthquake Prediction Examined via Models of Parkfield Earthquakes*, Proc. Nat. Acad. Sci. *93*, 3803–3810.

TULLIS, T. E. (1988), *Rock Friction Constitutive Behavior from Laboratory Experiments and its Implications for an Earthquake Prediction Field Monitoring Program*, Pure Appl. Geophys. *126*, 555–588.

WARD, S.N. (2000), *San Francisco Bay Area Earthquake Simulations, a Step Toward a Standard Physical Earthquake Model*, Bull. Seismol. Soc. Am. *90*, 370–386.

(Received February 20, 2001, revised June 11, 2001, accepted June 25, 2001)

B. Seismicity Change and its Physical Interpretation

Pure appl. geophys. 159 (2002) 2385–2412
0033–4553/02/102385–28 $ 1.50 + 0.20/0

© Birkhäuser Verlag, Basel, 2002

❚ **Pure and Applied Geophysics**

Accelerated Seismic Release and Related Aspects of Seismicity Patterns on Earthquake Faults

YEHUDA BEN-ZION[1] and VLADIMIR LYAKHOVSKY[2]

Abstract—Observational studies indicate that large earthquakes are sometimes preceded by phases of accelerated seismic release (ASR) characterized by cumulative Benioff strain following a power law time-to-failure relation with a term $(t_f - t)^m$, where t_f is the failure time of the large event and observed values of m are close to 0.3. We discuss properties of ASR and related aspects of seismicity patterns associated with several theoretical frameworks. The subcritical crack growth approach developed to describe deformation on a crack prior to the occurrence of dynamic rupture predicts great variability and low asymptotic values of the exponent m that are not compatible with observed ASR phases. Statistical physics studies assuming that system-size failures in a deforming region correspond to critical phase transitions predict establishment of long-range correlations of dynamic variables and power-law statistics before large events. Using stress and earthquake histories simulated by the model of BEN-ZION (1996) for a discrete fault with quenched heterogeneities in a 3-D elastic half space, we show that large model earthquakes are associated with nonrepeating cyclical establishment and destruction of long-range stress correlations, accompanied by nonstationary cumulative Benioff strain release. We then analyze results associated with a regional lithospheric model consisting of a seismogenic upper crust governed by the damage rheology of LYAKHOVSKY *et al.* (1997) over a viscoelastic substrate. We demonstrate analytically for a simplified 1-D case that the employed damage rheology leads to a singular power-law equation for strain proportional to $(t_f - t)^{-1/3}$, and a nonsingular power-law relation for cumulative Benioff strain proportional to $(t_f - t)^{1/3}$. A simple approximate generalization of the latter for regional cumulative Benioff strain is obtained by adding to the result a linear function of time representing a stationary background release. To go beyond the analytical expectations, we examine results generated by various realizations of the regional lithospheric model producing seismicity following the characteristic frequency-size statistics, Gutenberg-Richter power-law distribution, and mode switching activity. We find that phases of ASR exist only when the seismicity preceding a given large event has broad frequency-size statistics. In such cases the simulated ASR phases can be fitted well by the singular analytical relation with $m = -1/3$, the nonsingular equation with $m = 0.2$, and the generalized version of the latter including a linear term with $m = 1/3$. The obtained good fits with all three relations highlight the difficulty of deriving reliable information on functional forms and parameter values from such data sets. The activation process in the simulated ASR phases is found to be accommodated both by increasing rates of moderate events and increasing average event size, with the former starting a few years earlier than the latter. The lack of ASR in portions of the seismicity not having broad frequency-size statistics may explain why some large earthquakes are preceded by ASR and other are not. The results suggest that observations of moderate and large events contain two complementary end-member predictive signals on the time of future large earthquakes. In portions of seismicity following the characteristic earthquake distribution, such information exists directly in the associated quasi-periodic temporal distribution of large events. In portions of seismicity having broad frequency-size statistics with

---

[1] Department of Earth Sciences, Univ. of Southern CA, Los Angeles, CA, 90089-0740, U.S.A.
E-mail: benzion@terra.usc.edu
[2] Geological Survey of Israel, 30 Malkhei Israel, Jerusalem 95501, Israel. E-mail: vladi@geos.gsi.gov.il

random or clustered temporal distribution of large events, the ASR phases have predictive information. The extent to which natural seismicity may be understood in terms of these end-member cases remains to be clarified. Continuing studies of evolving stress and other dynamic variables in model calculations combined with advanced analyses of simulated and observed seismicity patterns may lead to improvements in existing forecasting strategies.

Key words: Continuum mechanics, damage rheology, heterogeneous faults, seismicity patterns, large earthquake cycles.

1. Introduction

Large earthquakes are often, but not always, preceded by a period during which the surrounding region experiences a phase of accelerated seismic release (ASR). This may be manifested as higher overall seismicity rates or elevated rates of low magnitude seismicity (e.g., PAPAZACHOS, 1973; JONES and MOLNAR, 1979; SHAW et al., 1992), increasing earthquake magnitude and/or number of moderate-size events with time (e.g., MOGI, 1969, 1981; ELLSWORTH et al., 1981; LINDH, 1990; KNOPOFF et al., 1996), and/or higher values of several functions of those (e.g., VARNES, 1989; SYKES and JAUMÉ, 1990; KEILIS-BOROK and KOSSOBOKOV, 1990; BUFE and VARNES, 1993; PRESS and ALLEN, 1995). The various forms of ASR activities occur in a broad region surrounding the following large earthquake ruptures. A recent review of features associated with observed ASR phases can be found in JAUMÉ and SYKES (1999).

BUFE and VARNES (1993) analyzed a few cumulative measures of seismic release during a period of about 150 years in northern California in terms of a power-law time-to-failure relation

$$\sum M_0^\zeta(t) = A + B(t_f - t)^m \ , \tag{1a}$$

where t is time, A, B, and m are adjustable parameters, t_f is failure time of a relatively large earthquake terminating a phase of ASR, and M_0^ζ is seismic moment for $\zeta = 1$, event count for $\zeta = 0$, and Benioff strain for $\zeta = 1/2$. They concluded that the cumulative Benioff strain in the space-time domain they study follows the power-law relation (1a) better than either a cumulative event count or cumulative seismic moment. SYKES and JAUMÉ (1990) fitted cumulative moment release in a similar space-time domain to that used by BUFE and VARNES (1993) with an exponential, rather than a power law function.

Subsequent observational analyses of ASR before relatively large earthquakes (BUFE et al., 1994; SORNETTE and SAMMIS, 1995; VARNES and BUFE, 1996; BOWMAN et al., 1998; BREHM and BRAILE, 1998; ROBINSON, 2000) focused on power-law time-to-failure fits of cumulative Benioff strain

$$\sum M_0^{1/2}(t) = A + B(t_f - t)^m \ . \tag{1b}$$

In retrospective- and forward-prediction applications of (1b) (e.g., BUFE *et al.*, 1994; SORNETTE and SAMMIS, 1995; BREHM and BRAILE, 1999; ROBINSON, 2000), values of t_f and A found by fitting procedures give estimates of the time and magnitude of the events culminating the ASR phases. It is well known (e.g., KANAMORI and ANDERSON, 1975) that event count N, earthquake magnitude M, and seismic moment M_0 of regional seismicity satisfy the scaling relations $Log_{10}N \sim -M$ and $Log_{10}M_0 \sim 1.5$ M. Thus applications of (1a) with $\zeta = 1$ and $\zeta = 0$ give dominating weights to the largest and smallest events within the analysis, respectively, while fractional values of ζ provide filters that modify the relative contributions of events in different magnitude ranges. With the above scaling relations, the contribution from each magnitude unit is approximately the same for $\zeta = 2/3$, and consequently the choice $\zeta = 1/2$ in (1b) associated with the Benioff strain gives a somewhat higher weight to smaller events (e.g., SAMMIS *et al.*, 1996).

Table 1 summarizes information on seismic regions, culminating large earthquakes, and best-fitting or fixed-used values of the exponent *m* in observational studies of ASR employing Benioff strain. We note that the number N of data points used to estimate the parameters of (1b) from observed ASR phases is typically less than 30. Synthetic data tests indicate that the uncertainty of estimated power-law parameters, approximately proportional to $N^{1/2}$, is rather high for such small data sets (Y. Huang, pers. comm., 2000). In addition, there are other possible errors in the best-fitting and used *m* values (and other ASR parameters) due to ambiguities associated with the identification of ASR phases, selection of involved spatio-temporal domains, and various other analysis issues (e.g., VERE-JONES *et al.*, 2001). Nevertheless, the results summarized in Table 1 form the current phenomenological basis of ASR and as such they are used in the present work. The relatively large earthquakes terminating the observed ASR phases range in size from M < 4 (BREHM and BRAILE, 1998) to M > 8 (BUFE *et al.*, 1994; BOWMAN *et al.*, 1998). Figures 1a and 1b show the distributions of all and best-fitting *m* values of Table 1, calculated with the kernel density method (SILVERMAN, 1986). The two distributions are peaked at *m* values of 0.29 and 0.28, respectively. Figures 1c and 1d show the mean, standard deviation, and median of all and best-fitting *m* values as a function of the magnitude cutoff M_{cut} of the relatively large events terminating the observed ASR phases. In both Figures 1c and 1d, the mean and median have a flat maximum region near $m = 0.3$ for $5 \leq M_{cut} \leq 7.5$ and they fall for smaller and larger M_{cut}.

In the following sections we present analytical and numerical results on ASR phases and related properties of seismicity patterns based on a number of different theoretical models. In Section 2.1 we discuss expectations associated with subcritical crack growth and conclude that this framework provides an inadequate explanation for observed ASR phases. In Section 2.2 we review statistical physics studies based on the assumption that large events in a deforming region correspond to phase transitions. Generic expectations in this framework include progressive establishment of long-range correlations of dynamic variables and

Table 1

Reported best-fitting and fixed-used values of the exponent m of the power-law time-to-failure equation (1b) in observational studies of accelerated seismic release

References	Seismic region	M large	m_{free}	m_{used}	Notes
Bufe and Varnes [1993] (BV 93)	Branches of the San Andreas fault system north of the creeping zone	7.9 6.9 6.8	0.32 0.30 0.34		
Bufe et al. [1994] (BNV 94)	Segments of subduction zone in the Alaska-Aleutian region	7.5 7.4 7.8 7.8 8.5		0.3 0.3 0.3 0.3 0.3	
Sornette and Sammis [1995] (SS95)	Loma Prieta Kommandorski Island	6.9 > 8.0	0.35, 0.34* 0.26, 0.28*		* denotes values obtained using equation (6), where the power law (1b) is augmented by log-periodic oscillations. Same sequences were also studied by BV93 and BNV94
Varnes and Bufe [1996] (VB96)	Virgin Islands	4.8 4.8		0.2 0.3	
Bowman et al. [1998]	California	7.5 7.3 7.0 6.7 6.7 6.6 6.6 6.5 7.7 5.6	0.3 0.18 0.28 0.18 0.1 0.13 0.43 0.55 0.49 0.12		Some sequences also studied by BV93, SS95, and VB96
	Assam Virgin Islands	8.6 4.8	0.22 0.11		
Brehm and Brail [1998]	New Madrid seismic zone	6.2 5.5 4.3 3.6 3.8 4.3 3.6 3.5 3.6 5.2 4.1 4.3 3.8	0.13 0.27 0.25 0.27 0.20 0.27 0.13 0.30 0.35 0.16 0.23 0.12 0.30		

Table 1

Continued

References	Seismic region	M large	m_{free}	m_{used}	Notes
Brehm and Brail [1998]	New Madrid seismic zone	4.3	0.20		
		4.8	0.16		
		3.9	0.27		
		3.5	0.18		
		4.2	0.24		
		3.5	0.47		
Robinson [2000]	New Zealand	7.0	0.31		
		6.7	0.29		
		6.7	0.46		

asymptotic power-law relations during the evolution leading to critical or spinodal phase transitions (e.g., SORNETTE and SAMMIS, 1995; SALEUR *et al.*, 1996; RUNDLE *et al.*, 2000b). In Section 2.3 we show that large earthquakes in the model of BEN-ZION (1996) for a discrete heterogeneous strike-slip fault in a 3-D elastic half space are associated with non-repeating cyclical establishment and destruction of long-range stress correlations accompanied by non-stationary cumulative Benioff strain. In Section 2.4 we demonstrate analytically that a 1-D version of the damage rheology of LYAKHOVSKY *et al.* (1997) leads to a power-law time-to-failure relation for strain with $m = -1/3$, and a corresponding power law for cumulative Benioff strain with $m = 1/3$. To derive an approximate expectation for regional deformation we add to the latter result a linear function of time representing release from background seismicity. Properties of evolving seismicity patterns in a regional model consisting of a seismogenic upper crust governed by the damage rheology of LYAKHOVSKY *et al.* (1997) over a viscoelastic substrate are discussed in Section 2.5. The results indicate that power-law build-up of cumulative Benioff strain exists only when the seismicity preceding the large earthquakes has broad frequency-size (FS) statistics. In such cases, the simulated ASR phases can be fitted well by all three forgoing analytical results. The ASR phases in the model simulations are accommodated both by increasing average rate of moderate events and increasing average earthquake size, with the former beginning a few years earlier. A brief discussion of the results including suggestions for future studies and implications for forecasting large event times is given in Section 3.

2. Analysis

Several theoretical frameworks may be used to explain the origin and parameters of the time-to-failure power-law relation of cumulative Benioff strain. These include

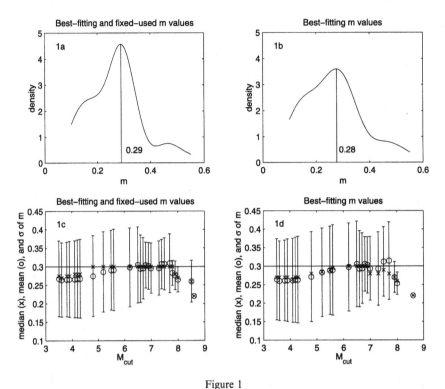

Figure 1
Statistics of best-fitting and fixed-used m values in the observations summarized in Table 1. (a) Density (smooth histogram) of all the m values of Table 1 calculated with the kernel density method. (b) Same as (a) for the best-fitting values only. (c) Median (x), mean (o), and standard deviation (vertical error bars) for all m values of Table 1 as a function of magnitude cutoff of the events terminating the reported accelerated seismic release (ASR) phases. (d) Same as (c) for the best-fitting m values only.

sub-critical crack growth, phase transition, and brittle deformation of heterogeneous faults in continuum solids. The first approach provides a deterministic description of failure at a single (relatively small, microscopic or mesoscopic) scale based on continuum mechanics. The second gives macroscopic statistical physics results for multi-scale failures near conditions of a global phase transition. The third, which we develop further in this work, incorporates elements of the other two.

2.1. Subcritical Crack Growth

In solids with microcracks, inclusions, and other flaws, the internal stress field is highly non-uniform. Such materials subjected to long-term loading show significant rates of macroscopic crack extension at nominal (macroscopic) values of stress intensity factor K significantly lower than the critical value for brittle failure. This phenomenon is known as subcritical crack growth (SWANSON, 1984; ATKINSON and MEREDITH, 1987; INGRAFFEA, 1987; COX and SCHOLZ, 1988). Experimental rates of subcritical crack growth are most commonly represented by a power-law equation (CHARLES, 1958; PARIS and ERDÖGAN, 1963)

$$\frac{dL}{dt} = C \cdot K^p \ , \tag{2}$$

where L is crack length and C and p are material parameters depending on confining pressure, temperature, and other conditions. The latter is referred to as crack index or stress corrosion index. Observed values of p in laboratory experiments vary greatly during crack evolution (SWANSON, 1984, 1987; MEREDITH and ATKINSON, 1985; ATKINSON and MEREDITH, 1987). In slow early deformation phases, p typically ranges from 2 to 5. Then p may decrease slightly in a regime apparently controlled by transport rates of reactive species at the crack tip. At some critical crack length there is a transition to dynamic rupture and p increases rapidly to values between 10 and 100. LYAKHOVSKY (2001) demonstrates that calculated rates of quasi-static crack propagation in a solid governed by the damage rheology of LYAKHOVSKY et al. (1997) fit the above observations of subcritical crack growth.

DAS and SCHOLZ (1981), VARNES (1989), BUFE and VARNES (1993), MAIN (1999) and others use the square-root relation between crack intensity factor and crack length to convert (2) into a differential equation for crack growth in the form

$$\frac{dL}{dt} = C^* L^{p/2} \ , \tag{3}$$

where C^* depends on the applied load (assumed constant), pressure, temperature and material properties. Integrating (3) and writing the solution for a crack growing from an initial length L_0 gives

$$L = [L_0^{(2-p)/2} - (p-2)C^* t/2]^{2/(2-p)} \ . \tag{4}$$

For $p > 2$, the crack length diverges when the quantity in the bracket becomes zero. Using this to define a failure time (e.g., DAS and SCHOLZ, 1981) and substituting back to (4) and (3) lead to a power-law relation between the crack length and time-to-failure. To connect the result with (1a) and (1b), we use a scaling relation $M_0 \sim L^\eta$ between seismic moment and crack dimension. The values of the exponent η range from 3 for a smooth classical crack (e.g., KANAMORI and ANDERSON, 1975) to 2 for a disordered fractal-like rupture (FISHER et al., 1997). Accounting for the square-root conversion from seismic moment to Benioff strain, the relation between the exponent m in (1b) and crack index p is

$$m = \eta/(p-2) \ . \tag{5}$$

Using (5) and observed values of p, we estimate the values of m expected to be associated with different stages of the subcritical crack growth process. During the initial slow stage of crack growth ($p = 2 - 5$), m varies from an unbounded value to 1 or 2/3 depending on whether η is 3 or 2. This regime probably represents very low seismic activity below the levels associated with observed ASR phases. During the transitional regime before dynamic rupture, p increases, m decreases, and both vary greatly. For $p > 32$ (or 22), m drops below 0.1 for $\eta = 3$ (or 2). The predicted high variability of m in the transition to dynamic rupture and low asymptotic value before large-scale failure are not compatible with the values of m estimated from observed ASR phases (Table 1 and Fig. 1).

The framework of subcritical crack growth was developed originally as an empirical description of laboratory observations involving primarily the evolution of a single crack from an early quasi-static deformation at small size scales to an unstable dynamic rupture at a critical length. This may be referred to as a deterministic continuum mechanics approach at a relatively small single scale. In contrast, the ASR phenomenology is associated with a network of faults at a variety of scales in a broad region around the eventual rupture that terminates an ASR phase. It thus appears that subcritical crack growth before a large dynamic failure on a fault does not provide a satisfactory explanation of ASR, both conceptually and in terms of stability and asymptotic value of the m exponent.

2.2. Phase Transition

An alternative approach to understanding ASR based on statistical physics of critical phenomena was developed by SORNETTE and SAMMIS (1995), SAMMIS et al. (1996), and SALEUR et al. (1996). The basic underlying assumption in this approach is that large earthquakes represent phase transitions in some extended spatial domain where stress is correlated and close to a critical level for a system-size failure. With additional assumptions and appropriate theoretical developments, this framework provides (in contrast to the subcritical crack approach) asymptotic statistical results for multi-scale failures in the space-time regions near the largest failure event. In this view, phases of ASR are associated with a progressive occurrence of increasingly larger events, due to progressive establishment of stress correlations over larger portions of a given seismogenic domain. The latter may be defined as a region where stress interaction with the tectonic loading and seismicity patterns are dominated by large earthquakes on a major through-going fault zone.

BEN-ZION and SAMMIS (2001) review a variety of observational and theoretical works on the character of fault zones and suggest that large-scale tectonic deformation in given crustal domains is indeed dominated by relatively regular, major through-going fault zones. This situation holds in the model of BEN-ZION (1996) for a single disordered strike-slip fault system in a 3D elastic half space discussed in Section 2.3, and the regional lithospheric model of BEN-ZION et al. (1999) and LYAKHOVSKY et al. (2001) discussed in Section 2.5. In practice, however, it is not yet clear how to divide the seismogenic crust into dominating fault zones and associated surrounding domains in the sense discussed above. BOWMAN et al. (1998),

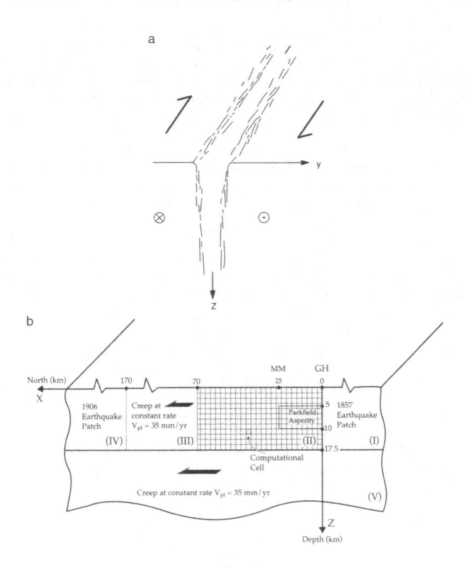

Figure 2
(a) A schematic representation of a 3-D disordered strike-slip fault structure (from BEN-ZION, 1996). (b) A planar representation of a disordered fault zone by a 2-D heterogeneous fault embedded in a 3-D elastic half-space. Each fault location $(x, y = 0, z)$ represents deformation in a volume centered on the line (x, y, z). The geometric disorder is modeled as disorder in strength properties of the planar fault. On regions I, II, IV, V, boundary conditions are specified. Region II is a computational grid where spatio-temporal evolution of stress and slip are calculated. The shown model configuration is tailored for the central San Andreas Fault. GH and MM mark approximate positions of Gold Hill and Middle Mountain (from BEN-ZION, 1996).

BREHM and BRAILE (1998), and ROBINSON (2000) identified about 30 crustal domains that were associated with phases of ASR before past large earthquakes, using search procedures that maximize the fit between observed cumulative Benioff strain and the time-to-failure power-law relation (1b). ZOLLER et al. (2001) identified about ten such domains (most of which overlap with and are somewhat larger than those of BOWMAN et al. (1998)) by optimizing the fit between growing correlations of hypocenter locations and power-law time-to-failure relation. It is important to examine in future works whether the regions defined in those studies based on past seismicity will preserve their intended functionality in the context of new ASR phases.

SORNETTE and SAMMIS (1995), SAMMIS et al. (1996), and SALEUR et al. (1996) applied the renormalization group theory to show that cumulative Benioff strain may represent the scaling regime of a critical phase transition, and that a complex-valued critical exponent produces log periodic correction to the right side of (1b),

$$\sum M_0^{1/2}(t) = A + B(t_f - t)^m [1 + C \cos(2\pi \log(t_f - t)/\log(\lambda) + \psi)] , \qquad (6)$$

with C, λ, and ψ being three additional free parameters. They also showed (SALEUR et al., 1996) that complex critical exponents might be generated by an underlying structure with discrete scale invariance (i.e., scale invariance at specific discrete magnifications) or a Euclidean heterogeneous system in which discrete scale invariance is produced by the dynamics. HUANG et al. (2000) demonstrated that log periodic oscillations in cumulative Benioff strain are also the expected generic outcome of routine processing of data containing an underlying power-law structure as in (1a) and (1b) with superimposed white noise. The integration to obtain cumulative Benioff strain transforms the original white noise to a correlated signal that is manifested as log periodic oscillations around the background power-law buildup. SORNETTE (1992) found that the mean field value of the exponent m associated with a critical phase transition is $m = 1/2$. This is close to the upper range rather than the mean or median of the observed m values with a peak around 0.3 (Table 1 and Fig. 1). RUNDLE et al. (2000b) used scaling arguments to show that power-law time-to-failure buildup of cumulative Benioff strain may represent the scaling regime of a spinodal phase transition, with an exponent $m = 1/4$ close to the observed values.

The statistical physics analyses provide powerful tools for studying possible types and statistical properties of event patterns. However, they cannot be used to calculate details of stress and displacement fields in a deforming solid. In the following sections we discuss results based on deterministic models of heterogeneous faults with many degrees of freedom in continuum solids. These heterogeneous models account for both complex evolving seismicity patterns with multi-scale failures and detailed deformation fields of the associated individual events.

2.3. Seismicity Patterns on Discrete Fault Systems with Quenched Heterogeneities

BEN-ZION (1996) simulated seismicity patterns for various cases of a 2-D discrete fault system in a 3-D elastic solid, based on earlier works of BEN-ZION and RICE

(1993, 1995). The model incorporates long-range elasticity, classical static/kinetic friction, power law creep, realistic boundary conditions, and various types of quenched heterogeneities. The work attempts to clarify the seismic response, on time scales of a few hundreds of years, of large individual fault systems, which have various levels of geometric disorder in the spatial distribution of brittle properties. A basic assumption of the model is that over such time scales the first-order deformational processes in a long and narrow 3-D fault zone (Fig. 2a) can be mapped onto a 2-D planar fault in a 3-D elastic solid (Fig. 2b). As discussed in Section 2.5, this is supported by good overall agreement between seismicity patterns generated by various cases of quenched heterogeneities in the planar model, and corresponding cases in the regional lithospheric model of LYAKHOVSKY *et al.* (2001) and BEN-ZION *et al.* (1999) with evolving non-planar structures. However, a rigorous mapping of various forms of 3-D geometric disorder onto corresponding planar representations remains an important unresolved issue.

The results of BEN-ZION (1996, Figs. 8–11) show that during gradual tectonic loading, small and intermediate earthquakes produce stress-roughening over their size scales, which collectively smooth the longer wavelength components of stress and develop long-range stress correlations on the fault. The pattern is reversed during large ruptures of size approaching the system dimension, which reduce the stress level and smooth the fluctuations along the large rupture area, while creating large stress concentrations near its boundary and increasing considerably the stress outside it. These system-size events reroughen the long wavelength stress field on the fault, destroy the long-range correlation in the system, and set the beginning of a new large earthquake cycle. ENEVA and BEN-ZION (1999) analyzed the progressive establishment and destruction of stress correlations in the simulated results using fluctuations of a stress-based order parameter. Following FERGUSON (1997) who evaluated slip-deficit fluctuations in a slider-block model, the analysis employs a stress fluctuation variable $F(t)$ defined as

$$F(t) = \sqrt{\frac{1}{K}\sum_{i=1}^{K}(\tau_i(t) - \bar{\tau}_i)^2} \ , \qquad (7)$$

where $\tau_i(t)$ is the stress at cell i and time t, $\bar{\tau}_i$ is the temporal average of stress at cell i during the deformation history, and K is the total number of computational cells. The variable $F(t)$ gives the temporal evolution of the RMS stress fluctuations along the fault. Increasing fluctuations correspond to increasing spatial correlation and increasing proximity to a global critical state.

Figure 3 top gives $F(t)$ of (7) for 150 years of deformation history simulated by the model realization of BEN-ZION (1996) with a fractal distribution of strength heterogeneities. The results show clearly the existence of non-repeating cyclic establishment and destruction of stress correlations on the fault. A large cycle begins when each large event (vertical lines in Fig. 3 top) destroys the long-range correlation

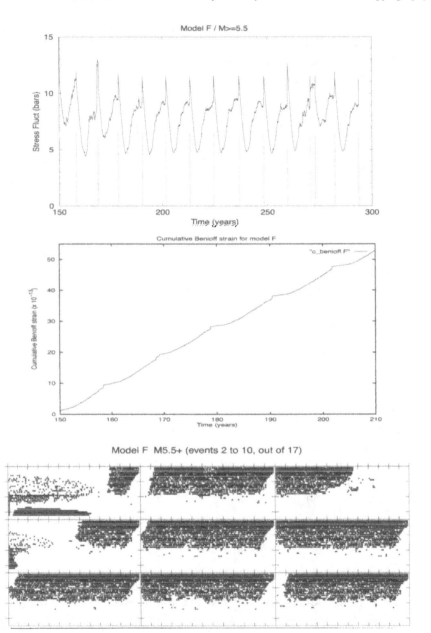

Figure 3
Top panel: Fluctuations of a stress-based order parameter for the model realization of BEN-ZION (1996) with fractal distribution of strength heterogeneities. The vertical lines show the times of 17 events with $M \geq 5.5$ that occur in the calculated history. Middle panel: Cumulative Benioff strain in the first 60 years of model evolution with 5 large earthquake cycles. Bottom panel: Rupture areas of events 2–10 of the 17 $M \geq 5.5$ earthquakes shown in the top panel. The event number increases from left to right and top to bottom (i.e., event 2 is on the left of the top raw and event 10 is on the right of the bottom raw).

of stress fluctuations. The combined occurrences of the subsequent small to intermediate size events increase the spatial correlation of fluctuations on the fault. The correlation length reaches a maximum value at about 3/4 the cycle time, when the ongoing model events establish "stress bridges" across the entire system. Then the correlation length fluctuates around the maximum value, sometimes with one or more clear drops, until one small event cascades to become the next large earthquake. This destroys the long-range correlation and starts a new cycle. The middle panel in Figure 3 shows the cumulative Benioff strain release on the fault for the first 60 years with five large earthquake cycles. As seen, there is clear overall correspondence between the cumulative Benioff strain, increasing stress correlation, and large earthquake cycle on the fault. However, since the model of Figure 2 does not simulate regional seismicity, the nonlinearities in the cumulative Benioff strain are produced primarily by drops after rather than build up before large events. Figure 3 bottom shows rupture areas of 9 large events that set the cycles duration. The areas are not necessarily contiguous, since the model incorporates long-range elasticity, however all these ruptures produce system-size stress changes that punctuate the relatively gradual preceding and following evolving correlations.

The simulated establishment and destruction of stress correlations within large earthquake cycles represent a nonrepeating cyclical approach to and retreat from criticality, or intermittent criticality, in general agreement with the cellular automata simulations of SAMMIS and SMITH (1999). Intermittent criticality associated with the occurrence of large earthquakes is supported by observed power law distributions of moment vs. time (ASR phases), number vs. moment (Gutenberg-Richter frequency-size statistics), and number vs. time (Omori law) before and after large earthquakes. We note that FISHER et al. (1997) demonstrated analytically and numerically that the model of BEN-ZION (1996) has an underlying critical point of a second-order phase transition. The critical point is associated with specific values of tuning parameters (dynamic weakening and conservation of stress transfer during failure events), and therefore it is "standard-ordinary-criticality" rather than "self-organized-criticality." In addition, the dynamics of the fault system of Figure 2 have a clear cyclical component, whereas self-organized-criticality describes stationary criticality where the only deviations from power-law distributions are statistical fluctuations (e.g., JENSEN, 1998). Tuning parameters and cyclical components are also present in the dynamics of the regional model for coupled evolution of earthquakes and faults discussed in the following two sections.

2.4. Continuum-mechanics-based Damage Rheology Model

BEN-ZION et al. (1999) and LYAKHOVSKY et al. (2001) used the damage rheology model of LYAKHOVSKY et al. (1997) in a regional lithospheric framework employed in the next section. The damage rheology of LYAKHOVSKY et al. (1997) provides a continuum-mechanics-based formulation for evolving non-linear properties of rocks under conditions of irreversible deformation. The framework adds to the Lamé parameters of linear Hookean elasticity λ and μ, a third parameter γ to account for the asymmetry of rock deformation under compression and tension conditions, and makes the moduli functions of an evolving damage state variable α. The damage variable α represents the local microcrack density as a function of the deformation history. An undamaged solid with $\alpha = 0$ is the ideal linear elastic material governed by the usual Hooke's law ($\gamma = 0$ for $\alpha = 0$). At the other extreme, a material with $\alpha = \alpha_c \leq 1$ is densely cracked and can not support any load. The damage rheology model of LYAKHOVSKY et al. (1997) calculates the instantaneous values of the elastic moduli for all intermediate states of the damage parameter ($0 < \alpha < \alpha_c$), based on the balance equations of energy and entropy and the above generalization of linear elasticity.

A full derivation of the governing equations and comparisons of model predictions with friction, fracture, and acoustic emission rock mechanics experiments (used both to validate the formulation and to constrain model parameters) are given by LYAKHOVSKY et al. (1997). The final equation for damage evolution, used below to derive a power-law time-to-failure relation, is

$$d\alpha/dt = CI_2(\xi - \xi_0) \, , \tag{8}$$

where $\xi = I_1/\sqrt{I_2}$, $I_1 = \varepsilon_{kk}$ and $I_2 = \varepsilon_{ij}\varepsilon_{ij}$ are the first and second invariants of the strain tensor ε_{ij}, the coefficient C describes the rate of damage evolution for a given deformation, and the critical strain parameter ξ_0 is qualitatively similar to internal friction in Mohr-Coulomb yielding criteria. A state of strain $\xi > \xi_0$ leads to material degradation (weakening of instantaneous elastic moduli) with a rate proportional to the second strain invariant multiplied by $(\xi - \xi_0)$. Similarly, a state of strain $\xi < \xi_0$ results in material strengthening (healing of instantaneous elastic moduli) proportional to the same factors. The damage rate coefficient C is constant during material degradation ($\xi > \xi_0$) and an exponential function of α during healing ($\xi < \xi_0$). The latter produces logarithmic healing in agreement with laboratory rate and state friction experiments (e.g., DIETERICH, 1972; SCHOLZ, 1990; MARONE, 1998).

BUFE and VARNES (1993) discuss general connections between ASR and damage mechanics. Below we show that a 1-D version of the damage rheology of LYAKHOVSKY et al. (1997) leads with a straightforward analytical derivation to power-law time-to-failure relation for strain with exponent $m = -1/3$, and corresponding power-law relation for cumulative Benioff strain release with $m = 1/3$. For 1-D deformation, equation (8) becomes

$$d\alpha/dt = C\varepsilon^2 \, , \tag{9}$$

where ε is the current strain. The stress-strain relation in this case is

$$\sigma = E_0(1 - \alpha)\varepsilon \, , \tag{10}$$

where $E_0(1 - \alpha)$ is the effective elastic modulus of a 1-D damaged material with E_0 being the initial modulus of the undamaged solid. Assuming constant stress σ and integrating (9) using (10) gives

$$\alpha = 1 - \{1 - (3C\sigma^2/E_0^2)t\}^{1/3} \ . \tag{11}$$

Substituting (11) back into (10) leads to strain accumulation in the power-law form

$$\varepsilon = \sigma/E_0\{1 - (3C\sigma^2/E_0^2)t\}^{-1/3} \ . \tag{12}$$

Using in (12) $t_f = E_0^2/3C\sigma^2$, defined by setting $\alpha = 1$ in (11), and changing constants gives

$$\varepsilon(t) = \sigma/E_0(1 - t/t_f)^{-1/3} = \sigma/E_0(\Delta t/t_f)^{-1/3} \tag{13}$$

with $\Delta t = t_f - t$. Equation (13) with negative exponent and strain singularity at the final failure time provides an appropriate physical expression for analyzing evolving deformation preceding a system-size event. However, analysis of observed ASR phases to date have focused on a nonsingular power-law time-to-failure equation of cumulative Benioff strain release with a positive exponent. Such an expression can be readily derived from the previous results. Using (10)–(13), the strain energy is

$$U(t) = (1/2)\sigma\varepsilon = (\sigma^2/2E_0)(\Delta t/t_f)^{-1/3} \ , \tag{14}$$

the energy and moment releases are proportional to

$$-\partial U/\partial t \sim -(\Delta t/t_f)^{-4/3} \ , \tag{15}$$

and the cumulative Benioff strain release is proportional to

$$-\int (\partial U/\partial t)^{1/2} dt \sim (\Delta t/t_f)^{1/3} \ . \tag{16}$$

Thus the 1-D version of our damage rheology predicts a power-law time-to-failure relation for cumulative Benioff strain with an exponent $m = 1/3$, close to the observed values.

Application of equation (16) to data of regional deformation containing many "damage degrees of freedom" requires modifications. The simplest generalization of the above result for such cases may be obtained by adding to the right side of (16) a linear function representing a stationary release associated with background regional seismicity (see also MAIN (1999)). With this, the expected cumulative Benioff strain in regional deformation has the form

$$\sum M_0^{1/2}(t) = A_1 + A_2 t + A_3(\Delta t/t_f)^{1/3} \ , \tag{17}$$

where A_1, A_2, A_3 are constants. In the next section we use equations (13), (16), and (17) to fit phases of ASR simulated by a regional lithospheric model with a

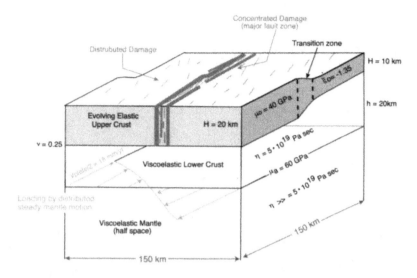

Figure 4

Geometry and parameters for a regional lithospheric model for coupled evolution of earthquakes and faults. The crust consists of a brittle upper layer governed by damage rheology over a viscoelastic lower crust driven by steady mantle motion from below. H and h mark the thickness of the upper and lower crust layers, respectively. Parameters μ, ξ, η, and v denote rigidity, critical strain coefficient, viscosity, and Poisson's constant, respectively. The boundary conditions are constant stress at the left and right edges and periodic repeats at the front and rear faces. The simulations leading to Figures 5 and 6 are done with a uniform crustal thickness H = 15 km. The simulations leading to Figures 7–10 are done with variable crustal thickness as indicated in the figure.

seismogenic upper crust governed by the damage rheology of LYAKHOVSKY *et al.* (1997).

2.5. Coupled Evolution of Earthquakes and Faults in a Regional Lithospheric Model

LYAKHOVSKY *et al.* (2001) and BEN-ZION *et al.* (1999) studied properties of crustal deformation associated with coupled evolution of earthquakes and faults using a version of the model shown in Figure 4 with a uniform upper crust thickness. The model consists of a seismogenic upper crust layer governed by the damage rheology of LYAKHOVSKY *et al.* (1997) over a layered Maxwell viscoelastic substrate. The evolving damage in the seismogenic layer simulates the creation and healing of fault systems as a function of the deformation history. The upper crust is coupled viscoelastically to the substrate where steady plate motion drives the deformation. The calculations employ vertically averaged variables of the thin sheet approximation for the viscous component of motion, and a Green function for a 3-D elastic half-space for the instantaneous component of deformation. Because of the thin sheet approximation, each model earthquake breaks the entire seismogenic zone so the smallest simulated event has a magnitude of about M = 6. In this sense the model is 2-D; however, stress transfer calculations are done, as mentioned above, with 3-D elasticity. We refer to this combined framework as a 2.5-D hybrid model. The formulation accounts in an internally consistent manner for evolving deformation fields, evolving fault structures, and spatio-temporal seismicity patterns. Simplified simulations with a prescribed narrow damage zone in an otherwise damage-free plate generate earthquake cycles on a large strike-slip fault with distinct inter-, pre-, co-, and post-seismic periods. LYAKHOVSKY et al. (2001) established that model evolution during each period is controlled by a subset of parameters that can be constrained by seismological, geodetic, and other geophysical data. Parameter values that are compatible with observations associated with the San Andreas fault are indicated in Figure 4 and used in the simulations discussed below.

Model realizations with the large-scale parameters of Figure 4 and random initial damage distribution produce large crustal faults and subsidiary branches with complex geometries. The parameter-space studies of BEN-ZION et al. (1999) and LYAKHOVSKY et al. (2001) with random initial damage and uniform upper crust thickness indicate that the results may be divided into three different dynamic regimes controlled by the ratio of time scale for damage healing τ_H to time scale for tectonic loading τ_L. The former characterizes the time for strength recovering after the occurrence of a brittle event and the latter the time for stress recovering at a failed location. High ratio of τ_H/τ_L leads to the development of geometrically regular fault systems and FS event statistics compatible with the characteristic earthquake (CE) distribution. In such cases, the event statistics are similar to those simulated by planar model realizations of BEN-ZION and RICE (1993, 1995) and BEN-ZION (1996) with relatively regular quenched heterogeneities. Conversely, low ratio of τ_H/τ_L leads to the development of a network of disordered fault systems, and power-law Gutenberg-Richter (GR) distribution. In these cases, the event statistics are similar to those simulated by model realizations of Ben-Zion and Rice with highly disordered quenched heterogeneities. For intermediate ratios of τ_H/τ_L, the results exhibit alternating overall switching of response, from periods of intense seismic activity and CE statistics to periods of low seismic activity and GR statistics. DAHMEN et al. (1998) demonstrated analytically and numerically that a similar mode switching behavior exists in the planar model of Ben-Zion and Rice for a range of dynamic weakening and conservation of stress transfer parameters.

Figure 5 shows cumulative Benioff strain simulated by the regional lithospheric model with a uniform upper crust thickness H = 15 km for cases producing the CE distribution, mode switching activity, and GR statistics. In the top panel with the CE distribution, the deviations from linear cumulative Benioff strain before the large events are abrupt and cannot be approximated well with a power-law relation. In the middle panel, there is a sharp transition in the cumulative Benioff strain when the mode of seismic release switches around 32 yr. from a period with relatively low release and GR distribution to a period with relatively high release and CE distribution. In the bottom panel with seismicity having GR statistics, there is a clear phase of ASR before the largest event. The cumulative Benioff strain in the ASR

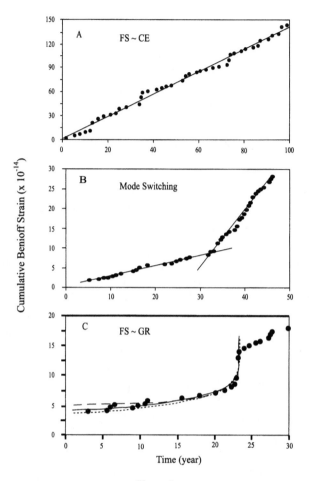

Figure 5

Cumulative Benioff strain for model realizations with uniform crustal thickness and three different ratios of healing time scale τ_H to loading time scale τ_L leading to different frequency-size (FS) distributions. (A) High ratio of τ_H/τ_L with characteristic earthquake distribution. (B) Intermediate ratio of τ_H/τ_L with mode switching activity. (C) Low ratio of τ_H/τ_L with power law FS statistics. The cumulative Benioff strain exhibits ASR before large events only when the earthquakes preceding the large event have a broad FS distribution. The ASR phase in the bottom panel is fitted with three power-law time-to-failure relations represented by $A_1 + A_2 t + A_3 (t_f - t)^m$. The long dash line corresponds to eq. (13) and is given by $m = -1/3, A_1 = 1.4 \ 10^{15}, A_2 = 0$, and $A_3 = 8 \ 10^{13}$. The solid line corresponds to eq. (16) and is given by $m = 0.2, A_1 = 1.6 \ 10^{15}, A_2 = 0$, and $A_3 = -1.0 \ 10^{14}$. The short dash corresponds to eq. (17) and is given by $m = 1/3, A_1 = 1.6 \ 10^{15}, A_2 = 1.2 \ 10^{12}$, and $A_3 = -8.2 \ 10^{13}$.

phase of the bottom panel can be fitted well by the singular power-law time-to-failure equation (13) with $m = -1/3$ (long dash curve), the nonsingular relation (16) with $m = 0.2$ (solid curve), and the generalized equation (17) with $m = 1/3$ (short dash curve). The good fits generated by the three different functions highlight the non-uniqueness associated with fitting such data, and stress the need for a careful

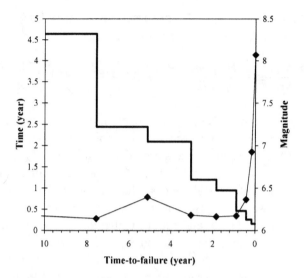

Figure 6

Average time interval between events (thick gray line) and average event magnitude (diamonds and thin line) during the ASR phase in the bottom panel of Figure 5.

estimation procedure when deriving functional forms and parameter values from similar data sets. VERE-JONES *et al.* (2001) provide several other examples of nonuniqueness and parameter estimation issues associated with ASR.

It is interesting to examine whether the activation process during the ASR phases involves increasing event size with time, increasing seismicity rates, or both. Figure 6 gives the average time interval between earthquakes and average event size in the ten years prior to the time t_f of the large event culminating the ASR phase in the bottom panel of Figure 5. The results show that the simulated ASR phase begins at about 8 years before t_f with increasing rates of moderate events (recall that the minimum simulated magnitude is M $\approx$ 6), and the rates continue to accelerate as the time of the culminating large earthquake is approached. This initial form of activation is followed a few years later, at about 1 year before t_f in the example shown, by a shorter phase of increasing event sizes. Similar trends are found in another set of simulations discussed below.

To verify the conclusions associated with the previous simulations we modify the model to include a variable crustal thickness in the direction parallel to the plate motion. In this version (Fig. 4), the thickness of the brittle upper crust changes from 10 km to 20 km through a narrow smooth transition zone. The rheological parameters are the same as before and the ratio of τ_H/τ_L is low enough to produce GR event statistics in the previous configuration with a uniform crustal thickness. To account for the variable thickness of the upper crust, we include in the governing equations of the 2.5-D hybrid model an additional force term proportional to the gradient of the upper crust thickness, as is commonly done in studies of lithospheric deformation (e.g., ARTUSHKOV, 1973; SONDER and ENGLAND, 1989). This amounts to replacing equation (13) of LYAKHOVSKY et al. (2001) with

$$H \iint_S G_{kn} \frac{\partial}{\partial t} \frac{\partial \sigma_{nm}}{\partial x_m} dS + \frac{Hh}{\eta} \left(\frac{\partial \sigma_{km}}{\partial x_m} + \frac{\rho_{uc}(\rho_{lc} - \rho_{uc})}{2\rho_{lc}} g \nabla H^2 \right) = \frac{\partial u_k}{\partial t} - V_{plate}^{(k)}, \quad (18)$$

where ρ_{uc} and ρ_{lc} are densities of the upper and lower crust layers, g is the gravitational acceleration, G is Green function for a 3-D elastic half space, u is displacement, and the other parameters are defined in Figure 4.

In our case with $\rho_{lc} - \rho_{uc} = 0.2$ gr/cm^3 and a mild difference between the thickness of the different crustal blocks, the additional force in (18) plays an insignificant role. However, the two different upper crust layers can store different amounts of elastic strain energy and produce different maximum earthquakes. This leads to another type of mode switching activity with space-time separation between event population with broad FS statistics and event population following the CE distribution (Fig. 7). During certain time intervals, seismicity not including the largest possible events occurs primarily in the thinner crustal block with FS statistics following the GR distribution. Occasionally, ruptures break through the transition zone initiating time intervals in which the thicker block participates in the earthquake activity. The seismicity in these time intervals includes clusters of the largest possible earthquakes and the associated FS statistics follows the CE distribution. During these latter periods, the cumulative Benioff strain experiences large abrupt jumps before the strongest events (Fig. 8) that cannot be fitted well by a power-law time-to-failure equation. In contrast, during the time intervals in which seismicity with broad FS statistics occurs in the thinner crustal block, the largest events are preceded by ASR phases (Fig. 9a) with a gradual buildup of activity. The cumulative Benioff strain in the ASR phases can be fitted well (Fig. 9b), as in Figure 5, by the singular analytical relation (13) with $m = -1/3$ (long dash curve), the non-singular equation (16) with $m = 0.2$ (solid curve), and the generalized relation (17) with $m = 1/3$ (short dash curve). Figure 10 displays the average time interval and average event size during the ASR phase of Figure 9b. As in Figure 6, the ASR phase begins with increasing rates of moderate events, here around 5 years before t_f, followed at about 3 years before t_f by increasing event sizes.

3. Discussion

We examined properties of ASR and related aspects of seismicity patterns associated with a number of theoretical frameworks. The studies continue our previous investigations of collective behavior of earthquakes and faults based on several different model categories. These include the discrete fault system with quenched heterogeneities of Section 2.3, the regional lithospheric model of Section

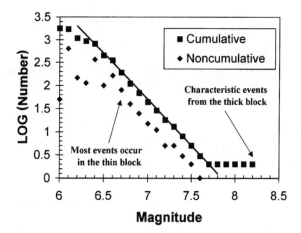

Figure 7
Frequency-size event statistics in model simulations with a variable upper crust thickness. Activity in the thinner block follows approximately a power-law distribution. Activity in the thicker block includes clusters of large events following the characteristic earthquake distribution.

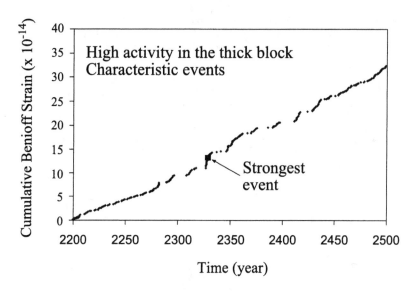

Figure 8
Cumulative Benioff strain release during a time interval with high activity in the thicker upper crust block. ASR phases are not observed.

2.5, and a smooth homogeneous fault in a continuum solid (e.g., BEN-ZION and RICE, 1997; LAPUSTA et al., 2000). The results from all these models indicate the existence of three basic dynamic regimes. The first is associated with strong fault heterogeneities, power-law FS statistics of earthquakes, and random or clustered temporal statistics of intermediate and large events. The second is associated with

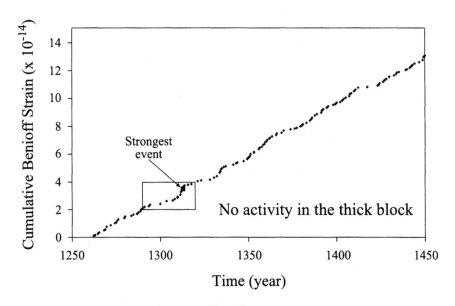

Figure 9a

Cumulative Benioff strain release for a time interval without activity in the thicker upper crust block. ASR phases are observed. The box marks an ASR phase that is fitted in Figure 9b.

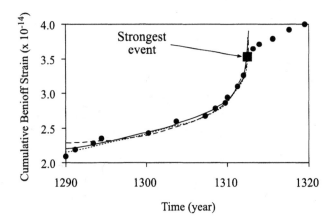

Figure 9b

Power-law time-to-failure fits to the ASR phase marked in Figure 9a with $A_1 + A_2t + A_3(t_f - t)^m$. The long corresponds to eq. (13) and is given by $m = -1/3$, $A_1 = 1.8 \ 10^{14}$, $A_2 = 0$, and $A_3 = 1.4 \ 10^{14}$. The solid line corresponds to eq. (16) and is given by $m = 0.2$, $A_1 = 4.3 \ 10^{14}$, $A_2 = 0$, and $A_3 = -1.1 \ 10^{14}$. The short dash line corresponds to eq. (17) and is given by $m = 1/3$, $A_1 = 3.6 \ 10^{14}$, $A_2 = -1.9 \ 10^{11}$, and $A_3 = -4.7 \ 10^{13}$.

homogeneous or relatively regular faults, FS statistics compatible with the characteristic earthquake distribution, and quasi-periodic temporal occurrence of large events. For a range of parameters, there is a third regime in which the response

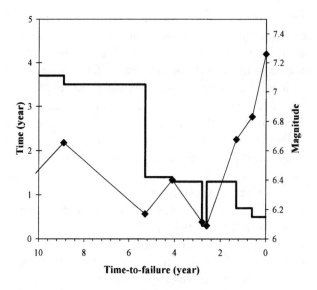

Figure 10
Average time interval between events (thick gray line) and average event magnitude (diamonds and thin line) in the ASR phase of Figure 9b

switches back and forth between the forgoing two modes of behavior. As discussed in our previous works, the model attributes and simulated patterns associated with the different dynamic regimes are compatible with friction, fracture, and other data of rock mechanics experiments, and *in situ* observations spanning wide ranges of space and time scales. The latter include fault trace maps and compiled earthquake statistics (WESNOUSKY, 1994; STIRLING *et al.*, 1996), high-resolution microseismicity patterns (e.g., JOHNSON and MCEVILLY, 1995; NADEAU *et al.*, 1995), and long paleoseismic records (e.g., GERSON *et al.*, 1993; MARCO *et al.*, 1996; ROCKWELL *et al.*, 2001; AMIT *et al.*, 2001).

In the present work we focus on connections between ASR and other properties of seismicity patterns. Various observational studies (e.g., BUFE and VARNES, 1993; BOWMAN *et al.*, 1998; BREHM and BRAILE, 1998; ROBINSON, 2000) fitted cumulative Benioff strain to power-law time-to-failure equation with exponent values *m* close to 0.3 (Table 1 and Fig. 1). Examining the subcritical crack growth process, we find that during the transition from stable slip to dynamic failure this framework predicts great variability and low asymptotic values of *m* that are not compatible with observations. Statistical physics results (e.g., SORNETTE and SAMMIS, 1995; SALEUR *et al.*, 1996; RUNDLE *et al.*, 2000b) provide scaling relations for deformation leading to global failures, assuming that those correspond to critical or spinodal phase transitions. Generic expectations in these models include progressive establishment of long-range correlations and asymptotic power-law relations during the final stages of the evolution.

One fundamental obstacle of progress toward a better understanding of earthquake processes is the fact that the evolution of stress and other governing dynamic variables on natural faults cannot be directly observed. We demonstrate that large earthquakes in the discrete heterogeneous fault model of BEN-ZION (1996) are associated with nonrepeating cyclical establishment and destruction of long-range stress correlations accompanied by non-stationary cumulative Benioff strain. These results are compatible with the idea that large earthquake cycles involve an approach to and retreat from criticality. Since stress evolution is not directly observable, it is important to find seismicity-based parameters that can be used as surrogate variables for stress and other dynamic variables of interest. This may be done by continuing studies of the type discussed in Section 2.3, combined with advanced analysis (e.g., PEPKE et al., 1994; ENEVA and BEN-ZION, 1997; RUNDLE et al., 2000a) of the associated seismicity parameters.

Damage mechanics provides a framework that can be used to model brittle deformation in a region with evolving material properties. We show analytically that a 1-D version of the damage rheology of LYAKHOVSKY et al. (1997) leads to a singular power-law time-to-failure relation with $m = -1/3$ for strain preceding catastrophic failure, and a corresponding nonsingular equation for cumulative Benioff strain release with $m = 1/3$. An approximate generalization of the latter result for regional deformation is obtained by adding a linear function of time representing a stationary background cumulative Benioff strain release.

Having these analytical expectations, we analyze results obtained by numerical simulations employing various realizations of a regional lithospheric model with a seismogenic upper crust governed by the damage rheology of LYAKHOVSKY et al. (1997). The simulations indicate that ASR phases exist only when the events before a given large earthquake have broad FS statistics. This appears to be a clear necessary condition; however it is not a sufficient one. That is because the existence of ASR requires not only a broad distribution of event sizes but also a failure sequencing within the distribution that produces increasing seismic release. The ASR phases in the simulated data are associated with both increasing rates of moderate events and increasing average event size; with the former starting a few years before the latter (Figs. 6 and 10). The simulated rate increase of moderate events is compatible with the observations of ELLSWORTH et al. (1981), LINDH (1990), SYKES and JAUMÉ (1990), and KNOPOFF et al. (1996), among others. The simulated increase of average earthquake size is compatible with the observations of JAUMÉ and SYKES (1999) and JAUMÉ (2000). The earlier occurrence in the simulations of the former type of activation and generality of the other results should be examined further in future studies.

The cumulative Benioff strain in the simulated ASR phases can be fitted well by the singular, nonsingular, and generalized nonsingular functions discussed above with exponent values close to those of the analytical derivations and observational results. The good fit obtained with all three functions emphasizes the non-uniqueness associated with fitting such data sets and the need for better estimation procedures (see also VERE-JONES et al., 2001). It is also important to develop better procedures for selecting space-time domains associated with ASR phases, and to examine whether the "zonings" of previous observational works (e.g., BOWMAN et al., 1998; BREHM and BRAILE, 1998; ROBINSON, 2000; ZOLLER et al., 2001) remain valid for future ASR phases.

The results of this and other works point to the existence of two complementary endmember predictive signals in patterns of moderate and large events associated with the first two dynamic regimes summarized at the beginning of the discussion. On regular fault systems with FS event statistics compatible with the CE distribution, there is predictive information in the associated quasi-periodic temporal distribution of large events. On highly disordered fault systems with power-law FS statistics and random or clustered temporal distribution of large events, phases of ASR before large events have predictive information. Mode switching behavior, mixed populations of faults, and various forms of transients are some examples among many possible complicating factors. A continuing multi-disciplinary research employing a combination of numerical simulations, analytical work, and advanced analysis of synthetic and observed data can clarify the generality of the discussed results and may lead to recognition of additional predictive signals. Further progress in prediction studies will also require more rigorous hypothesis testing and data analysis (e.g., KAGAN, 1999).

Acknowledgments

The results of Figure 3 are taken from work in progress of YBZ with Mariana Eneva and Yunfeng Liu. Yueqiang Huang helped to prepare Figure 1. We thank anonymous reviewers for comments. YBZ acknowledges support from the National Science Foundation (grant EAR-9725358) and the Southern California Earthquake Center (based on NSF cooperative agreement EAR-8920136 and USGS cooperative agreement 14-08-0001-A0899). VL acknowledges support from the United States – Israel Binational Science Foundation (BSF), Jerusalem Israel (grant No. 9800198).

REFERENCES

AMIT, R., ZILBERMAN, E., PORAT, N., and ENZEL, Y. (2002), *Paleoseismic Evidence for Time-dependency of Seismic Response on a Fault System – The Southern Arava Valley, Red Sea Rift, Israel*, Geol. Soc. Am. Bull. *114*, 192–206.

ARTUSHKOV, E. V. (1973), *Stresses in the Lithosphere Caused by Crustal Thickness Inhomogeneities*, J. Geophys. Res. *78*, 7675–7708.

ATKINSON, B. K. and MEREDITH, P. G. *The theory of subcritical crack growth with applications to minerals and rocks*. In: *Fracture Mechanics of Rock* (B. K. Atkinson, ed.), (Academic Press, San Diego 1987), pp. 110–166.

BEN-ZION, Y. (1996), *Stress, Slip and Earthquakes in Models of Complex Single-fault Systems Incorporating Brittle and Creep Deformations*, J. Geophys. Res. *101*, 5677–5706.

BEN-ZION, Y., DAHMEN, K., LYAKHOVSKY, V., ERTAS, D., and AGNON, A. (1999), *Self-driven Mode Switching of Earthquake Activity on a Fault System*, Earth Planet. Sci. Lett. *172/1–2*, 11–21.

BEN-ZION, Y. and RICE, J. R. (1993), *Earthquake Failure Sequences along a Cellular Fault Zone in a Three-dimensional Elastic Solid Containing Asperity and Nonasperity Regions*, J. Geophys. Res. *98*, 14,109–14,131.

BEN-ZION, Y. and RICE, J. R. (1995), *Slip Patterns and Earthquake Populations along Different Classes of Faults in Elastic Solids*, J. Geophys. Res. *100*, 12,959–12,983.

BEN-ZION, Y. and RICE, J. R. (1997), *Dynamic Simulations of Slip on a Smooth Fault in an Elastic Solid*, J. Geophys. Res. *102*, 17,771–17,784.

BEN-ZION, Y. and SAMMIS, C. G. (2001), *Characterization of Fault Zones*, Pure Appl. Geophys., in press.

BOWMAN, D. D., OUILLON, G., SAMMIS, C. G., SORNETTE, A., and SORNETTE, D. (1998), *An Observational Test of the Critical Earthquake Concept*, J. Geophys. Res. *103*, 24,359–24,372.

BREHM, D. J. and BRAILE, L. W. (1998), *Intermediate-term Earthquake Prediction Using Precursory Events in the New Madrid Seismic Zone*, Bull. Seismol. Soc. Am. *88*, 564–580.

BUFE, C. G. and VARNES, D. J. (1993), *Predictive Modeling of the Seismic Cycle of the Greater San Francisco Bay Region*, J. Geophys. Res. *98*, 9871–9883.

BUFE, C. G., NISHENKO, S. P., and VARNES, D. J. (1994), *Seismicity Trends and Potential for Large Earthquakes in the Alaska-Aleutian Region*, Pure Appl. Geophys. *142*, 83–99.

CHARLES, R. J. (1958), *Static Fatigue of Glass*, J. Appl. Physics *29*, 1549–1560.

COX, S. J. D. and SCHOLZ, C. H. (1988), *Rupture Initiation in Shear Fracture of Rocks: An Experimental Study*, J. Geophys. Res. *93*, 3307–3320.

DAHMEN, K., ERTAS, D., and BEN-ZION, Y. (1998), *Gutenberg Richter and Characteristic Earthquake Behavior in Simple Mean-field Models of Heterogeneous Faults*, Phys. Rev. E *58*, 1494–1501.

DAS, S. and SCHOLZ, C. H. (1981), *Theory of Time-dependent Rupture in the Earth*, J. Geophys. Res. *86*, 6039–6051.

DIETERICH, J. H. (1972), *Time-dependent Friction in Rocks*. J. Geophys. Res. *77*, 3690–3697.

ELLSWORTH, W. L., LINDH, A. G., PRESCOTT, W. H., and HERD, D. J. *The 1906 San Francisco Earthquake and the seismic cycle*. In *Earthquake Prediction: An Internation Review*, Maurice Ewing Ser., vol. 44, (eds. D.W. Simpson and P.G. Richards), (AGU, Washington, D.C. 1981) pp. 126–140.

ENEVA, M. and BEN-ZION, Y. (1997), *Application of Pattern Recognition Techniques to Earthquake Catalogs Generated by Models of Segmented Fault Systems in Three-dimensional Elastic Solids*, J. Geophys. Res. *102*, 24,513–24,528.

ENEVA, M. and BEN-ZION, Y. (1999), *Criticality in Stress/Slip Distribution and Seismicity Patterns in Fault Models*, EOS Trans. Amer. Geophys. Union *80*, S331.

FERGUSON, C. D. (1997), *Numerical Investigations of an Earthquake Fault Based on a Cellular Automaton, Slider-block Model*, Ph.D. Dissertation, Boston University.

FISHER, D. S., DAHMEN, K., RAMANATHAN, S., and BEN-ZION, Y. (1997), *Statistics of Earthquakes in Simple Models of Heterogeneous Faults*, Phys. Rev. Lett. *78*, 4885–4888.

GERSON, R., GROSSMAN, S., AMIT, R., and GREENBAUM, N. (1993), *Indicators of Faulting Events and Periods of Quiescence in Desert Alluvial Fans*, Earth Surface Processes and Landforms *18*, 181–202.

HUANG, Y., SALEUR, H., JOHANSEN, A., LEE, M., and SORNETTE, D. (2000), *Artifactual Log-periodicity in Finite Size Data: Relevance for Earthquake Aftershocks*, J. Geophys. Res. *105*, 25,451–25,471.

INGRAFFEA, A. R. *Theory of crack initiation and propagation in rock*. In *Fracture Mechanics of Rock* (B. K. Atkinson, ed.) (Academic Press, San Diego 1987) pp. 76–110.

JAUMÉ, S. C. (2000), *Changes in earthquake size-frequency distributions underlying accelerating seismic moment/energy release*. In Geophys. Mono. Series 120, *GeoComplexity and the Physics of Earthquakes* (J. B. Rundle, D. L. Turcotte, and W. Klein, eds.) 199–210 (American Geophysical Union, Washington D. C).

JAUMÉ, S. C. and SYKES, L. R. (1999), *Evolving Toward a Critical Point: A Review of Accelerating Seismic Moment/Energy Release Prior to Large and Great Earthquakes*, Pure Appl. Geophys. *155*, 279–305.

JENSEN, H. J. *Self-organized Criticality* (Cambridge University Press 1998).

JOHNSON, P. and MCEVILLY, T. V. (1995), *Parkfield Seismicity: Fluid-driven?* J. Geophys. Res. *100*, 12,937–12,950.

JONES, L. M. and MOLNAR, P. (1979), *Some Characteristics of Foreshocks and their Possible Relationship to Earthquake Prediction and Premonitory Slip on Faults*, J. Geophys. Res. *84*, 3596–3608.

KAGAN, Y. Y. (1999), *Is Earthquake Seismology a Hard, Quantitative Science?*, Pure Appl. Geophys. *155*, 233–258.

KANAMORI, H. and ANDERSON, D. L. (1975), *Theoretical Basis of Some Empirical Relations in Seismology*, Bull. Seismol. Soc. Am. *65*, 1073–1095.

KEILIS-BOROK, V. I. and KOSSOBOKOV, V. G. (1990), *Premonitory Activation of Earthquake Flow: Algorithm M8*, Phys. Earth Planet. Inter. *61*, 73–83.

KNOPOFF, L., LEVSHINA, T., KEILIS-BOROK, V. I., and MATTONI, C. (1996), *Increased long-range Intermediate-magnitude Earthquake Activity prior to Strong Earthquakes in California*, J. Geophys. Res. *101*, 5779–5796.

LAPUSTA, N., RICE, J. R., BEN-ZION, Y., and ZHENG, G. (2000), *Elastodynamic Analysis for Slow Tectonic Loading with Spontaneous Rupture Episodes on Faults with Rate- and State-dependent Friction*, J. Geophys. Res. *105*, 23,765–23,789.

LINDH, A. G. (1990), *The Seismic Cycle Pursued*, Nature *348*, 580–581.

LYAKHOVSKY, V. (2001), *Scaling of Fracture Length and Distributed Damage*, Geophys. J. Int. *144*, 114–122.

LYAKHOVSKY, V., BEN-ZION, Y., and AGNON, A. (1997), *Distributed Damage, Faulting, and Friction*, J. Geophys. Res. *102*, 27,635–27,649.

LYAKHOVSKY, V., BEN-ZION, Y., and AGNON, A. (2001), *Earthquake Cycle, Fault Zones, and Seismicity Patterns in a Rheologically Layered Lithosphere*, J. Geophys. Res. *106*, 4103–4120.

MAIN, I. G. (1999), *Applicability of Time-to-failure Analysis to Accelerated Strain before Earthquakes and Volcanic Eruptions*, Geophys. J. Int. *139*, F1–F6.

MARCO, S., STEIN, M., AGNON, A., and RON, H. (1996), *Long-term Earthquake Clustering: A 50,000 Year Paleoseismic Record in the Dead Sea Graben*, J. Geophys. Res. *101*, 6179–6192.

MARONE, C. (1998), *Laboratory-derived Friction Laws and their Application to Seismic Faulting*, Annu. Rev. Earth Planet. Sci. *26*, 643–649.

MEREDITH, P. G. and ATKINSON, B. K. (1985), *Fracture Toughness and Subcritical Crack Growth during High-temperature Tensile Deformation of Westerly Granite and Black Gabbro*, Tectonophysics *39*, 33–51.

MOGI, K. (1969), *Some Features of Recent Seismic Activity in and near Japan {2}: Activity Before and After Great Earthquakes*, Bull. Eq. Res. Inst. Univ. Tokyo *47*, 395–417.

MOGI, K. *Seismicity in Japan and long-term earthquake forecasting.* In *Earthquake Prediction: An Internation Review*, Maurice Ewing Ser., vol. 4 (eds. D.W. Simpson and P.G. Richards) pp. 43–51, (AGU, Washington, D.C. 1981).

NADEAU, R. M., FOXALL, W., and MCEVILLY, T. V. (1995), *Clustering and Periodic Recurrence of Microearthquakes on the San Andreas Fault at Parkfield, California*, Science *267*, 503–507.

PAPAZACHOS, B. C. (1973), *The Time Distribution and Prediction of Reservoir-associated Foreshock and its Importance to the Prediction of the Principal Shock*, Bull. Seismol. Soc. Am. *63*, 1973–1978.

PARIS, P. C. and ERDOGAN, F. (1963), *A Critical Analysis of Crack Propagation Laws*, J. Basic Engineering, ASME Transactions, Series D *85*, 528–534.

PEPKE, S. L., CARLSON, J. M., and SHAW, B. E. (1994), *Prediction of Large Events on a Dynamical Model of a Fault*, J. Geophys. Res. *99*, 6769–6788.

PRESS, F. and ALLEN, C. (1995), *Patterns of Seismic Release in the Southern California Region*, J. Geophys. Res. *100*, 6421–6430.

ROBINSON, R. (2000), *A Test of the Precursory Accelerating Moment Release Model on Some Recent New Zealand Earthquakes*, Geophys. J. Int. *140*, 568–576.

ROCKWELL, T. K., LINDVALL, S., HERZBERG, M., MURBACH, D., DAWSON, T., and BERGER, G. (2000), *Paleoseismology of the Johnson Valley, Kickcapoo and Homestead Valley Faults of the Eastern California Shear Zone*, Bull. Seismol. Soc. Am. *90*, 1200–1236.

RUNDLE, J. B., KLEIN, W., TIAMPO, K., and GROSS, S. (2000a), *Linear Patterns Dynamics in Nonlinear Threshold System*, Phys. Rev. E *61*, 2418–2431.

RUNDLE, J. B., KLEIN, W., TURCOTTE, D. L., and MALAMUD, B. D. (2000b), *Precursory Seismic Activation and Critical-point Phenomena*, Pure Appl. Geophys. *157*, 2165–2182.

SALEUR, H., SAMMIS, C. G., and SORNETTE, D. (1996), *Discrete Scale Invariance, Complex Fractal Dimensions, and Log-periodic Fluctuations in Seismicity*, J. Geophys. Res. *101*, 17,661–17,677.

SAMMIS, C. G. and SMITH, S. W. (1999), *Seismic Cycles and the Evolution of Stress Correlation in Cellular Automaton Models of Finite Fault Networks*, Pure Appl. Geophys. *155*, 307–334.

SAMMIS, C. G., SORNETTE, D., and SALEUR, H. *Complexity and earthquake forecasting.* In *Reduction and Predictability of Natural Disasters*, SFI Studies in the Sciences of Complexity, vol. XXV, (eds. J. B. Rundle, W. Klein, and D. L. Turcotte) pp. 143-156 (Addison-Wesley, Reading, Mass. 1996).

SCHOLZ, C. H. *The Mechanics of Earthquakes and Faulting* (Cambridge Press 1990).

SHAW, B. E., CARLSON, J. M., and LANGER, J. S. (1992), *Patterns of Seismic Activity Preceding Large Earthquakes*, J. Geophys. Res. *97*, 479–488.

SILVERMAN, B. W. *Density estimation for statistics and data analysis.* In *Monographs on Statistics and Applied Probability* (eds. D. R. Cox, D. V. Hinkley, D. Rubin and B. W. Silverman), 26 (Chapman and Hall, New York 1986).

SONDER, L. J. and ENGLAND, P. C. (1989), *Effects of a temperature-dependent Rheology on Large-scale Continental Extension*, J. Geophys. Res. *94*, 7603–7619.

SORNETTE, D. (1992), *Mean-field Solution of a Block-spring Model of Earthquakes*, J. Phys. I France *2*, 2089–2096.

SORNETTE, D. and SAMMIS, C. G. (1995), *Complex Critical Exponent from Renormalization Group Theory of Earthquakes: Implications for Earthquake Predictions*, J. Phys. I France, 5, 607–619.

STIRLING, M. W., WESNOUSKY, S. G., and SHIMAZAKI, K. (1996), *Fault Trace Complexity, Cumulative Slip, and the Shape of the Magnitude-frequency Distribution for Strike-slip Faults: A Global Survey*, Geophys. J. Int. *124*, 833–868.

SWANSON, P. L. (1984), Subcritical Crack Growth and Other Time and Environment-dependent Behavior in Crustal Rocks, J. Geophys. Res. *89*, 4137–4152.

SWANSON, P. L. (1987), *Tensile Fracture Resistance Mechanisms in Brittle Polycrystals: An Ultrasonics and in situ Microscopy Investigation*, J. Geophys. Res. *92*, 8015–8036.

SYKES, L. R. and JAUMÉ, S. C. (1990), *Seismic Activity on Neighboring Faults as a Long-term Precursor to Large Earthquakes in the San Francisco Bay Region*, Nature 348, 595–599.

VARNES, D. J. (1989), *Predicting Earthquake by Analyzing Accelerating Precursory Seismic Activity*, Pure Appl. Geophys. *130*, 661–686.

VARNES, D. J. and BUFE, C. G. (1996), *The Cyclic and Fractal Seismic Series Preceding an* $M_b = 4.8$ *Earthquake on 1980 February 14 near the Virgin Islands*, Geophys. J. Int. *124*, 149–158.

VERE-JONES, D., ROBINSON, R., and YANG, W. (2001), *Remarks on the Accelerated Moment Release Model: Problems of Model Formulation, Simulation and Estimation*, Geophys. J. Int. *144*, 517–531.

WESNOUSKY, S. G. (1994), *The Gutenberg-Richter or Characteristic Earthquake Distribution, which is it?*, Bull. Seismol. Soc. Am. *84*, 1940–1959.

ZOLLER, G., HAINZL, S., and KURTHS, J. (2001), *Observation of Growing Correlation Length as an Indicator for Critical Point Behavior prior to Large Earthquakes*, J. Geophys. Res. *106*, 2167–2175.

(Received February 20, 2001, revised June 11, 2001, accepted June 15, 2001)

To access this journal online:
http://www.birkhauser.ch

Pure appl. geophys. 159 (2002) 2413–2427
0033–4553/02/102413–15 $ 1.50 + 0.20/0

© Birkhäuser Verlag, Basel, 2002

❙ **Pure and Applied Geophysics**

Stress Correlation Function Evolution in Lattice Solid Elasto-dynamic Models of Shear and Fracture Zones and Earthquake Prediction

Peter Mora[1] and David Place[2]

Abstract — It has been argued that power-law time-to-failure fits for cumulative Benioff strain and an evolution in size-frequency statistics in the lead-up to large earthquakes are evidence that the crust behaves as a Critical Point (CP) system. If so, intermediate-term earthquake prediction is possible. However, this hypothesis has not been proven. If the crust does behave as a CP system, stress correlation lengths should grow in the lead-up to large events through the action of small to moderate ruptures and drop sharply once a large event occurs. However this evolution in stress correlation lengths cannot be observed directly. Here we show, using the lattice solid model to describe discontinuous elasto-dynamic systems subjected to shear and compression, that it is for possible correlation lengths to exhibit CP-type evolution. In the case of a granular system subjected to shear, this evolution occurs in the lead-up to the largest event and is accompanied by an increasing rate of moderate-sized events and power-law acceleration of Benioff strain release. In the case of an intact sample system subjected to compression, the evolution occurs only after a mature fracture system has developed. The results support the existence of a physical mechanism for intermediate-term earthquake forecasting and suggest this mechanism is fault-system dependent. This offers an explanation of why accelerating Benioff strain release is not observed prior to all large earthquakes. The results prove the existence of an underlying evolution in discontinuous elasto-dynamic systems which is capable of providing a basis for forecasting catastrophic failure and earthquakes.

Key words: Numerical simulation, stress correlation function evolution, critical point hypothesis for earthquakes, earthquake prediction, lattice solid model.

Introduction

The power-law size-frequency statistics of earthquakes has been cited as evidence that the earth's crust is in a self-organized critical state, and hence, that earthquake prediction is impossible (GELLER *et al.*, 1997). On the other hand, observations of earthquake size scaling (e.g., KNOPOFF, 2000; AKI, 2000; WU, 2000) suggest that the SOC hypothesis is inadequate and that prediction may be possible. Furthermore, observations of accelerating Benioff strain release (BUFE and VARNES, 1993; BOWMAN *et al.*, 1998) and an evolution in earthquake statistics prior to large events

[1,2] QUAKES, Department of Earth Sciences, The University of Queensland, St Lucia, Brisbane QLD 4072, Australia. E-mails: mora@quakes.uq.edu.au; place@quakes.uq.edu.au

(JAUMÉ, 2000) are cited as evidence that the crust is not perpetually in a critical state, but acts as a critical-point system in which case it approaches and retreats from criticality. In this view, an earthquake cycle is proposed to proceed as follows. Initially, small earthquakes occur and redistribute stress locally. This process gradually allows long-range stress correlations to be established which are argued to be a precondition for a rupture to runaway to the largest event size (RUNDLE et al., 1999). Once this occurs, the large earthquake will destroy the long-range stress correlations and the cycle will repeat. Consideration of the earth's crust as a critical system has led various researchers to draw on statistical physics to model seismicity patterns and to propose that a large earthquake may be viewed as a critical point (SORNETTE and SAMMIS, 1995). Under this hypothesis, regional seismic energy release should fit a power-law time-to-failure function such as

$$\varepsilon(t) = A + B(t - t_f)^c \ , \tag{1}$$

in the lead-up to a large event, where t_f is the time of the critical point and c is the power-law exponent. Energy release measures such as cumulative seismic moment and Benioff strain have been observed to fit well with such a power-law time-to-failure function in a number of cases for real data (BOWMAN et al., 1998) as well as synthetic data obtained from cellular automaton (CA) models of seismicity (SAMMIS and SMITH, 1999; JAUMÉ et al., 2000) and a 2-D elasto-dynamic model for a granular system (MORA et al., 2000). However, such observations provide only indirect evidence that the earth or model systems are critical point systems. Recent observations of growing correlation lengths in seismicity data prior to large earthquakes provide additional support for the critical point hypothesis for earthquakes (ZOELLER et al., 2001). Studies of different classes of CA models indicate that the stress field may evolve like that of critical point system or self-organized critical systems, depending on the rules used in the CA model to approximate the rupture and stress redistribution processes (SAMMIS and SMITH, 1999; WEATHERLEY et al., 2000, 2002). Here we study the evolution in the stress correlation function in a physically based numerical model which accurately simulates the elasto-dynamics of a stress transfer and rupture in simplified 2-D systems (MORA and PLACE, 1994; PLACE and MORA, 1999). Numerical experiments are conducted for the case of a 2-D granular region subjected to shear and an intact block subjected to compression, resulting in the development of a fracture system. These models can be considered as analogs to crustal fault systems.

The Lattice Solid Model

The lattice solid model (LSM) is a particle-based numerical model that was developed to simulate the nonlinear dynamics of earthquakes (MORA, 1992; MORA and PLACE, 1993, 1994, 2000; PLACE and MORA, 1999, 2000, 2001). The model simulates the elasto-dynamics of a system made up of model material that may fracture and undergo frictional forces along any fracture surface. Model material has elastic moduli $\lambda = \mu$ and consists of bonded groups of particles whose equations of motion are integrated taking into account all elastic and frictional forces. The lattice solid model is similar to the discrete element method (CUNDALL and STRACK, 1979) to model granular assemblies but involves a different computational approach (PLACE and MORA, 1999) and is being developed to model the physical processes underlying the dynamics of fault zones and earthquake nucleation. It is presently capable of simulating physical processes such as friction (PLACE and MORA, 1999), fracture (MORA and PLACE, 1993; Place et al., 2002), granular dynamics and thermal effects including thermo-mechanical and thermo-porous feedback (ABE et al., 2000; MORA et al., 2000). The LSM has been applied to the study of the heat-flow paradox (MORA and PLACE, 1998, 1999), and rock fracture and localization phenomena (PLACE and MORA, 2000, 2001; PLACE et al., 2002). Intact regions of model material can be shown to obey the equations of motion of continuous elastic media in the macroscopic limit – the elastic wave equation (ABE et al., 2000), and the model realistically simulates rupture and seismic wave propagation (MORA and PLACE, 1994). Recently, shear experiments involving a granular system exhibited power-law acceleration in Benioff strain release and evolving frequency-magnitude statistics in the lead-up to large events (MORA et al., 2000), suggesting the lattice solid model may be used to probe the Critical Point question. The lattice solid computational approach is well described elsewhere and will not be reviewed here (see MORA and PLACE, 1994; PLACE and MORA, 1999). In the following, numerical experiments of a granular system subjected to shear, and of an initially intact block subjected to compression are described.

Shear Experiment

Numerical experiments were conducted using a granular model subjected to shear (Fig. 1) using the same configuration as in previous experiments to study fault zone dynamics and localization phenomena (MORA and PLACE, 1999; PLACE and MORA, 2000; MORA et al., 2000). The model consists of a central granular region sandwiched between elastic blocks which are attached to rigid driving plates at their outer boundaries. The granular region is composed of groupings of particles representing pieces or blocks of rock (see PLACE and MORA, 2000 for a description of the initialization). Touching blocks interact through a simple frictional force with magnitude proportional to the normal force not exceeding that required to stop slip during a time-step (PLACE and MORA, 1999). The shearing is achieved by moving the driving plates at a constant rate of $0.00005 \times V_p$ while maintaining a normal stress of 150 MPa on them. Boundary conditions are circular in the x-direction.

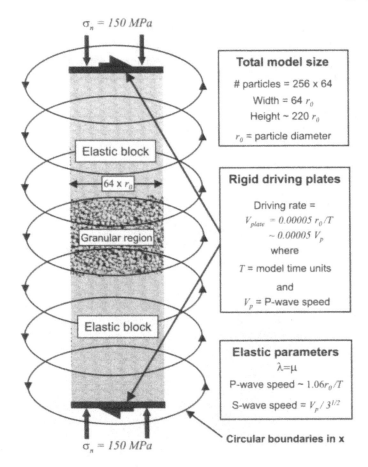

$\sigma_n = 150\ MPa$

Total model size

\# particles = 256 x 64

Width = 64 r_0

Height ~ 220 r_0

r_0 = particle diameter

Elastic block

64 x r_0

Rigid driving plates

Driving rate =
$V_{plate} = 0.00005\ r_0/T$
$\sim 0.00005\ V_p$
where
T = model time units
and
V_p = P-wave speed

Granular region

Elastic block

Elastic parameters
$\lambda = \mu$
P-wave speed ~ $1.06 r_0/T$
S-wave speed = $V_p / 3^{1/2}$

Circular boundaries in x

$\sigma_n = 150\ MPa$

Figure 1
The experimental setup for the shear experiment. A normal stress of $\sigma_n = 150\ MPa$ was maintained on the upper and lower rigid driving plates.

Since the model is fully elasto-dynamic and allows ruptures to occur along any internal surface within the granular region, the numerical model can be considered as a simplified model for an interacting fault system. Figure 2 shows the deviatoric stress $(\sigma_1 - \sigma_2)$ in the central granular region. The deviatoric stress was calculated from the forces acting on each particle and is interpolated between voids for graphical display purposes. The deviatoric stress shows a complex pattern and has a filamentary appearance that reflects paths of high stress. Movies showing the evolution of the deviatoric stress indicate that large events typically involve stress dropping suddenly on high stress paths and being redistributed elsewhere.

Kinetic energy during the simulation is plotted in Figure 3a. Each spike represents a synthetic earthquake event with its height proportional to the seismic energy release. Each spike consists of a rapid rise as a rupture event nucleates and

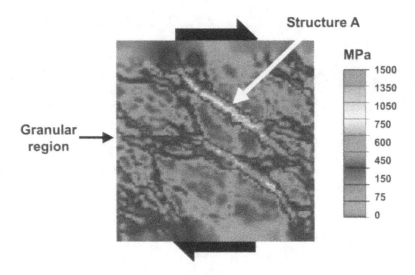

Figure 2
Deviatoric stress in the central granular region of the shear experiment.

elastic strain energy is converted to kinetic energy (seismic waves), followed by a more gradual decay once the rupture stops as the radiated waves are gradually damped through the action of an artificial viscosity. The artificial viscosity was introduced into the numerical model to avoid buildup of kinetic energy in the closed system and does not substantially affect the dynamics of rupture if the viscosity is sufficiently small and is carefully chosen (see MORA and PLACE, 1994). As the seismic waves propagate away from the rupture they redistribute the stress within the model. The largest event is labeled A and the largest subsequent event is labeled B. Of all events, only A ruptured a global structure that cut across the entire granular region – i.e., it involved redistribution of stress from a global structure that cut across the entire granular region (this global structure grew from structure A shown in Fig. 2). Hence, event A will be referred to as a global rupture event in the following. Event B ruptured a part of the same structure (A) a second time. This reruptured segment of structure A spanned about 2/3 the length of the original structure. The cumulative Benioff strain (cumulative square-root of seismic energy) is shown in Figure 3b. This plot shows that large events are often preceded by a period of accelerating Benioff strain release. Previous work (MORA et al., 2000) has demonstrated that the sequences prior to events A and B as well as sequences within these fit well to a power-law time-to-failure criterion with log-periodic fluctuations (SORNETTE and SAMMIS, 1995) given by

$$\varepsilon(t) = A + B(t - t_f)^c [1 + D\cos(2\pi \log(t_f - t) / \log(\lambda) + \Psi)] . \qquad (2)$$

Figure 4 (left) shows a best fit power-law time-to-failure function of the cumulative Benioff strain for sequence A. In this sequence, the best-fit power-law exponent was

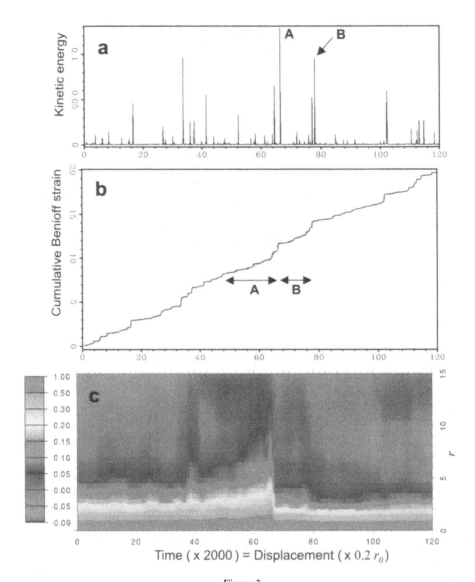

Figure 3
a: Kinetic energy. b: Cumulative Benioff strain. c: Scalar stress correlation function. Notes: (1) *P* waves traverse the model in the *x* direction in about 60 units of model time. (2) The correlation function was calculated every 2000 units of time which was the rate at which the stress field was saved to disc. (3) Displacement denotes the horizontal displacement between the rigid driving plates. (4) The correlation function has its maximum value at $r = 0.0$ and decreases monotonically as r increases.

$c = 0.37$. The RMS error of the power law fit divided by the RMS error of the linear fit was 0.51 and 0.39, respectively for the cases without and with log-periodic fluctuations. Figure 4 (right) shows interval frequency-magnitude plots for events in the first and last half of sequence A. This plot shows that there is an increased rate of

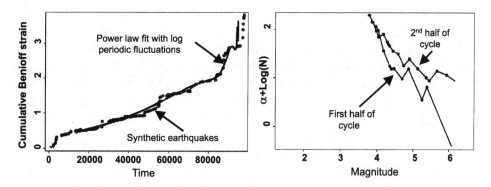

Figure 4

Left: Cumulative Benioff strain release for the simulated earthquake cycle A along with the power-law fit with log-periodic fluctuations. Time here is measured from the start of sequence A. Right: Interval frequency-magnitude plot where α = constant and the total number of earthquakes in this sequence is 227.

moderate to large events which occurs in the lead-up to the global rupture event labeled A. The earlier study (MORA *et al.*, 2000) presented plots showing the evolution of cumulative frequency magnitude statistics for sequence A and B as well as subsequences within these. The results demonstrated that whereas sequence A exhibits an increased rate of moderate earthquakes in the lead-up to the large event, sequence B exhibits an increased rate at all magnitudes in the lead-up to its terminating event.

Here, our purpose is not to expand on the previous results but to study the evolution of the stress correlation function. We define the *scalar stress correlation function* as the following function of scalar variable r

$$C(r) \;=\; \langle \triangle\sigma(\mathbf{x})\triangle\sigma(\mathbf{x}+\mathbf{r})\rangle / \langle \triangle\sigma(\mathbf{x})\triangle\sigma(\mathbf{x})\rangle \;, \tag{3}$$

where $\triangle\sigma = \sigma - \bar{\sigma}$ and $\sigma = \sigma_1 - \sigma_2$ denotes the deviatoric stress, $\bar{\sigma}$ the mean deviatoric stress, and σ_1 and σ_2 the two principle stresses. The averaging in Equation (3) is taken over all positions $\mathbf{x}$ and $\mathbf{x} + \mathbf{r}$ within the granular region, and all vectors $\mathbf{r}$ of length $r = |\mathbf{r}|$.

The denominator in Equation (3) $\langle \triangle\sigma(\mathbf{x})\triangle\sigma(\mathbf{x})\rangle$ is the normalizing factor specified in standard definitions of correlation coefficients in statistics texts and correlation functions in statistical physics texts. Its effect is to normalize $C(0)$ to unity (i.e., a value and itself is perfectly correlated so the correlation coefficient must be unity). Correlation functions decrease with r and may become negative if a field is negatively correlated at certain distances. This is easy to see in 1-D. For example, a sine wave is perfectly correlated over a length scale of 2π (i.e., $C(2\pi) = 1$) and perfectly negatively correlated over a length scale of π (i.e., $C(\pi) = -1$). The correlation length L is defined as the value of r at which the correlation function reaches $1/e$ (i.e., $C(L) = 1/e$).

Figure 3c depicts the scalar stress correlation function defined by Equation (3) plotted as a function of time. The main feature is a trend of increasing correlation length in the lead-up to event A and a dramatic drop in correlation length immediately after this event. One also observes fluctuations in correlation length superimposed on this trend. These results show that small ruptures in the lead-up to the global rupture A act to build up the correlation length and that the global rupture event causes the correlation length to dramatically drop. Sequence B also seems to display evidence for a similar trend although the effect is less clear than for sequence A. This is possibly due to event A representing a global catastrophe whereas event B may represent a local catastrophe. If so, then one would expect that the growth in correlation length prior to event B would become clearer if the correlation function were calculated within a subregion corresponding to the critical region for this event (c.f., YIN et al., 2002). Preliminary calculations of the correlation function around rupture B show a slightly clearer growth in correlation lengths prior to the event, followed by a somewhat sharper drop once it occurs. However, the difference was not dramatic, thus further research beyond the scope of the present study would be required to probe the possibility of correlation function evolution within subregions prior to local catastrophes.

To investigate further the evolution in the stress field, we also calculated the *vector stress correlation function*

$$C_2(\mathbf{r}) = \langle \triangle\sigma(\mathbf{x})\triangle\sigma(\mathbf{x}+\mathbf{r})\rangle / \langle \triangle\sigma(\mathbf{x})\triangle\sigma(\mathbf{x})\rangle \,, \tag{4}$$

which defines the correlation between deviatoric stress values separated by vector $\mathbf{r}$. Here, the averaging is taken for all points separated by vector $\mathbf{r}$ rather than for all points separated by scalar distance r as in Equation (3), and thus retains directional information. Figure 5 exhibits snapshots of the stress field and the vector correlation function. The stress field and vector correlation function snapshots show that the stress field is anisotropic, and that high stress tends to be correlated along a preferred angle. This is the typical angle of high stress paths through the granular system. As with the scalar correlation function, one observes a pattern of progressive increase in correlation length in the lead-up to event A (progressive stretching of the yellow band in $C_2(\mathbf{x})$ between $t = 0$ and $t = 66$). One clearly sees that correlations in deviatoric stress exist at all length scales in the model immediately prior to the global rupture event (i.e., the hot colored band stretches diagonally across the entire correlation function field in snapshot $t = 66$). Between $t = 66$ and $t = 67$, the hot colored band compacts suddenly indicating that the global rupture event has destroyed long range stress correlations in the model. The stress correlation evolution can be understood from the stress field snapshots. The stress field progressively becomes smoother (snapshots $t = 0$ to $t = 66$) and suddenly becomes rougher between $t = 66$ and $t = 67$ once the global rupture occurs.

The results to date have shown that an elasto-dynamic 2-D numerical model of a granular system subjected to shear, small simulated ruptures can progressively act to

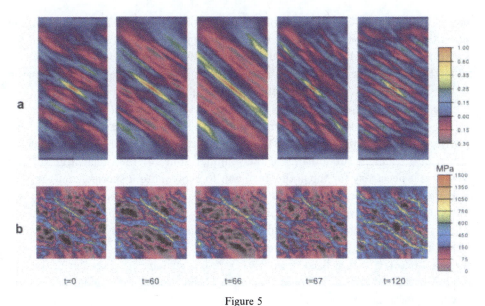

Figure 5

a: Stress correlation function $C_2(\mathbf{x})$. The origin is located in the center of the plot where the correlation function reaches unity (i.e., $C_2(\mathbf{0}) = 1$ depicted as red. b: Stress field $\triangle\sigma(\mathbf{x})$ Note: Snapshot times $t = n$ are in units of model time $\times$ 2000.

build-up the stress correlation length, and that a large global rupture event is capable of causing a dramatic drop in the stress correlation length. This underlying evolution in the stress correlation function is consistent with that of critical point systems. This suggests catastrophic failure of such granular systems may be predictable in some cases if it is possible to observe the evolution of stress correlation length, or to infer it through, for example, accelerating rates of acoustic emissions, as has been proposed previously.

Further work is required to study the conditions under which the stress correlation function evolves in the predictable manner seen prior to the global rupture event here, and to determine the extent to which the granular model provides a good analog for a crustal fault system. One difference between the granular model and a crustal fault system is the relatively rapid evolution of internal geometric structure that can occur as grains slide, roll or jostle past one another during ruptures. In contrast, crustal fault systems can be considered as stable since the evolution effect is minor over the time frame between large events. During the simulation, the total displacement between the upper and lower driving plates is $24 \times r_0$ where r_0 is the diameter of particles used to construct grains within the granular region. Thus, during the period when the stress correlation function shows the critical-point-like evolution, the displacement between driving plates was relatively small ($\sim 4r_0$) and the evolution in the internal configuration of the granular region was minor (i.e., the total displacement of $4r_0$ was accommodated by many small ruptures along different internal surfaces within the granular region rather than by a major change of the internal configuration). This suggests that stress is becoming correlated through ruptures occurring within a relatively stable structure rather than the stress correlation function tracking major internal configuration changes as the system evolves to a state where a global structure exists.

The fact that the same pattern of growth in the stress correlation length was not clearly observed prior to other large but non-global ruptures raises interesting questions. Does the predictable evolution only occur for global rupture events or does it also occur in subdomains of the granular zone for the other large rupture events? Is it necessary for the granular region to evolve into a specific configuration that allows critical-point-like evolution of the stress correlation function? If so, crustal fault systems with certain geometries may exhibit critical-point-like evolution whereas others may not. This offers a potential explanation of why accelerating energy release is not observed prior to all large earthquakes. Alternatively, it is possible that even with a fixed geometry, there is a switching between critical point and self-organized critical behavior such as has been observed in cellular automaton models (WEATHERLEY et al., 2002).

Compression Experiment

The shear experiment presented above involved a pre-existing granular layer. In the following, we initialize a model with random particles bonded by elastic brittle bonds and subject it to uniaxial compression in the y direction (see PLACE and MORA, 2001 and PLACE et al., 2002 for a description of the random lattice solid model). Snapshots of the simulation are shown in Figure 6 and illustrate that after about Time step $\times$ 10 = 2000, a well-developed fracture system has developed.

A plot of stress as a function of time is shown in Figure 6 (top curve). Since the driving rate on the upper and lower edges was constant, this plot depicts the stress-strain curve. Sharp drops in stress correlate with spikes in the kinetic energy plot (Ke in Fig. 6) and indicate rupture events. Early in the simulation, these correspond to the development of new fractures but later the rupture events may also involve slip along the existing fracture surfaces.

Figure 6 presents the evolution of the scalar stress correlation function with time. One observes that early in the simulation (Time step $\times$ 10 < 2000), there is no clear evolution. However, later in the simulation there are examples where the stress correlation function appears to evolve in a manner consistent with the CP hypothesis. Namely, prior to the large events labeled A, B and C the correlation length grows followed by a sharp drop when a large event occurs. This is most clearly visible in the sequence after event A in the lead-up towards event B. Figure 7 shows the vector stress correlation and deviatoric stress in the lead-up to event B and just after this event. The evolution of the deviatoric stress (Fig. 7) clearly indicates the

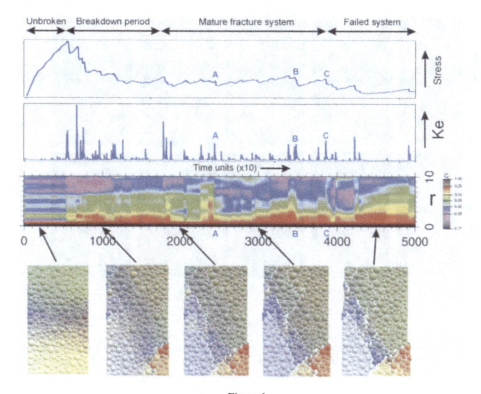

Figure 6

From top to bottom, the stress versus time (equivalent to stress-strain curve), the kinetic energy, the scalar stress correlation function using the same color scale as in Figure 3 and snapshots. The plots use the same time axis. The snapshots on the bottom depicts the evolution of fracturing through time (colors from red to blue and green to yellow represent the vertical and horizontal particle displacement). The three larger events later in the simulation are labeled A, B and C.

stress pattern is evolving towards a highly correlated state in the lead-up towards the main event. After the event, the stress pattern is reset to a more random state which correlates with the sharp drop in correlation length.

The snapshots (Fig. 6, bottom) show the evolution of the fracture system. One observes that when the block is intact and undergoing breakdown, there is no clear evolution in the stress correlation function ($t < 2000$). Once a fracture system is well developed ($t = 2000$ to $t = 4000$), there is evidence for CP-like evolution of the correlation function. When the block has almost failed entirely along a dominant fracture ($t > 4000$), there is no longer an obvious CP-like evolution of the correlation function. This suggests a well-developed system of interacting faults is necessary for the crust to behave as a CP system. If the crust acts as an intact system or has only one dominant fault that is relatively unaffected by neighboring faults, then the region may not behave as a CP system. This offers an explanation that accelerating seismic moment release (AMR) is not universally observed prior to large earthquakes. For

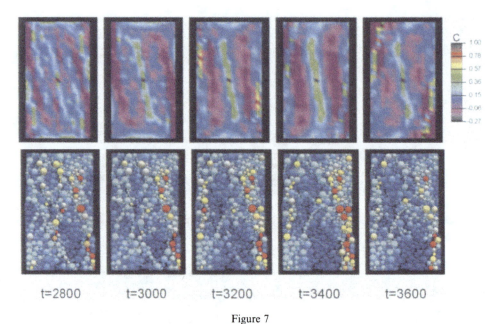

Figure 7
Evolution of the vector stress correlation function in the lead-up to event B at $t = 3500$ and just after this event. The top row of images depicts the evolution of the vector stress correlation function using the color scheme displayed on the right. The bottom row presents the deviatoric stress (colors from red to blue depict high to low values) at different points in time.

example, prior to the 1988 Tennant Creek earthquake in Northern Australia there was no observation of AMR whereas prior to the 1989 Newcastle earthquake in Australia, there was (YIN *et al.*, 2002). Thus, the Tennant Creek region may act as an isolated fault in an otherwise largely intact crust whereas the Newcastle region may act as an interacting fault region. This interpretation is consistent with the lack of seismicity in the Tennant Creek region prior to the 1988 earthquake sequence in contrast to the observed distributed seismicity in the Newcastle region prior to the Newcastle earthquake.

Discussion

The shear experiment sequence reflected an evolution in the stress correlation function consistent with the Critical Point Hypothesis for earthquakes in the lead-up to a global rupture event. In the case of the compression experiment, this evolution became evident only after establishment of a well-developed fracture system. These results suggest that well developed systems of interacting faults may behave as CP systems but that regions with few interacting faults will not. If so, earthquake forecasting will rely on our ability to characterize the crustal fault system and related system dynamics of any given region. This possibility is also supported by Cellular Automaton research (WEATHERLEY et al., 2002) which suggests that even simplified crust-like model systems can exhibit behavior ranging from critical point to self-organized critical, depending on model characteristics such as effective stress interaction range.

The results presented in this paper provide direct evidence of stress correlation evolution in the model – and hence a physical mechanism for forecasting catastrophic failures in discontinuous elasto-dynamic model systems. It does not prove the same physical mechanism exists in the earth. However, recent observational studies (YIN et al., 2002) provide compelling evidence that a CP or CP-like mechanism operates in the crust. This study showed that the critical region size – as estimated from two completely different observations – scales with magnitude as predicted by the CP hypothesis (i.e., correlations must exist in a larger region for the crust to be prepared for a larger earthquake). In this work, the critical region size was estimated from accelerating moment release observations (AMR) by calculating the fit parameter to the cumulative Benioff strain release $C = (RMS\ error\ power\text{-}law\ time\text{-}to\text{-}failure\ fit)/$ $(RMS\ error\ linear\ fit)$ using data within different radii from the mainshock. The radius that minimizes C provides an estimate of the critical region size based on AMR. Similarly, the radius that maximizes the Load-Unload Response Ratio (LURR) prior to large earthquakes provides an estimate of the critical region size based on LURR. The common critical region size to magnitude scaling according to both AMR and LURR observations strongly supports a common underlying physical mechanism for these observations and is consistent with the CP hypothesis. If so, LURR may provide a means to detect when the crust has reached the critical state when the crust is prepared for a large earthquake, AMR may indicate the lead-up towards this critical state, and the critical region size for AMR and LURR may provide an estimate of mainshock magnitude. Numerical simulations (MORA et al., 2002) of LURR using the lattice solid model support the notion that LURR provides a means to detect the critically sensitive state prior to catastrophe.

The combination of the stress correlation function evolution results presented in this paper, new observational results of critical region – magnitude scaling (YIN et al., 2002) and growth in seismicity data correlation lengths (ZOELLER et al., 2001), and simulation studies of LURR (MORA et al., 2002) provide compelling evidence for CP or CP-like behavior in the crust. This gives impetus to research aiming to develop a comprehensive understanding of the physics of crustal fault systems with the ultimate goal of earthquake forecasting.

Conclusions

The numerical experiments demonstrate that at least under certain conditions, it is possible for there to be an evolution in the stress field of an elasto-dynamic system that is consistent with the critical point hypothesis for earthquakes. This proves the possibility of existence of a physical mechanism in discontinuous elasto-dynamic systems that can allow forecasting of catastrophic failure or earthquakes. Conversely, the results disprove that such a system needs to be in a perpetual critical state in which the large events can occur at any time, and hence, remove one of the principle arguments against earthquake prediction. A compression experiment of intact model material resulting in progressive development of a fracture system demonstrates that the evolution in correlation lengths occurs only once a well developed fracture system has evolved. This suggests that intermediate-term earthquake forecasting is achievable, at least for certain well developed fault systems.

Acknowledgments

This research was funded by the Australian Research Council and The University of Queensland. Computations were made on QUAKES' 12 processor SGI Origin 2000 and the Australian Solid Earth Simulator (ASES) thematic parallel supercomputer facility (phase I – 16 processor SGI Origin 3800) funded by the ARC, The University of Queensland, CSIRO, University of Western Australia and Silicon Graphics. We express our appreciation to the reviewers whose helpful comments improved the manuscript, and to an anonymous reviewer for drawing our attention to the recent paper by ZOELLER et al., (2001).

REFERENCES

ABE, S., MORA, P., and PLACE, D. (2000), *Extension of the Lattice Solid Model to Incorporate Temperature Related Effects*, Pure Appl. Geophys. *157*, 1867–1887.

AKI, K. (2000), *Scale Dependence in Earthquake Processes and Seismogenic Structures*, Pure Appl. Geophys., *157*, 2249–2258.

BOWMAN, D. D., OUILLON, G., SAMMIS, C. G., SORNETTE, A., and SORNETTE, D. (1998), *An Observational Test of the Critical Earthquake Hypothesis*, J. Geophys. Res. *103*, 24,359–24,372.

BUFE, C. G., and VARNES, D. J. (1993), *Predictive Modeling of the Seismic Cycle in the Greater San Francisco Bay region*, J. Geophys. Res. *98*, 9,871–9,833.

CUNDALL, P. A. and STRACK, O. D. L. (1979), *A Discrete Numerical Model for Granular Assemblies*, Géotectonique, *29*, 47–65.

GELLER, R. J., JACKSON, D. D., KAGAN, Y. Y., and MULARGIA, F. (1997), *Earthquakes Cannot be Predicted*, Science *275*, 1,616.

JAUMÉ, S. C. (2000), *Changes in earthquake size-frequency distributions underlying accelerating seismic moment/energy release prior to large and great earthquakes*. In *GeoComplexity and the Physics of Earthquakes* (Geophysical Monograph series; no. 120) (eds Rundle, J.B., Turcotte, D.L. and Klein, W.) pp 199–210 (Am. Geophys. Union, Washington, DC 2000).

JAUMÉ, S. C., WEATHERLEY, D., and MORA, P. (2000), *Accelerating Seismic Energy Release and Evolution of Event Time and Size Statistics: Results from two Heterogeneous Cellular Automaton Models*, Pure appl. Geophys. *157*, 2209–2226.

JAUMÉ, S. C. and SYKES, L. (1999), *Evolving Towards a Critical Point: a Review of Accelerating Seismic Moment/Energy Release Prior to Large and Great Earthquakes*, Pure appl. Geophys *155*, 279–305.

KNOPOFF, L. (2000), *The Magnitude Distribution of Declustered Earthquakes in Southern California*, Proc. Nat. Acad. Sci. *97*, 11,880–11,884.

MORA, P. (1992), *A Lattice Solid Model for Rock Rheology and Tectonics*, The Seismic Simulation Project Tech. Rep. 4, 3–28 (Institut de Physique du Globe, Paris).

MORA, P. and PLACE, D. (1993), *A Lattice Solid Model for the Nonlinear Dynamics of Earthquakes*, Int. J. Mod. Phys. C *4*, 1059–1074.

MORA, P. and PLACE, D. (1994), *Simulation of the Frictional Stick-slip Instability*, Pure Appl. Geophys. *143*, 61–87.

MORA, P. and PLACE, D. (1998), *Numerical Simulation of Earthquake Faults with Gouge: Towards a Comprehensive Explanation of the Heat Flow Paradox*, J. Geophys. Res. 103/B9, 21,067–21,089.

MORA, P. and PLACE, D. (1999), *The Weakness of Earthquake Faults*, Geophys. Res. Lett. *26*, 123–126.

MORA, P., PLACE, D., ABE, S., and JAUMÉ, S. (2000), *Lattice solid simulation of the physics of earthquakes: The model, results and directions*. In GeoComplexity and the Physics of Earthquakes (Geophysical Monograph series; no. 120) (eds. Rundle, J.B., Turcotte, D.L. and Klein, W.) pp 105–125 (American Geophys. Union, Washington, DC 2000).

MORA, P., WANG, Y. C., YIN, C., PLACE, D., and YIN, X. C. (2002), *Simulation of the Load-unload Response Ratio and Critical Sensitivity in the Lattice Solid Model*, Pure Appl. Geophys. *159*, 2525–2536.

PLACE, D. and MORA, P. (1999), *A Lattice Solid Model to Simulate the Physics of Rocks and Earthquakes: Incorporation of Friction*, J. Comp. Phys. *150*, 1–41.

PLACE, D. and MORA, P. (2000), *Numerical Simulation of Localisation Phenomena in a Fault Zone*, Pure Appl. Geophys. *157*, 1821–1845.

PLACE, D. and MORA, P. (2001), *A random lattice solid model for simulation of fault zone dynamics and fracture processes*. In Bifurcation and Localisation Theory for Soils and Rocks'99 (eds Mühlhaus H-B., Dyskin A.V. and Pasternak, E.) (AA Balkema, Rotterdam/Brookfield 2001).

PLACE, D., LOMBARD, F., MORA, P., and ABE, S. (2002), *Simulation of the Micro-physics of Rocks Using LSMearth*, Pure Appl. Geophys. *159*, 1911–1932.

RUNDLE, J. B., KLEIN, W., and GROSS, S. (1999), *A Physical Basis for Statistical Patterns in Complex Earthquake Populations: Models, Predictions and Tests*, Pure Appl. Geophys. *155*, 575–607.

SAMMIS, C. G. and SMITH, S. W. (1999), *Seismic Cycles and the Evolution of Stress Correlations in Cellular Automaton Models of Fault Networks*, Pure Appl. Geophys. *155*, 307–334.

SORNETTE, D. and SAMMIS, C. G. (1995), *Complex Critical Exponents from Renormalization Group Theory of Earthquakes: Implications for Earthquake Prediction*, J. Phys. I. 5, 607–619.

WEATHERLEY, D., JAUMÉ, S. C., and MORA, P. (2000), *Evolution of Stress Deficit and Changing Rates of Seismicity in Cellular Automaton Models of Earthquake Faults*, Pure appl. Geophys. *157*, 2183–2207.

WEATHERLEY, D., MORA, P., and XIA, M. (2002), *Long-range Automaton Models of Earthquakes: Power-law Accelerations, Correlation Evolution, and Mode Switching*, Pure Appl. Geophys. *159*, 2469–2490.

WU, Z.L. (2000), *Frequency-size Distribution of Global Seismicity seen from Broadband Radiated Energy*, Geophys. J. Int. *142*, 59–66.

YIN, X-C, MORA, P., PENG, K., WANG, Y., and WEATHERLEY, D. (2002), *Load-Unload Response Ratio, Accelerating Moment/Energy Release, Critical Region Scaling, and Earthquake Prediction*, Pure Appl. Geophys. *159*, 2511–2523.

ZOELLER, G., HAINZL, S., and KURTHS, J. (2001), *Observation of Growing Correlation Length as an Indicator for Critical Point Behaviour Prior to Large Earthquakes*, J. Geophys. Res. *106*, 2,167–2,175.

(Received February 20, 2001, revised June 11, 2001, accepted June 15, 2001)

To access this journal online:
http://www.birkhauser.ch

Pure appl. geophys. 159 (2002) 2429–2467
0033–4553/02/102429–39 $ 1.50 + 0.20/0

© Birkhäuser Verlag, Basel, 2002

Pattern Dynamics and Forecast Methods in Seismically Active Regions

Kristy F. Tiampo,[1] John B. Rundle,[2]
Seth A. McGinnis,[1] and William Klein[3]

Abstract—Large, extended fault systems such as those in California demonstrate complex space-time seismicity patterns, which include repetitive events, precursory activity and quiescence, and aftershock sequences. Although the characteristics of these patterns can be qualitatively described, a systematic quantitative analysis remains elusive. Our research suggests that a new pattern dynamics methodology can be used to define a unique, finite set of seismicity patterns for a given fault system. In addition, while a long-sought goal of earthquake research has been the reliable forecasting of these events, very little progress has been made in developing a successful, consistent methodology. In this report, we document the discovery of systematic space-time variations in seismicity from southern California using a new technique. Here we present examples of this analysis technique on data obtained *prior* to events in seismically active areas that show coherent regions associated with the future occurrence of major earthquakes in the same areas. These results strongly support the hypothesis that seismic activity is highly correlated across many space and time scales within large volumes of the earth's crust.

Key words: Fault system dynamics, pattern dynamics, mathematical methods in geophysics, seismicity.

Introduction

While the historic earthquake record is not complete, it has long been recognized that earthquake mainshocks occur at quasi-periodic intervals and that, for some parts of the world, average recurrence intervals are well defined (KANAMORI, 1981). In addition, both temporal and spatial clustering is evident in the data, with the result that neither the recurrent nature of the mainshocks, nor the observed phenomena such as foreshocks, aftershocks, seismic gaps, or mainshock triggering, is compatible with a Poisson probability function (JONES and HAUKSSON, 1997; KAGAN and

[1] CIRES, University of Colorado, Boulder, CO U.S.A.
E-mails: kristy@fractal.colorado.edu; sethmc@turcotte.colorado.edu
[2] Department of Physics, Colorado Center for Chaos and Complexity, CIRES, University of Colorado, Boulder, CO, 80309, U.S.A, and Distinguished Visiting Scientist, Jet Propulsion Laboratory, Pasadena, CA, 91125, U.S.A. E-mail: rundle@cires.colorado.edu
[3] Department of Physics, Boston University, Boston, MA USA and Center for Nonlinear Science, Los Alamos National Laboratory, Los Alamos, NM U.S.A. E-mail: klein@buphyc.bu.edu

JACKSON, 1992; SAVAGE, 1993; DIETERICH, 1994; GRANT and SIEH, 1994; RUNDLE and KLEIN, 1995; TURCOTTE, 1997; MAIN, 1999b). Much of the recent geophysical research associated with earthquakes themselves has centered on investigating these spatial and temporal patterns in local and regional seismicity data (KANAMORI, 1981). Notable examples include characteristic earthquakes (SWAN et al., 1980; ELLSWORTH and COLE, 1997), repeating earthquakes (BAKUN et al., 1986; MARONE et al., 1995), seismic gaps (HABERMAN, 1981; HOUSE et al., 1981; KAGAN, 1981; KAGAN and JACKSON, 1992; WYSS and WIEMER, 1999), well-defined recurrence intervals (BAKUN and McEVILLY, 1984; LYZENGA et al., 1991; SAVAGE, 1993), Mogi donuts (MOGI, 1969; MOGI, 1977), temporal clustering (FROHLICH, 1987; PRESS and ALLEN, 1995; DODGE et al., 1996; ENEVA and BEN-ZION, 1997; JONES and HAUKSSON, 1997; RUNDLE et al., 1997; HUANG et al., 1998), 'slow' earthquakes (LINDE et al., 1996; McGUIRE et al., 1996; KERR, 1998), precursory quiescence (YAMASHITA and KNOPOFF, 1989; WYSS et al., 1996; KATO et al., 1997; WYSS et al., 2000), aftershock sequences (GROSS and KISSLINGER, 1994; NANJO et al., 1998), earthquake triggering over large distances (STEIN et al., 1992; HILL et al., 1993,1995; KING et al., 1994; DENG and SYKES, 1996; GOMBERG, 1996; STARK and DAVIS, 1996; POLLITZ and SACKS, 1997; STEIN, 1999), scaling relations (RUNDLE, 1989; PACHECO et al., 1992; ROMANOWICZ and RUNDLE, 1993; RUNDLE, 1993; SALEUR et al., 1995; RUNDLE et al., 1999), and time-to-failure analyses (BUFE and VARNES, 1993; BOWMAN et al., 1998; GROSS and RUNDLE, 1998; BREHM and BRAILE, 1999; JAUMÉ and SYKES, 1999; MAIN, 1999a). Although much of this work represents important attempts to describe these characteristic patterns using empirical probability density functions, none of these observations or methodologies systematically identifies all possible seismicity patterns. The quantification of all possible space-time patterns would seem to be a necessary first step in the process of identifying which patterns are precursory to large events, leading to the possible development of new approaches in forecast methodology. Yet, as can be seen in Figure 1, a plot of relative southern California seismicity during the time period 1932–1991, the identification and quantification of these patterns is no easy matter.

Recent large earthquakes include the M~7.4 event that struck Izmit, Turkey in August of 1999, the M~7.6 Taiwan earthquake that occurred in September of 1999, and the M~7.1 Hector Mine, California earthquake of October 1999. Many similar examples have been documented over the course of time (RICHTER, 1958; SCHOLZ, 1990), yet despite the fact that the largest of these events span distances exceeding 500 km, no reliable precursors have been detected with any repeatability (KANAMORI, 1981; GELLER et al., 1997). It is difficult for most scientists to understand why events of this magnitude are not preceded by at least some causal process. While various patterns of seismic activity centered on the source region have been proposed, as noted above, these efforts to identify the premonitory signals have focused predominantly on local regions near the earthquake source. As a result, these techniques often require intensive and expensive monitoring efforts and have been largely unsuccessful

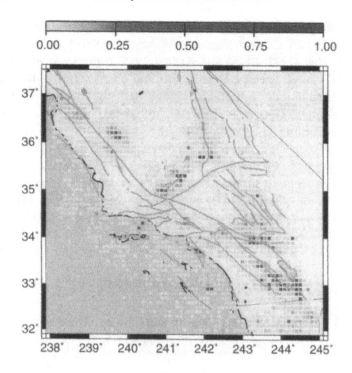

Figure 1
Seismicity for southern California, 1932–1991, normalized to the maximum number of events for the period.

(KANAMORI, 1981). Since these hypothesized patterns are localized on the eventual source region, the fact that one must know or suspect where the event will occur before they can be applied is a major drawback to their implementation.

In this report, we discuss a new pattern dynamics methodology that can be used to define a unique, finite set of seismicity patterns for a given fault system. Similar in nature to the empirical orthogonal functions historically employed in the analysis of atmospheric and oceanographic phenomena (PREISENDORFER, 1988), this method derives the eigenvalues and eigenstates from the diagonalization of a correlation matrix using a Karhunen-Loeve expansion (FUKUNAGA, 1970). This pattern dynamics technique has been successfully applied to the study of numerically modeled seismicity for fault networks similar in character and extent to those found in California (RUNDLE et al., 2000a). We implement this same methodology in order to analyze historical seismicity in California and derive space-time eigenvalue patterns for the San Andreas fault system. The significant eigenstates for this relatively short period of time can be directly correlated with the known California faults and associated events (TIAMPO et al., 2000).

In this work we present a method for identifying these areas of increased probability of an event. The success of the Karhunen-Loeve decompositions discussed above, coupled with recent observational evidence, suggests that earthquake faults are characterized by strongly correlated space-time dynamics (BUFE and VARNES, 1993; PRESS and ALLEN, 1995; BOWMAN *et al.*, 1998; TIAMPO *et al.*, 2000). We have extended these results and observations of their application to numerical simulations of earthquake fault systems, to the development of a method for identifying areas of increased probability of an event, $\Delta \mathbf{P}$. Realistic numerical simulations of earthquakes also suggest that space-time pattern structures are non-local in character, a consequence of strong correlations in the underlying dynamics (RUNDLE, 1988; RUNDLE *et al.*, 2000a). The procedure described below is based upon the idea that seismic activity corresponds geometrically to the rotation of a pattern state vector in the high-dimensional correlation space spanned by the eigenvectors of an equal-time correlation operator (RUNDLE *et al.*, 2000b; TIAMPO *et al.*, 2000).

Background and Theory

Earthquake fault systems are now thought to be an example of a complex nonlinear system (BAK *et al.*, 1987; RUNDLE and KLEIN, 1995). Interactions among a spatial network of fault segments are mediated by means of a potential that allows stresses to be redistributed to other segments following slip on any particular segment. For faults embedded in a linear elastic host, this potential is a stress Green's function whose exact form can be calculated from the equations of linear elasticity, once the current geometry of the fault system is specified. A persistent driving force, arising from plate tectonic motions, increases stress on the fault segments. Once the stresses reach a threshold characterizing the limit of stability of the fault, a sudden slip event results. The slipping segment can also trigger slip at other locations on the fault surface whose stress levels are near the failure threshold as the event begins. In this manner, earthquakes occur that result from the interactions and nonlinear nature of the stress thresholds.

The Karhunen-Loeve method, a linear decomposition technique in which a dynamical system is decomposed into a complete set of orthonormal subspaces, has been applied to a number of other complex nonlinear systems over the last fifty years, including the ocean-atmosphere interface, turbulence, meteorology, biometrics, statistics, and even solid earth geophysics (HOTELLING, 1993; FUKUNAGA, 1970; AUBREY and EMERY, 1983; PREISENDORFER, 1988; SAVAGE, 1988; PENLAND, 1989; VAUTARD and GHIL, 1989; GARCIA and PENLAND, 1991; PENLAND and MAGORIAN, 1993; PENLAND and SARDESHMUKH, 1995; HOLMES *et al.*, 1996; MOGHADDAM *et al.*, 1998). The notable success of this method in analyzing the ocean-atmosphere interface and such features as the El Niño Southern Oscillation (ENSO), a nonlinear system whose underlying physics is governed by the Navier-Stokes equation, suggested its application to the analysis of the earthquake fault system (NORTH, 1984; PREISEN-

DORFER, 1988; PENLAND and MAGORIAN, 1993; PENLAND and SARDESHMUKH, 1995). Building on these methods for analyzing nonlinear threshold systems, space-time seismicity patterns can be identified in both numerical simulations using realistic earthquake models for southern California (BUFE and VARNES, 1993; BOWMAN et al., 1998; GROSS and RUNDLE, 1998; BREHM and BRAILE, 1999; JAUMÉ and SYKES, 1999; RUNDLE et al., 2000a) and actual historic seismicity records (TIAMPO et al., 1999, 2000). In this paper we apply this Karuhunen-Loeve expansion (KLE) technique (FUKUNAGA, 1970; HOLMES et al., 1996) to the analysis of observed seismicity data from southern California in order to identify basis patterns for all possible space-time seismicity configurations. These basis states represent a complete, orthonormal set of eigenvectors and associated eigenvalues, obtained from the diagonalization of the correlation operators computed for the regional historic seismicity data, and, as such, can be used to reconstitute the data for various subset time periods of the entire data set.

Variables in many dynamical systems can be characterized by a phase function that involves both amplitude and phase angle (MORI and KURAMOTO, 1998). Our simulations have suggested that seismicity can be described by pure phase dynamics (MORI and KURAMOTO, 1998; RUNDLE et al., 2000a, 2000b), in which the important changes in seismicity are associated primarily with rotations of the vector phase function in a high-dimensional correlation space (FUKUNAGA, 1970; HOLMES et al., 1998; RUNDLE et al., 2000a). Changes in the amplitude of the phase function are unimportant, or not relevant. Examples of pure phase dynamical systems in the classical world include weak turbulence in fluids and reaction-diffusion systems (MORI and KURAMOTO, 1998). Another nonclassical example is a quantum system in which the wave function is the phase function. By mapping our problem into the mathematics of quantum mechanics, we are treating the underlying stress-strain dynamics as the hidden variables and the seismicity patterns as the wave functions (RUNDLE and KLEIN, 1995; RUNDLE et al., 2000a, 2000b).

Observations and numerical simulations suggest that space-time patterns of seismic activity directly reflect the existence of space-time correlations in the underlying stress and strain fields (RUNDLE et al., 2000a, 2000b). A spatially coherent, uniformly high-level of stress on a fault is a necessary precondition for the occurrence of a large earthquake. Recently, several groups have found that spatial coherence in the stress field is reflected in a similar coherence in the seismic activity (RUNDLE, 1988; BUFE and VARNES, 1993; KNOPOFF et al., 1996; MAIN, 1999a; SALEUR et al., 1996; BREHM and BRAILE, 1998; RUNDLE et al., 2000a, 2000b). It should therefore be possible to compute the increase in probability of observing such an anomalous correlation, ΔP, directly from the observed seismicity data. Using the fact that seismicity is an example of phase dynamics, it follows that ΔP can be calculated from the square of the phase function for the associated pattern state vector (FUKUNAGA, 1970; HOLMES et al., 1998; RUNDLE et al., 2000a). To emphasize the connection to phase dynamics, we call the function ΔP the Phase Dynamical Probability Change (PDPC).

Finally, we note that the space-time correlations or patterns that lead to a uniformly high stress field on the fault represent emergent space-time structures, which evidently form and evolve over time intervals of years preceding the mainshock. Longer time intervals and larger correlated areas should be associated with larger mainshocks.

Data

The primary seismicity data set for southern California employed in this analysis is the entire Caltech catalog from 1932 through December of 1999, obtained from the Southern California Earthquake Center (SCEC) database, with all blast events specifically removed from the catalog [http://www.scecdc.scec.org.]. Relevant data include location, in latitude and longitude, and the time the event occurred. Seismic events between $-115°$ and $-122°$ longitude and $32°$ and $37°$ latitude were selected, and all quality events were acquired. Separate analyses were performed for the entire data set, consisting of all events of magnitude greater than or equal to 0.0, and on another data set in which only those events of magnitude greater than or equal to 3.0 were included in the binning process described below. The time periods evaluated spanned 1932 to 1991, and 1932 through August of 1999. In both cases, the seismicity was binned into squares of $0.1°$ latitude and $0.1°$ longitude to a side, and a time series constructed for each location square, boxes of approximately 11 km to a side. Each time step is given an initial value of 1.0 if one or more events occur in that time period, or a value of 0.0 otherwise. Subsequently, the mean for each time series is removed from the data.

For the time period 1932 through August of 1999, the seismicity was analyzed using the entire data set, including the entire areal extent and events of all quality. The time interval for this decomposition was increased to one day, so that the total number of time steps is approximately 24,333. In addition, all locations from the entire database, and all quality events, were included, even those where no event occurred for the more than 67 years. The number of location time series affected by the seismicity, p, therefore is 3621. For the time period 1932 to 1991, the time interval for the analysis was again one day, so that the total number of time steps is approximately 21,535. Again, all locations, and all quality events were included, even those where no event occurred for the entire 59 years. The total number of location time series remains constant at 3621.

Method

Karhunen-Loeve Decomposition

Pattern evolution and prediction in nonlinear systems is complicated by nonlinear mode coupling and noise, however understanding such patterns, which are the surface expression of the underlying dynamics, is critical to understanding and perhaps characterizing the physics which control the system. Karhunen-Loeve expansion (KLE) methods can be used to define a unique, complete pattern basis set for a given dynamical system (FUKUNAGA, 1970; NORTH, 1984; PENLAND, 1989; HOLMES et al., 1996). For driven threshold systems, an adaptation of these KLE methods can be employed to characterize both the space-time patterns of threshold transitions, i.e., "firings," as well as the underlying, usually unobservable Markov variables that define the dynamics (FUKUNAGA, 1970; HOLMES et al., 1996; RUNDLE et al., 2000). In either case, the patterns are defined by the eigenstates and eigenvalues of one of an appropriately constructed family of correlation operators.

Earthquake fault systems are examples of driven nonlinear threshold systems, comprised of interacting spatial networks of statistically identical, nonlinear units that are subjected to a persistent driving force (SCHOLZ, 1990; RUNDLE and KLEIN, 1995; FISHER et al., 1997; RUNDLE et al., 1997; FERGUSON et al., 1999). Numerous examples of such systems exist, including neural networks (HERTZ et al., 1990; HERZ and HOPFIELD, 1995), sandpiles (BAK et al., 1987), and superconductors (FISHER et al., 1997), of which earthquakes are but another example. Such systems re composed of cells which fire, or fail, when the driving force causes the force or potential, $\sigma(\mathbf{x},t)$, on a cell at location $\mathbf{x}$ and time t, to reach a predefined threshold value σ^F. The behavior of these systems is determined by parameters such as threshold values, residual stresses, quenched disorder and noise (BAK et al., 1987; RUNDLE and KLEIN, 1995; FISHER et al., 1997). Complex spatial and temporal firing patterns result that are difficult to analyze deterministically (OUCHI, 1993; NIJHOUT, 1997). In the case of an earthquake fault system, the driving force is tectonic plate motion, and the internal potential is the stress on each fault cell or patch. The firing, or failure of each patch results in an increase in the internal state variable $s(\mathbf{x},t)$ and a decrease in the cell potential to some residual value σ^R. The interactions between the cells, or fault patches, may be excitatory, bringing another closer to failure, or inhibitory, in which the failure of one cell can move neighboring cells further from failure. The spatial and temporal firing patterns, $\psi(\mathbf{x},t)$, of these driven threshold systems are complex and often difficult to understand and interpret from a deterministic perspective, as these patterns are emergent processes that develop from the obscure underlying structures, parameters, and dynamics of a multidimensional nonlinear system (OUCHI, 1993; NIJHOUT, 1997).

Analysis of a number of these driven threshold systems often is complicated by the fact that the underlying dynamics and the state variables which control the physics of the system, $s(\mathbf{x},t)$, are unknown and difficult to observe. The earthquake fault system is no exception. While it is not only probable, but essential, that space-time patterns and correlations exist in the variables and interactions which control earthquake dynamics, $s(\mathbf{x},t)$ and $\sigma(\mathbf{x},t)$, from which the observable surface patterns and correlations, $\psi(\mathbf{x},t)$ arise, those true dynamical patterns are difficult or impossible to observe within the earth (SCHOLZ, 1990; TURCOTTE, 1997). The schematic shown in Figure 2 illustrates the physical problem. As the state variable $s(\mathbf{x},t)$ at a particular location $\mathbf{x}$ evolves in time under the deterministic dynamics $\mathbf{D_t}$ to a value $s(\mathbf{x},t + \Delta t)$, the force or potential $\sigma(\mathbf{x},t)$, also evolves to $\sigma(\mathbf{x},t + \Delta t)$. While the values of $s(\mathbf{x},t)$ and $\sigma(\mathbf{x},t)$, along with the specifics of $\mathbf{D_t}$, are hidden from view below the dashed line of Figure 2, the firing patterns of $\psi(\mathbf{x},t)$ are observable. In the earth, there is no means at present to measure the stress and strain at every point in an earthquake fault system, or the constitutive parameters, which characterize the heterogeneous medium and its dynamics. However, the seismicity, which is the surface expression of its firing activity, can be located in both space and time with considerable accuracy (BAKUN and McEVILLY, 1984; SIEH *et al.*, 1989; HILL *et al.*, 1990). For example, the firing activity, $\psi(\mathbf{x},t)$, can be represented as a set of time series at all positions $\mathbf{x}$, where $\psi(\mathbf{x},t)=1$ if an event occurs in the time interval between t and $t + \Delta t$, and $\psi(\mathbf{x},t) = 0$ otherwise.

Observations and numerical simulations suggest that space-time patterns of seismic activity directly reflect the existence of space-time correlations in the underlying stress and strain fields (PRESS and ALLEN, 1995). A spatially coherent, uniformly high level of stress on a fault is a necessary precondition for the occurrence of a large earthquake. Recently, several groups have found that spatial coherence in

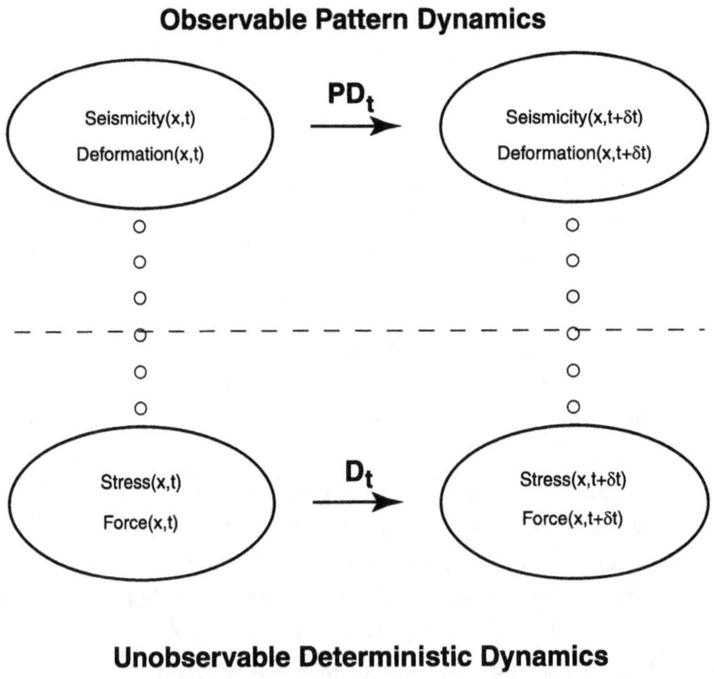

Figure 2
Schematic diagram of threshold systems.

the stress field is reflected in a similar coherence in the seismic activity (BUFE and VARNES, 1993; BOWMAN et al., 1998; RUNDLE et al., 2000a).

RUNDLE et al. (2000) extended the standard KLE methods to include the construction of pattern states that can be used to forecast events in time, in much the same manner as EOF analysis is used to predict El Niño events in meteorology (PREISENDORFER, 1988; PENLAND; GARCIA and PENLAND, 1991). This procedure involves constructing a correlation operator, $C(x_i,x_j)$, for the sites that contain the spatial relationship of slip events over time. $C(x_i,x_j)$ is decomposed into the orthonormal spatial eigenmodes for the nonlinear threshold system, e_j, and their associated time series, $a_j(t)$.

The Karhunen-Loeve expansion is obtained from the p time series that record the deformation history at particular locations in space. Each time series, $y(x_s,t_i) = y_i^s$, $s = 1,\ldots, p$, consists of n time steps, $i = 1,\ldots, n$. The goal is to construct a time series for each of numerous locations that records, for a given short period of time, whether an earthquake occurred at that location (value = 1) or did not occur (value = 0). If, for example, the time interval was decimated into units of days, the result would be a time series of 365 time steps for every year of data, with either a zero or a one at each time step. These time series are incorporated into a matrix, $\mathbf{T}$, consisting of time series of the same measurement for p different locations, i.e.,

$$\mathbf{T} = [\bar{y}_1, \bar{y}_2, \ldots, \bar{y}_p] = \begin{bmatrix} y_1^1 & y_1^2 & \cdots & y_1^p \\ y_2^1 & y_2^2 & \cdots & y_2^p \\ \vdots & \vdots & \ddots & \vdots \\ y_n^1 & y_n^2 & \cdots & y_n^p \end{bmatrix}.$$

$\mathbf{T}$ is therefore an $n \times p$, matrix of real values (FUKUNAGA, 1970). The covariance matrix, $S(x_i,x_j)$, for these events is formed by multiplying $\mathbf{T}$ by $\mathbf{T}^T$, where S is a $p \times p$ real, symmetric matrix. The covariance matrix, $S(x_i,x_j)$, is converted to a correlation operator, $C(x_i,x_j)$, by dividing each element of $S(x_i,x_j)$, by the variance of each time series, $y(x_i,t)$ and $y(x_j,t)$, as follows:

$$\sigma_p = \sqrt{\frac{1}{n}\sum_{k=1}^{n}(y_k^p)^2},$$

and

$$\mathbf{C} = \begin{bmatrix} \frac{S_{11}}{\sigma_1\sigma_1} & \frac{S_{12}}{\sigma_1\sigma_2} & \cdots & \frac{S_{1p}}{\sigma_1\sigma_p} \\ \frac{S_{21}}{\sigma_2\sigma_1} & \frac{S_{22}}{\sigma_2\sigma_2} & \cdots & \frac{S_{2p}}{\sigma_2\sigma_p} \\ \vdots & \vdots & \ddots & \vdots \\ \frac{S_{p1}}{\sigma_p\sigma_1} & \frac{S_{p2}}{\sigma_p\sigma_2} & \cdots & \frac{S_{pp}}{\sigma_p\sigma_p} \end{bmatrix}.$$

This equal-time correlation operator, $C(x_i,x_j)$, is decomposed into its eigenvalues and eigenvectors in two parts. The first employs the trireduction technique to reduce the matrix $\mathbf{C}$ to a symmetric tridiagonal matrix, using a Householder reduction. The second part employs a ql algorithm to find the eigenvalues, λ_j^2, and eigenvectors, e_j, of the tridiagonal matrix (PRESS *et al.*, 1992).

These eigenstates thus represent the orthonormal basis vectors arranged in order of decreasing correlation, and reflect the relative importance of the various modes over the time interval of interest. Dividing each corresponding eigenvalue, λ_j^2, by the sum of the eigenvalues, yields that percent of the correlation accounted for by that particular mode. The associated orthonormal time series can be reconstructed by projecting the initial data set onto these basis vectors (PREISENDORFER, 1988; HOLMES *et al.*, 1996). The time dependent expansion coefficients, $a_j(t)$, which represent temporal eigenvectors, are reconstructed by multiplying the original data matrix by the eigenvectors, i.e.,

$$a_j(t_i) = \vec{e}^T \cdot T = \sum_{s=1}^{p} e_j y_i^s,$$

where $j,s = 1,\ldots,p$ and $i = 1,\ldots,n$. This eigenstate decomposition technique produces the orthonormal spatial eigenmodes for this nonlinear threshold system, e_j, and the associated principal component time series, $a_j(t)$. These principal component time series represent the signal associated with each particular eigenmode over time. For purposes of clarity, the spatial eigenvectors are designated "KLE modes" and the associated time series "Principal Component (PC)" vectors.

Phase Dynamical Probability Change

Our method is based on the idea that the time evolution of seismicity can be described by pure phase dynamics (MORI and KURAMOTO, 1998; RUNDLE *et al.*, 2000a, 2000b). We therefore construct a real-valued seismic phase function $\hat{\mathbf{S}}(x_i, t_0, t)$. For our analysis, the phase function $\hat{\mathbf{S}}(x_i, t_0, t)$ characterizes the seismic activity in southern California between 32° and 37° latitude, and −115° to −122° longitude. It should be noted that this southern California catalog *has not been declustered*, because the space-time clustering carries the information concerning the space-time correlations that we seek to identify. Since seismicity in active regions is a noisy function (KANAMORI, 1981), we work with temporal averages of seismic activity. The geographic area is partitioned into N = 3621 square regions approximately 11 km on a side, centered on a point x_i. Within each box, a time series is defined using the Caltech seismic catalog obtained from the online SCEC database. For southern California, the instrumental data begins in 1932 and extends to the present.

We define the activity rate $\psi_{obs}(x_i, t)$ as the number of earthquakes per unit time, of any size, within the box centered at x_i at time t. The geographic region that $\hat{\mathbf{S}}(x_i, t_0, t)$ represents is taken large enough so that seismic activity can be considered an incoherent superposition of phase functions. To construct the phase function $\hat{\mathbf{S}}(x_i, t_0, t)$, we define the time-averaged seismicity function $\mathbf{S}(x_i, t_0, t)$ over the interval $(t - t_0)$:

$$\mathbf{S}(x_i, t_0, t) = \frac{1}{(t - t_0)} \int_{t_0}^{t} \psi_{\text{obs}}(x_i, t)dt.$$

Since there are N numbers $\mathbf{S}(x_i, t_0, t)$, and if we assume t_0 to be a fixed time, we can then consider $\mathbf{S}(x_i, t_0, t)$ to be the ith component of a general, time-dependent vector evolving in an N-dimensional space. In previous work, we showed that this N-dimensional *correlation space* is spanned by the eigenvectors of an $N \times N$ correlation matrix (RUNDLE *et al.*, 2000a, 2000b).

Denoting spatial averages over the N boxes by $\langle \rangle$, the phase function $\hat{\mathbf{S}}(x_i, t_0, t)$ is then defined to be the mean-zero, unit-norm function obtained from $\mathbf{S}(x_i, t_0, t)$:

$$\hat{\mathbf{S}}(x_i, t_0, t) = \frac{\mathbf{S}(x_i, t_0, t) - \langle \mathbf{S}(x_i, t_0, t) \rangle}{\|\mathbf{S}(x_i, t_0, t)\|},$$

where $\|\mathbf{S}(x_i, t_0, t)\|$ is the norm over all spatial boxes. Note that

$$\|\mathbf{S}(x_i, t_0, t)\| = \{\langle (\mathbf{S}(x_i, t_0, t) - \langle \mathbf{S}(x_i, t_0, t) \rangle)^2 \rangle\}^{1/2}.$$

Based upon the assumption of pure phase dynamics (RUNDLE *et al.*, 2000a, 2000b), the important changes in seismicity will be given by $\Delta\hat{\mathbf{S}}(x_i, t_1, t_2) = \hat{\mathbf{S}}(x_i, t_0, t_2) - \hat{\mathbf{S}}(x_i, t_0, t_1)$, which is a pure rotation of the N-dimensional unit vector $\hat{\mathbf{S}}(x_i, t_0, t)$ in time.

In order to both remove the last free parameter in the system, the choice of base year, as well as to reduce the random noise component, we sum over all base years, t_0, from 1932 through t_2, such that

$$\Delta\hat{\mathbf{S}}(x_i, t_1, t_2) = \frac{\int_{1932}^{t_2} [\Delta\hat{\mathbf{S}}(x_i, t, t_2) - \Delta\hat{\mathbf{S}}(x_i, t, t_1)] \, dt}{\int_{1932}^{t_2} dt}.$$

In phase dynamical systems, probabilities are related to the square of the associated vector phase function (MORI and KURAMOTO, 1998; RUNDLE *et al.*, 2000b). Thus we find

$$\Delta\mathbf{P}(x_i, t_1, t_2) = \{\Delta\hat{\mathbf{S}}(x_i, t_1, t_2)\}^2 - \mu_P.$$

In other words, μ_P is the spatial mean of $\{\Delta\hat{\mathbf{S}}(x_i, t_1, t_2)\}^2$, and $\Delta\mathbf{P}(x_i, t_1, t_2)$ is the square of the value of $\Delta\hat{\mathbf{S}}(x_i, t_1, t_2)$ at each x_i minus that spatial mean. Note that the integral of $\Delta\mathbf{P}(x_i, t_1, t_2)$ over all N sites vanishes for all time intervals $[t_1, t_2]$, as it should, to conserve probability. Finally, note again that there are *no remaining free parameters in this method*. For any given catalog, up to the present, it is possible to compute the relative change in probability of an event over any given time period.

Results

The decomposition of nonlinear systems into their orthonormal eigenfunctions has been used successfully in the atmospheric sciences for many years (PREISENDOR-FER, 1988; PENLAND and MAGORIAN, 1993). The Karhunen-Loeve approach, the theoretical basis for EOF techniques, represents these space-time patterns as a set of eigenvectors, or normal modes, of an equal-time correlation function, their associated time series, and N total eigenfrequencies, where N is the total number of locations. The eigenvectors provide information pertaining to the spatial correlations of the patterns; the time series characterize each eigenvectors temporal pattern; the eigenfrequencies provide information regarding how often they occur in the data. In a number of applications, a complex, linear correlation operator is constructed in order to extrapolate future system behavior such as the El Niño southern oscillation (PENLAND and MAGORIAN, 1993). Here we apply this decomposition method to historical seismicity data in southern California in order to identify basis patterns for all possible space-time seismicity configurations. These basis states are a complete, orthonormal set of eigenvectors and associated eigenvalues that are obtained from the diagonalization of the correlation operators computed for this regional historic seismicity data.

Time Period 1932 through August, 1999

In our first analysis example, the time period starting in 1932 and continuing through the end August of 1999 was analyzed using the entire data set, including the entire areal extent and events of all quality. Figure 3a is a plot of the first 25 normalized eigenvalues, while Figure 3b is the first 1000 normalized eigenvalues, plotted on a log-log scale. Figure 4 shows the first two modes for southern California, for this data set. The absolute maximum value in each plot is normalized to one, where squares are positive and diamonds are negative, i.e., squares and diamonds are anticorrelated. The correct interpretation is that while a square location is "on," a diamond location is "off," and *vice versa*. The first mode is effectively a background hazard map, where small events are mutually correlated throughout southern California, while the second mode is the Landers event. The accompanying PC time series are shown in Figures 4a and 4c. The influence of spatial and temporal variations due to the density and completeness of network coverage is visible in the PC time series. For example, note the distinctive wave associated with the Landers sequence and its large numbers of aftershocks, punctuated by the occurrence of the Northridge event.

Figure 4d displays the second KLE mode. Here the region surrounding the 1992 Landers event is "on" (squares) whereas the rest of the southern San Andreas fault system is "off" (diamonds). The Coalinga earthquake is visible in this mode, and an apparent correlation between Landers and a set of events in eastern Nevada is revealed. As can be seen in Figures 5 and 6, a number of the KLE eigenpatterns are

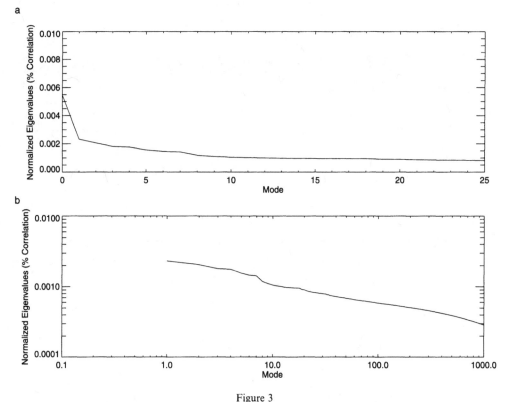

Figure 3
a) The first 25 normalized eigenvalues, for the time period 1932–1999. (b) The first 1000 normalized eigenvalues, plotted on a log-log scale.

lower order harmonics of the second mode. Interestingly, Figure 6b, KLE7, illustrates the correlations between the North Palm Springs event and other major southern California earthquakes, although with minimal correlation with the Landers sequence.

Typically, the higher order KLE modes display signal on shorter spatial and temporal scales than the initial, lower modes. This is illustrated in Figure 6. Figure 6c, KLE8, is a smaller scale harmonic of Figure 6a, KLE6, while KLE9, Figure 6d, shows the Landers sequence, essentially isolated, with the Joshua Tree earthquake anticorrelated with the Landers event to the north and Big Bear to the northwest.

Time Period: 1932 through December, 1991

One of the interesting questions which arises in studying the results above is how exactly the dominance of the Landers sequence, and its associated instrumentation, affects the eigenpatterns. This encouraged the removal of that event from the data set by cutting off the time series before its occurrence, at the end of 1991. For this time

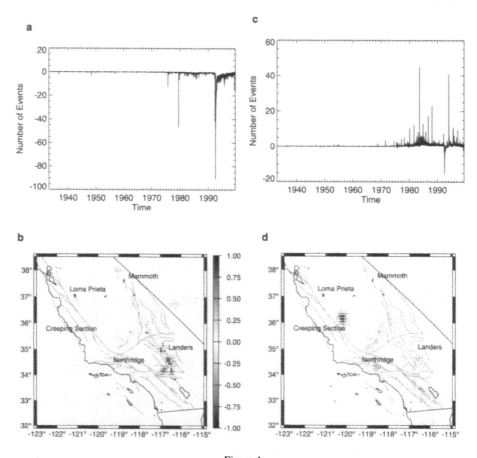

Figure 4

First two KLE modes for southern California seismicity, 1932–1999. Diamonds are negative, square boxes are positive. a) PC time series for first KLE mode; b) first KLE mode, normalized to maximum; c) PC time series for second KLE mode; and d) second KLE mode, also normalized to the maximum.

period, 1932 to 1991, the time series interval was again one day, so that the total number of time steps is approximately 21,535. Again, all locations from the entire database, and all quality events, were included in the decomposition, consequently the number of location time series is 3621. Figure 7a shows the first 25 normalized eigenvalues, while Figure 7b is the first 1000 normalized eigenvalues, plotted on a log-log scale. The eigenvalue plot is now smoother, without the large drop after the first mode that was a function of the large wave in the first PC time series generated by the Landers aftershocks (see Fig. 4b).

KLE modes one and two are shown in Figure 8, where KLE1 is the mode associated with background seismicity. As expected, there is no large signal for Landers visible in Figure 8b. In addition, many of what were the lower modes in the previous analysis have moved up in the eigenvalue ranking, replacing the large

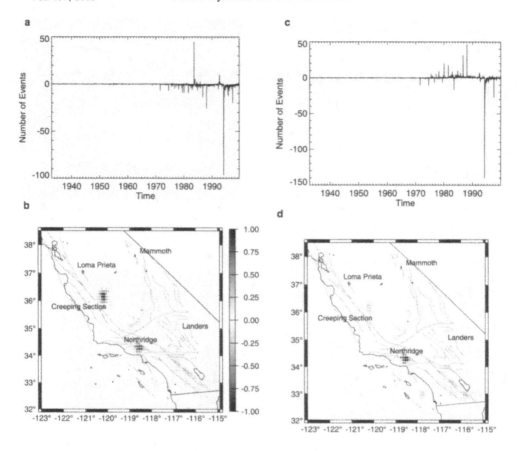

Figure 5
KLE modes three and four for southern California seismicity, 1932–1999. a) PC time series for third KLE mode; b) third KLE mode, normalized to maximum; c) PC time series for fourth KLE mode; and d) fourth KLE mode, also normalized to the maximum.

number of Landers harmonics. KLE2 (Fig. 8d) is the 1983 Coalinga earthquake, anticorrelated with the 1986 Oceanside and North Palm Spring events. The 1971 San Fernando event is a single diamond, correlated with the Coalinga earthquake.

Figure 9 again demonstrates both the addition of lower order harmonics and increasingly smaller spatial and temporal scales with increasing mode number. The fourth KLE mode shown in Figure 9a is the 1987 Superstition Hills and Elsinore Ranch events correlated with the Whittier Narrows earthquake and anticorrelated with the North Palm Springs and Oceanside events. Figure 9b, KLE mode five, also shows the North Palm Springs and Oceanside events correlated with each other. In KLE7, Point Mugu appears correlated with the 1971 San Fernando event. The Tejon Ranch earthquake of 1988, in addition to the 1979 Homestead Valley earthquake, south of Landers, is also correlated with these events, while Imperial Valley is

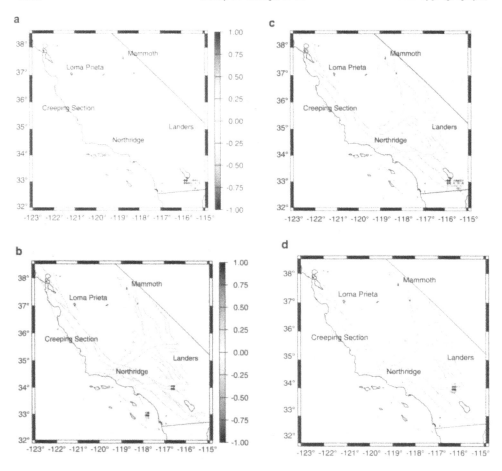

Figure 6
KLE modes six through nine for southern California seismicity, 1932–1999, each normalized to its maximum. a) Sixth KLE mode; b) seventh KLE mode; c) eighth KLE mode; and d) ninth KLE mode.

anticorrelated with these events. KLE8 again presents the 1979 Homestead Valley event, now anticorrelated with Point Mugu. Note the arcuate structure, which cuts across the faults at the location of the 1992 Landers earthquake. This is a feature of the local seismicity that has only been recognized in recent years with the occurrence of the 1992 Landers sequence and the 1999 Hector Mine earthquake, but which was clearly visible in this decomposition as early as 1991 (Fig. 9d). It should be noted again that *no seismicity data after December, 1992, are included in this analysis.*

Analysis of this same data set, 1932 through 1991, but with a magnitude cut of 3.0 applied to the data, yields the KLE modes shown in Figures 10 and 11. The application of a magnitude cut has allowed events from earlier in the data set to achieve a greater prominence. Interestingly, while the background mode has dropped out, the first mode is not Coalinga, the second mode in the entire data set (see Fig. 8b

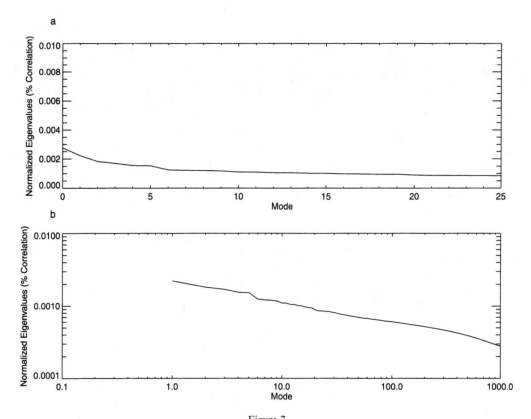

Figure 7
a) The first 25 normalized eigenvalues, for the time period 1932–1991. b) The first 1000 normalized eigenvalues, plotted on a log-log scale.

above), but the 1971 San Fernando event. The 1983 Coalinga earthquake is now the third KLE mode, behind the 1969 Avila Beach earthquake (see Fig. 10). The correlation between the 1979 Imperial Valley and 1987 Superstition Hills sequence is the fourth mode. The 1952 Kern County event has appeared as the seventh mode (Fig. 11a), while the eighth mode is also new to the decomposition – the 1968 Borrego Mountain earthquake (Fig. 11b). Figure 11 also shows KLE modes ten and eleven, now harmonics of the earlier modes, on smaller spatial and temporal scales. One interesting feature of KLE mode 11 is the anticorrelation between the 1983 Coalinga event and the historic location of the Parkfield earthquake to the west, which has not taken place in recent years as expected. This mode supports speculation that the Coalinga event delayed its occurrence.

Precursory Modes

The presence of both large- and small-scale correlations in the data, evident in the KLE decompositions shown above, prompted a study of the change in these

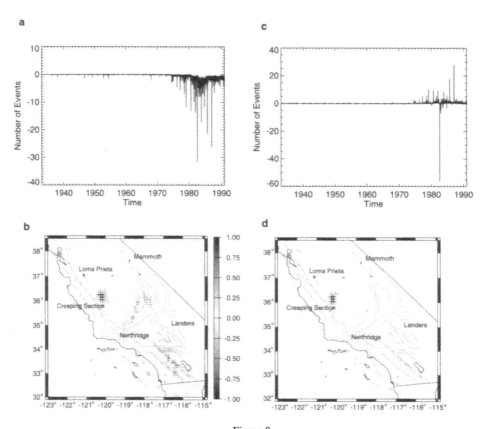

Figure 8
First two KLE modes for southern California seismicity, 1932–1991. a) PC time series for first KLE mode;
b) first KLE mode, normalized to maximum; c) PC time series for second KLE mode; and d) second KLE
mode, also normalized to the maximum.

modes for each year, in an attempt to identify modes which consistently appear over some identifiable time period prior to an event. While a complex rate correlation operator, $K(x_i, x_j)$, can be used to compute the probability of future events on a fault patch model producing events over time periods of thousands of years (RUNDLE *et al.*, 2000a), its application to historic seismicity data is limited. Neither the long-time periods nor the abundance of moderate to large events produced by numerical simultions are available in the actual data, nor is the same accuracy in time and space possible. Consequently, the following method was developed.

If the seismicity in a given year, **S** is known, and the eigenmodes, or eigenvectors e_i, are calculated using all seismicity data $(i = 1, \ldots N, N = 3621$ sites), then the eigenvectors are a complete, orthonormal set of basis functions, and any seismicity over that space can be decomposed into those eigenvectors.

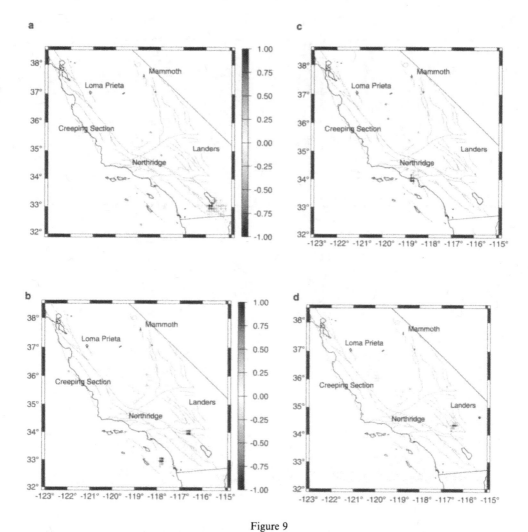

Figure 9
a) Fourth KLE mode; b) fifth KLE mode; c) seventh KLE mode; and d) eighth KLE mode for southern
California seismicity, 1932–1991, each normalized to its maximum.

$$S = \sum_{i=1}^{N} \alpha_i e_{ji},$$

where α_i are the eigenvalues for that particular year. The eigenvalues, α_i, are then
computed from

$$\alpha_i = \sum_{i=1}^{N} e_{ji} S_i.$$

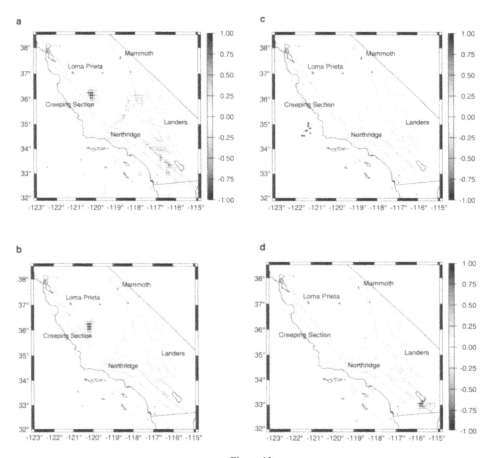

Figure 10
a) First KLE mode; b) second KLE mode; c) third KLE mode; and d) fourth KLE mode for southern California seismicity, magnitudes ≥ 3.0, 1932-1991, each normalized to its maximum.

Computing the α_i for any given year, given the KLE decompositions above, is a relatively simple process. The data set used was that described above, for the time period 1932 through 1991, with a magnitude cutoff of 3.0. The resulting $(\alpha_i)^2$, i.e., the power spectrum of the eigenmodes, for each year prior to the 1992 Landers sequence, are plotted in Figure 12. Note again, that no data after December 1991 are included in this analysis. Events that occur in the data set, for example the 1979 Imperial Valley event or the 1983 Coalinga earthquake, have signal in the corresponding eigenmodes, and would be expected to produce signal prior to those events. In addition, several of the higher order eigenmodes, on smaller spatial and temporal scales, appear to change substantially over the several years prior to 1992. Interestingly, a number of the modes that increase are those that include signal for the 1992 Landers earthquake sequence. For example, in Figure 13 is shown KLE modes 46, 49, 58, and 62, which, while noisy, contain significant correlations for the Joshua Tree, Landers and Big Bear events.

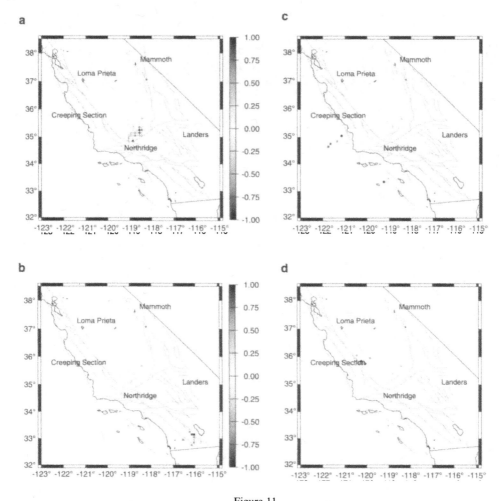

Figure 11
a) Seventh KLE mode; b) eighth KLE mode; c) tenth KLE mode; and d) eleventh KLE mode for southern California seismicity, magnitudes $\geq$ 3.0, 1932–1991, each normalized to its maximum.

Figure 14 depicts the $(\alpha_i)^2$, plotted for the seismicity over each year, summed sequentially from 1987 through 1991. Systematic changes now appear in the $(\alpha_i)^2$. These show the growth in the smaller scale Landers modes over time, as the fault system becomes increasingly correlated prior to the 1992 event. Again, it is important to remember that no data after December 31, 1991, is included in the analysis of either these eigenmodes or the associated $(\alpha_i)^2$.

Note that a multidimensional vector can be formed from the seismic activity by assuming that each of the locations on the grid of southern California is one of the dimensions, and that as the amount of seismicity at each location changes in time, the vector undergoes a rotation about some mean. If the seismicity

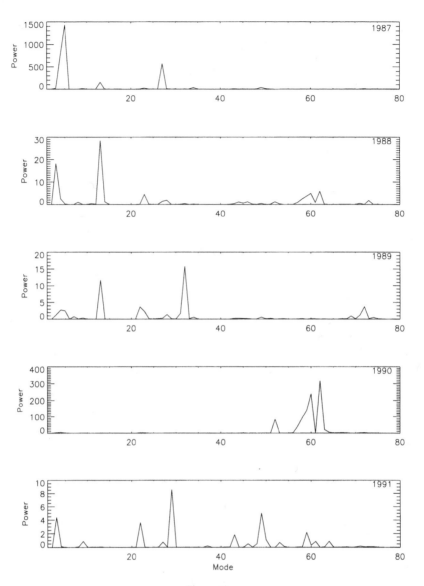

Figure 12

Eigenvalue power, $(\alpha_i)^2$, for the individual years 1987–1991, decomposed using eigenmodes derived using southern California seismicity, magnitudes ≥ 3.0, 1932 through 1991.

patterns, as described by the correlations in the $(\alpha_i)^2$, are undergoing a systematic change due to the growth of precursory modes, such as seen in Figures 12 and 14, then the vector is no longer experiencing a random walk about some mean. It undergoes a persistent rotation away from that mean that is quantifiable and can be converted into a probability of current and future events.

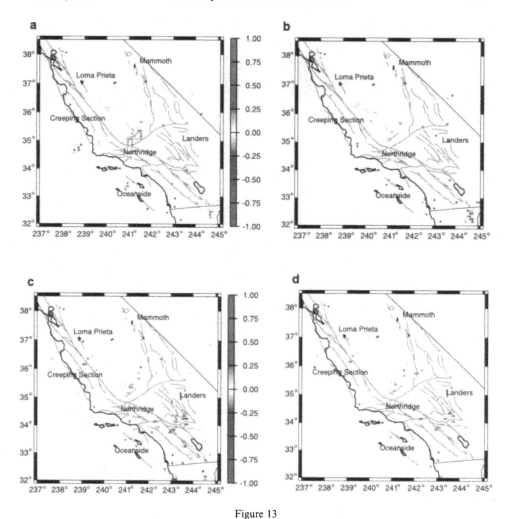

Figure 13
a) KLE mode 46; b) KLE mode 49; c) KLE mode 58; and d) KLE mode 62 for southern California seismicity, magnitudes ≥ 3.0, 1932–1991. The color scale is linear, blue to white to red, where blue is negative.

Probability of an Earthquake

Our simulations have suggested that the correlations in the seismicity represented by the KLE modes above can be described by phase dynamics. Phase dynamics is a method used in various branches of physics to describe the behavior of important parameters of the physical system (FUKUNAGA, 1970; MORI and KURAMOTO, 1998). Variables in many dynamical systems can be characterized by using this phase dynamical technique, represented as a phase function that involves both amplitude and phase angle. Changes in the amplitude of the phase function are unimportant, or not relevant. Examples of pure phase dynamical systems in the classical world

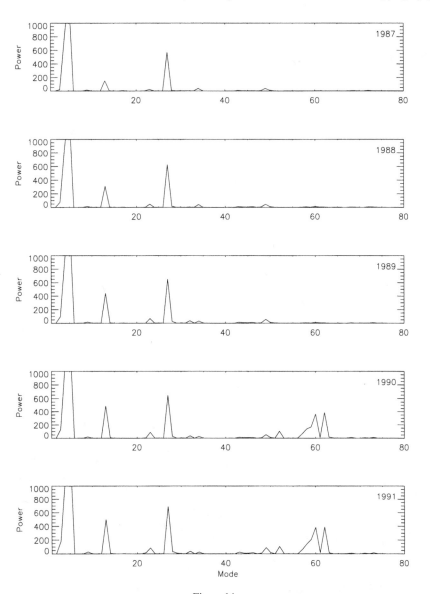

Figure 14

Eigenvalue power, $(\alpha_i)^2$, for the individual years *summed* from 1987 through 1991, decomposed using eigenmodes derived using southern California seismicity, magnitudes $\geq$ 3.0, 1932–1991.

include weak turbulence in fluids and reaction-diffusion systems. Another nonclassical example is a quantum system in which the wave function is the phase function.

In the case of earthquake fault systems, the important changes in seismicity are associated primarily with rotations of the vector phase function in a high-dimensional correlation space. Using the assumption that seismicity is an example of pure phase dynamics, it follows that $\Delta\mathbf{P}$ can be calculated from the square of the phase function for the associated pattern state vector (RUNDLE et al., 2000a). It should therefore be possible to compute the increase in probability of observing an anomalous correlation, $\Delta\mathbf{P}$, directly from the observed seismicity data and its rotation. To emphasize the connection to phase dynamics, we call the function $\Delta\mathbf{P}$ the Phase Dynamical Probability Change (PDPC) (TIAMPO et al., 2000).

The technique described here is not a model, rather it is a new method for processing seismicity data to reveal underlying space-time structure. The purpose of the remainder of this paper is to exhaustively test the method, without regard to its formulation, and to apply it to the forecast of future earthquakes.

The seismicity data employed in our analysis are the same as those used above. Using only the subset of these data at locations $\mathbf{x}$ in southern California and covering the period from January 1, 1932 through December 31, 1991, we compute the PDPC function $\Delta\mathbf{P} = \Delta\mathbf{P}(x_i, 1932, 1991)$ for detecting anomalous spatial correlations at sites in southern California prior to January 1, 1992. Note that we use only events having magnitude $M \geq 3$, to ensure completeness of the catalog. The hypothesis to be tested is that these anomalous correlated regions are associated with large mainshocks that occurred after January 1, 1992. Figure 15a showed the relative seismic activity for this period, $S(x_i, 1932, 1991)$, superimposed on a map of southern California. The intensity scale is logarithmic, where the scale value indicates the exponent, to the base ten, of the grayscale intensity value. It is clear that $S(x_i, 1932, 1991)$ is an unremarkable function. For example, there is little evidence of any phenomena precursory to the $M\sim7.3$ Landers, California event that occurred on June 28, 1992.

Figure 15b exhibits the PDPC anomalies in southern California for the time period 1978 to 1991. Note that no data after December of 1991 were used in this analysis. The triangles denote events of $M > 5$ which go off during this time period, while the open circles are events which occur after 1991. Note the frequent occurrence of large earthquakes at the locations of increased relative probability.

Figure 15b shows plots of all $\Delta\mathbf{P} > 0$, using only existing seismicity data acquired prior to January 1, 1992; six months before the occurrence of the Landers earthquake sequence. The increase in $\Delta\mathbf{P}$ above the background level, as measured by μ_P, the spatial mean of $\Delta\mathbf{P}$, should be interpreted as the formulation of a spatially coherent region of either anomalous activation or quiescence, associated with an increased chance of a major earthquake. The scale is again logarithmic, scaled to the largest value of $\Delta\mathbf{P}$ on any of Figure 15b. $\Delta\mathbf{P}$ indicates only a relative, not an absolute change in probability. The inverted triangles represent events that occurred during the time period covered by the plot, indicating locations associated with large events that are present in the data used to construct $\Delta\mathbf{P}$. At a minimum, the method should identify areas of increased $\Delta\mathbf{P}$ associated with the triangles, and it is clearly successful. Of most interest are regions of increased $\Delta\mathbf{P}$ with no nearby triangles. According to our hypothesis, these locations may represent sites for future large earthquakes.

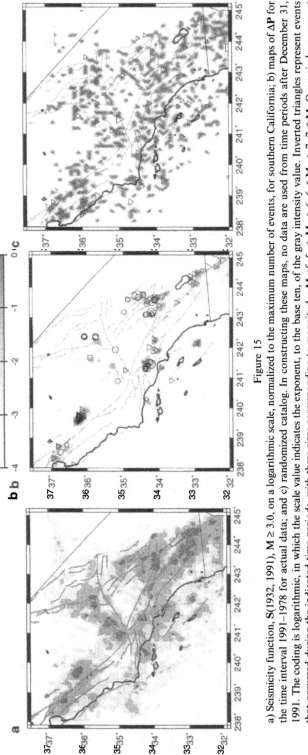

Figure 15

a) Seismicity function, **S**(1932, 1991), M ≥ 3.0, on a logarithmic scale, normalized to the maximum number of events, for southern California; b) maps of Δ**P** for the time interval 1991–1978 for actual data; and c) randomized catalog. In constructing these maps, no data are used from time periods after December 31, 1991. The coding is logarithmic, in which the scale value indicates the exponent, to the base ten, of the gray intensity value. Inverted triangles represent events that occurred during the indicated time periods, with three sizes corresponding to magnitudes M of: 5 < M < 6, 6 ≤ M < 7, 7 ≤ M. Open circles represent events that occur after January 1, 1992. Three increasing circle sizes again correspond to the same magnitude ranges as for the inverted triangles.

We then superimpose on this map the locations of mainshocks larger than 5.0 that occurred between January 1, 1992 and November 1, 1999, that is, the ~8 years *following* the time interval from which we computed the probabilities. Black circles denote these locations. One can observe an obvious correspondence between regions of increased probability and the location of the subsequent mainshocks, tending to support the results first observed in our simulations. Included are circles representing the 1992 Landers sequence, the 1994 Northridge earthquake, and the recent Hector Mine event.

Figure 15b shows that the recent large earthquakes occurring *after* January 1, 1992 are associated with areas of $\Delta P > 0$ that formed *prior* to January 1, 1992. In particular, a bright area has developed close to the epicenter of the Joshua Tree event, latitude 33.95°, longitude 243.7°, which occurred in April of 1992. The anomalous area north of that location corresponds to the Landers earthquake sequence of June 1992. The epicenter of the recent October 16, 1999, M~7.0 Hector Mine also occurred on the northernmost end of that anomaly.

There is clearly variability, particularly for smaller events, depending on the choice of time interval. Larger events tend to be associated with larger anomalous regions that form earlier and persist longer after the mainshock. Since earthquake fault dynamics are now believed to be associated with critical phenomena (RUNDLE and KLEIN, 1995; GELLER et al., 1997; KLEIN et al., 1997) we hypothesize that there may be a scaling relation between the area A of the correlated region and the time interval τ prior to the mainshock at which the anomalous correlation begins to form, such that $\tau \propto A^{\eta}$, where η is a critical exponent near 1. Since the linear size of our location grid boxes is approximately 11 km, one should not expect events significantly smaller than M~6, whose characteristic linear source dimension is 10 km, to be well resolved by our procedure. Yet even the smaller circles associated with M~5 – 6 events often seem to occur in proximity to these anomalous areas. One such example is the China Lake event that occurred at approximately latitude 36.75°, longitude 242.25°. Here we also should note that a number of the anomalies correspond to events with magnitude less than 5, but that display persistent moderated seismicity, such as the Durrwood Meadows swarm, highlighted with a black square, at 35.5° latitude, 241.75° longitude. This set of events, which included an earthquake of magnitude 4.9, occurred from late 1983 through mid-1984.

These results suggest that the anomalies are correlated with total seismic moment release. The large, intense anomaly associated with the Coalinga event in 1983, for example, was followed by numerous aftershocks of magnitude 5.0 or greater. While there are earthquakes of M > 5 which occur without the formation of an area of increased probability ("false negatives"), this may be the result of a lower total seismic moment release. In the future, we will investigate whether this technique can be extended to estimate the potential magnitude and time of occurrence of forecast events from both the sizes of the candidate source regions and the temporal persistence and duration.

Likelihood Tests

To test the hypothesis that the formation of correlated regions of anomalous activity can be identified by our method, and are related to future large events, we carried out thousands of likelihood ratio tests (BEVINGTON and ROBINSON, 1992; GROSS and RUNDLE, 1998) on the method using values of ΔP obtained from random seismicity catalogs that were used as null hypotheses. In the likelihood ratio test, a probability density function (PDF) is required. For our PDF we used a Gaussian distribution whose width is that of our original location grid, approximately 11 km, and whose peak value is given by $\Delta P + \mu_P$, because in a likelihood ratio test the value of probability at all sites must be positive. The log likelihood is then calculated for the circles, which identify the locations of future events, as shown in Figure 15b, and provides a measure of how well the colored regions predict the actual events as quantified by the locations of the circles.

Each random catalog was constructed from the instrumental catalog by using the same total number of events, but assigning occurrence times from a uniform probability distribution over the years 1932–1991, and distributing them uniformly over the original locations. This procedure produces a Poisson distribution of events in space with an exponential distribution of interevent times. Randomizing the catalog in this way destroys whatever coherent space-time structure may exist in the data, *thus effectively declustering the catalog*. While random "Poisson clustering" may remain, there will be no "Omori clustering" left in these catalogs.

Likelihood ratio tests were then performed on thousands of such random catalogs, comparing calculations of ΔP from the actual catalog with calculations of ΔP from each random catalog. Success of the prediction is scored by how well the circles are forecasted by the colored areas. The better the prediction, the larger the likelihood value. Figure 16 shows the distribution of likelihood values for five hundred of the random catalogs, for the same time period as shown in Figure 15b. Superimposed on this plot as a dashed line is the likelihood ratio value for Figure 15b, computed from the actual catalog. One can see that the value of likelihood computed from the actual catalog is always substantially better than likelihoods computed from the random catalogs.

For purposes of illustration, we compute ΔP for one random catalog whose likelihood value is close to the mean of the distribution in Figure 16, as shown in Figure 15c. One can see that there are many more anomalous areas in Figure 15c than in Figure 15b; these regions are far more broadly distributed in space, and contain many more dark areas. The specific likelihood value for Figure 15b equals -311.665, whereas that for Figure 15c equals -363.344. These values correspond to a likelihood ratio $e^{51.68} \approx 10^{22.44}$, indicating that the locations of increased probability obtained from the actual instrumental catalog are far more likely to be associated with the locations of the circles than those obtained from the

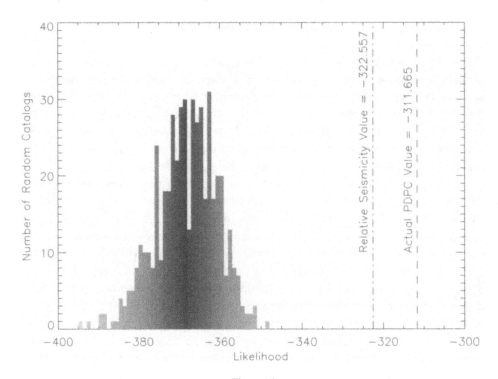

Figure 16

Histogram of likelihood values for one hundred random catalogs of southern California seismicity. In addition, the likelihood value for the actual catalog, as shown in Figure 15b, is superimposed on this plot as a dashed line, and the likelihood value for the relative southern California seismicity as shown in Figure 15a is plotted as the dash-dotted line.

random catalog. The physical reason for this large ratio is that the likelihood test invokes a penalty for predictions that are not sufficiently near the circles ("false positives"), and there are many more such locations in Figure 15c than in Figure 15b.

In addition, we computed the likelihood ratio using Figure 15a as the null hypothesis that may be taken to correspond to a map of relative hazard based upon recent instrumental seismic intensity. The likelihood value for this map is −322.557, equating to a likelihood ratio of $e^{11.22} \approx 10^{4.87}$. When compared to the computations for Figure 15b, this suggests that our method is considerably more successful at forecasting the locations of future major earthquakes than the current practice of relying on hazard maps to predict broad areas of increased seismic potential. From these statistical tests, we conclude that there are coherent, anomalous space-time correlation structures in the instrumental catalog that our method identifies, and that these correlations are effectively destroyed by the common practice of declustering seismicity catalogs.

Discussion

We emphasize that while our method may identify higher risk areas, there is no certainty at this time that every location of increased ΔP will be located near the site of a future large earthquake. There are numerous examples in Figure 15b where a colored area appears without the occurrence of a major earthquake between 1992 and 1999. One example appears near 34° latitude, 244.25° longitude. Further attempts at optimization of the method must focus on better spatial location of events and minimizing the numbers of both false positives and false negatives.

As discussed below, we have seen that regions with $\Delta P > 0$ may correspond either to anomalous seismic activity or anomalous quiescence. In some locations, a region that had been the site of a recent major earthquake may evidently indicate values of $\Delta P > 0$ that are linked with a future large event that will occur at a neighboring, although somewhat disjointed, location. The positive value of ΔP may appear only at the location of the past event, rather than at the neighboring location of the future event. In these cases, we have found that the neighboring aftershock zone of the past event is participating in the anomalous activity or quiescence that defines the future event. In addition, it is possible that anomalous activation or quiescence may appear in events having magnitudes less than the uniform cutoff we use, $M = 3$. While we have performed this analysis with other magnitude cuts and have found the results to be relatively insensitive between values of 2.5 and 4.0, this may not apply to other tectonic settings, based on the quality of the data available. The appropriateness of various magnitude cutoffs should therefore be tested in future work.

An example of these effects is the anomalous area at the location of the 1971 San Fernando earthquake, shown in Figure 15b, that is evidently associated with the coming 1994 Northridge event. This area is present *not* because it represents aftershock activity from the 1971 San Fernando event (anomalous seismic activity), rather it represents an area of relative anomalous quiescence. To further illustrate the point that the San Fernando region is anomalously quiescent prior to Northridge, we plot the change in the phase function $\Delta \hat{S}(x_i, 1978, 1991)$, as defined in the method section above, spanning the years 1978–1991. In Figure 17, a positive value of $\Delta \hat{S}$ represents anomalous seismic activity, and a negative value of $\Delta \hat{S}$ represents anomalous quiescence. The intense anomaly near Northridge-San Fernando seen in the plot of ΔP in Figure 15b is seen to be negative in Figure 17, indicating that the positive value of ΔP arises from anomalous quiescence in the San Fernando aftershock zone. In addition, it will be seen below that the construction of $\Delta P(x_i, 1978, 1991)$ actually subtracts the effect of any changes in seismicity prior to 1978, therefore San Fernando aftershock activity between 1971–1978 does not contribute to the calculation. Figure 17 also indicates that the 1983 Coalinga earthquake displays seismic activation, as would be expected since it occurred during this time period, while the Landers sequence is a mix of anomalous quiescence and activation. These results support the conclusion that the PDPC function does not simply identify areas associated with past

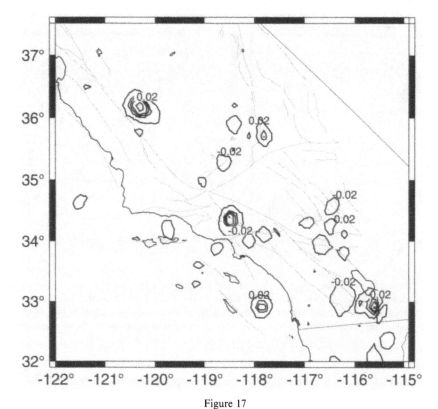

Figure 17

Contour plot of $\Delta\hat{S}(x_i, t_1, t_2)$, normalized to the maximum absolute value, for the time period 1991–1978.

events and their aftershock sequences, rather it quantifies the underlying stress coherence and correlations associated with the regional seismicity.

Finally, we calculate the PDPC for all of California over three additional time periods. Figure 18a shows the increased probability for the time period 1968 through 1978. Again, the intensity scale is as shown in Figure 15, and the inverted triangles represent those events that occur during the time period, while the open circles are those earthquakes that occur throughout the next ten years. Despite the sparser networks in place in California at that time, which affects the spatial completeness of the catalog, the 1979 Imperial Valley earthquake, M = 6.4, occurs near a bright set of anomalies, which includes anomalies for the 1987 Superstition Hills and Elmore Ranch events. Also clearly visible is the upcoming 1983 Coalinga event at 36.25° latitude, 239.75° longitude, a previously unanticipated hidden thrust event.

Figure 18b shows the PDPC encompassing 1978 through 1988. Clearly visible is another relatively unexpected event at the time, the 1989 Loma Prieta earthquake. The epicenter of the mainshock occurs within 10 kms of the nearest anomaly, which extends down through the aftershock zone. Again, the Durrwood Meadows swarm,

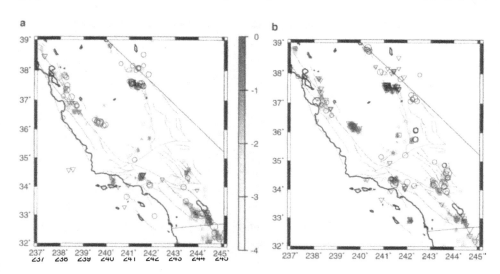

Figure 18

Maps of ΔP for a) for the time period 1978–1968, and for b) the time period 1988–1978. In constructing these maps, no data is used for time periods after the final year. The color scale is the same as in Figure 15. Inverted triangles represent events that occurred during the indicated time periods, as before. Open circles represent events that occur during the ten years after the referenced time period. Three increasing sizes again correspond to the same magnitude ranges as Figure 15.

at 35.5° latitude, 241.75° longitude, including one earthquake of magnitude 4.9, is plotted as a square box.

Finally, Figure 19 shows a forecast for roughly the next ten years, using a white to yellow to red logarithmic intensity scale. The PDPC analysis is performed for the time period 1989 through 1999. The persistent anomaly located at the northeast end of the White Wolf Fault, 35.25° latitude, 241.5° longitude, corresponds to an area of recurrent seismicity throughout the 1980s and 1990s, which may represent the birth of a newly forming structure, a blind strike-slip fault connecting the Kern County and Walker Pass, California, earthquakes (BAWDEN *et al.*, 1999). Or, the orange zones at 33° latitude, 244° longitude, may identify a silent earthquake that occurred in the mid-1990s at that location (Paul Vincent, personal communication). Note that the northwesternmost anomaly, at 38.75° latitude, 237.2° longitude, identifies the location of a series of events, M = 4, followed by a number of smaller events, which occurred in January of 2000.

Figure 19 identifies certain areas that may be destined for activity in the next 5–10 years. For example, the yellow area just to the north of the 1988 Loma Prieta event, along the southern Hayward fault, is one potential area of concern. However, in agreement with deformation work by BURGMANN *et al.* (2000), the northern section of the Hayward fault manifests a lesser potential for an independent event. In southern California, the area just to the north of the 1992 Landers sequence, as well as the site of the 1983 Coalinga earthquake, display a

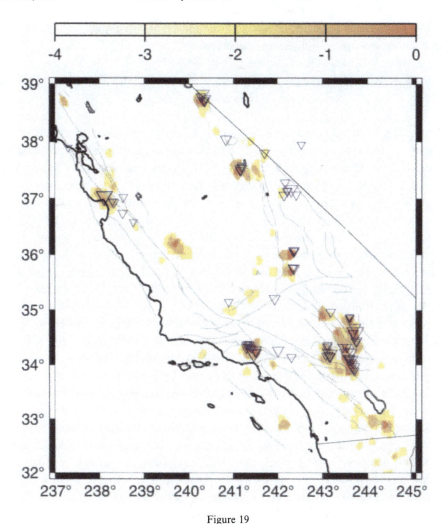

Figure 19

Map of ΔP for the time period 1999–1989. In constructing this map, no data are used for the year 2000. The color scale is still logarithmic, as in Figure 15. Inverted triangles represent events that occurred during the indicated time periods, as before. Open circles represent events that occur during the ten years after the referenced time period. Three increasing sizes again correspond to the same magnitude ranges as Figure 15.

large PDPC anomaly. Finally, several areas of relative probability increase are visible along the southern San Andreas and San Jacinto faults. We must point out that, for reasons already detailed, absolute probabilities cannot be attached to these anomalies at this time. In addition, these anomalies represent areas of varying seismic activity – it is possible that some or all of these locations may represent areas of either recurring moderate events of magnitude less than 5.0, or locations of silent earthquakes occurring between 1989 and 1999. Finally, the occurrence of one or more of these events could modify the underlying dynamics of the system, requiring the revision of subsequent forecasts.

Conclusions

This pattern dynamics approach that we have applied to historical seismicity data in southern California reveals a wealth of interesting spatial patterns. In particular, it provides a new methodology for classifying all of the possible seismicity patterns that can exist in terms of mutually orthogonal eigenstates. In fact, a number of the descriptive patterns cited earlier can be readily identified among the eigenstates depicted in Figures 4 through 12. For example, Figure 4d can be interpreted either as seismic activation within the Landers epicentral region, or else as quiescence near Landers coincident with seismic activity surrounding the region, i.e., a "Mogi donut."

Our results argue strongly for the development of realistic numerical simulations of fault systems such as those in southern California (RUNDLE, 1988; TIAMPO *et al.*, 1999; RUNDLE *et al.*, 2000a). Because the historic data set is incomplete worldwide, construction of such numerical simulations is necessary to more accurately define the most significant eigenpatterns, which can be applied to understanding the nature of the observed seismicity. For example, while a magnitude cutoff of 3.0 may be appropriate for southern California, it may not be applicable for other tectonic settings. Simulations can help to better define the sensitivity of the analysis to parameters such as this.

Moreover, computer simulations will be of critical importance for relating the observable pattern basis set for the seismicity data to the pattern basis set for stress, strains, and displacements through time. While seismicity is readily observable by standard methods, stress and strain within the earth are not. However, stress and strain are the primary dynamical variables, and are also the Markov variables in which the underlying nonlinear dynamics are most likely formulated. It will be most important to relate a readily observable, seismicity pattern basis set to the actual, unobservable dynamical pattern basis set, so that mode-shaping techniques can be applied to the underlying dynamics (FUKUNAGA, 1970; HOLMES *et al.*, 1996). In this manner it may be possible to characterize the spatially coarse-grained features of local and regional stress levels, coefficients of friction, failure and residual stress levels, and fault interactions. Finally, RUNDLE *et al.* (2000a), demonstrated, using simulations, that such methods can in principle forecast future events with accuracies considerably better than a standard Poisson process. With the development of a quantitative, readily reproducible technique for characterizing all possible seismicity patterns, these methods may allow us to test the hypothesis that large damaging earthquakes on a given subset of faults are preceded by one of a small set of characteristic precursory seismicity patterns. If this hypothesis is true, then it may well be possible to develop a quantitative method to forecast large, infrequent events using the patterns of seismic activation and quiescence associated with smaller, more frequent events on local and regional fault systems.

Finally, our results suggest that systematic variations in seismicity prior to recent southern California earthquakes can be observed. Our method employs data from existing seismic monitoring networks as well as a theoretical understanding obtained from numerical computer simulations to identify anomalous activity or quiescence in seismicity. These space-time patterns in the seismic activity directly reflect the existence of correlated structure in the underlying stress and strain fields, a necessary precondition for the occurrence of large earthquakes. While at this time we offer this forecast for scientific evaluation only, depending on the nature of future seismicity in the region, as well as ongoing modifications and extensions of the theory and technique, this procedure may prove a useful tool for forecasting seismic activity.

Acknowledgements

This work has been supported by CIRES and NASA student fellowships to KFT and SM; by US DOE grant DE-FG03-95ER14499 to JBR. Research by WK was supported by US DOE/OBES grant DE-FG02-95ER14498 and W-7405-ENG-6 at LANL. WK would like to acknowledge the hospitality and support of CNLS at LANL.

REFERENCES

AUBREY, D. G. and EMERY, K. O. (1983), *Eigenanalysis of Recent United States Sea Levels*, Continental Shelf Res. *2*, 21–33.

BAK, P. and TANG, C. (1989), *Earthquakes as Self-organized Critical Phenomena*, J. Geophys. Res. *94*, 15,635–15,637.

BAWDEN, G. W., MICHAEL, A. J., and KELLOGG, L. H. (1999), *Birth of a Fault: Connecting the Kern County and Walker Pass, California, Earthquakes*, Geology *27*, 601–601.

BAKUN, W. H., KING, G. C. P., and COCKERHAM, R. S. (1986), *Seismic slip, aseismic slip, and the mechanics of repeating earthquakes on the Calaveras fault, California*. In *Earthquake Source Mechanics*, AGU Monograph, AGU, Washington, D.C., 195–208.

BAKUN, W. H. and McEVILLY, T. V. (1984), *Recurrence Models and Parkfield, California, Earthquakes*, J. Geophys. Res. *89*, 3051–3058.

BOWMAN, D. D., OUILLON, G., SAMMIS, C. G., SORNETTE, A., and SORNETTE, D. (1998), *An Observational Test of the Critical Earthquake Concept*, J. Geophys. Res. *103*, 24,359–24,372.

BREHM, D. J. and BRAILE, L. W. (1999), *Intermediate-term Earthquake Prediction Using the Modified Time-to-failure Method in Southern California*, BSSA *89*, 275–293.

BUFE, C. G. and VARNES, D. J. (1993), *Predictive Modeling of the Seismic Cycle of the Greater San Francisco Bay Region*, J. Geophys. Res. *98*, 9871–9883.

BURGMANN, R. *et al.* (2000), *Earthquake Potential along the Northern Hayward Fault, California*, Science *289*, 1178–1182.

DENG, J. S. and SYKES, L. R. (1996), *Triggering of 1812 Santa Barbara Earthquake by a Great San Andreas Shock: Implications for Future Hazards in Southern California*, Geophys. Res. Lett. *23*, 1155–1158.

DIETERICH, J. H. (1994), *A Constitutive Law for Rate of Earthquake Production and an Application to Earthquake Clustering*, J. Geophys. Res. *99*, 2601–2618.

DODGE, D. A., BEROZA, G. C., and ELLSWORTH, W. L. (1996), *Detailed Observations of California Foreshock Sequences: Implications for the Earthquake Initiation Process*, J. Geophys. Res. *101*, 22,371–22,392.

ELLSWORTH, W. I. and COLE, A. T. (1997), *A Test of the Characteristic Earthquake Hypothesis for the San Andreas Fault in Central California*, Seis. Res. Lett. *68*, 298.

ELLSWORTH, W. I., COLE, A. T., and DIETZ, L. (1998), *Repeating Earthquakes and the Long-term Evolution of Seismicity on the San Andreas Fault near Bear Valley, California*, Seis. Res. Lett. *69*, 144.

ENEVA, M. and BEN-ZION, Y. (1997), *Techniques and Parameters to Analyze Seismicity Patterns Associated with Large Earthquakes*, J. Geophys. Res. *102*, 17,785–17,795.

EVISON, F. F. (1977), *Fluctuations of Seismicity before Major Earthquakes*, Nature *266*, 957–972.

FERGUSON, C., KLEIN, W., and RUNDLE, J. B. (1999), *Spinodals, Scaling and Ergodicity in a Model of an Earthquake Fault with Long-range Stress Transfer*, Phys. Rev. E. *60*, 1359–1373.

FISHER, D. S., DAHMEN, K., RAMANATHAN, S., and BEN-ZION, Y. (1997), *Statistics of Earthquakes in Simple Models of Heterogeneous Faults*, Phys. Rev. Lett. *78*, 4885–4888.

FROHLICH, C. (1987), *Aftershocks and Temporal Clustering of Deep Earthquakes*, J. Geophys. Res. *92*, 13,944.

FUKUNAGA, K., *Introduction to Statistical Pattern Recognition* (Academic Press, N.Y., 1970).

GARCIA, A. and PENLAND, C. (1991), *Fluctuating Hydrodynamics and Principal Oscillation Pattern Analysis*, J. Stat. Phys. *64*, 1121–1132.

GELLER, R. J., JACKSON, D. D., KAGAN, Y. Y., and MULARGIA, F. (1997), *Enhanced: Earthquakes Cannot Be Predicted*, Science *278*, 488–490.

GOMBERG, J. (1996), *Stress/Strain Changes and Triggered Seismicity Following the M_W 7.3 Landers, California, Earthquake*, J. Geophys. Res. *101*, 751–764.

GRANT, L. and SIEH, K. (1994), *Paleoseismic Evidence of Clustured Earthquakes on the San Andreas Fault in the Carrizo Plain*, J. Geophys. Res. *99*, 6819–6842.

GRAY, C. M. (1997), *Synchronous oscillations in neuronal systems: Mechanisms and functions*. In *Pattern Formation in the Physical and Biological Sciences*, Lecture Notes V, SFI, Addison-Wesley, Reading, MA, 93–134.

GROSS, S. J. and KISSLINGER, C. (1994), *Tests of Models of Aftershock Rate Decay*, BSSA *84*, 1571–1579.

GROSS, S. and RUNDLE, J. B. (1998), *A Systematic Test of Time-to-failure Analysis*, Geophys. J. Int. *133*, 57–64.

HABERMAN, R. E. (1981), *Precursory seismicity patterns: Stalking the mature seismic gap*. In *Earthquake Prediction: An International Review*, AGU Monograph, AGU, Washington, D.C., 29–42.

HERTZ, J., KORGH, A., and PALMER, R. G. (1990), *Introduction to the Theory of Neural Computation*, Lecture Notes I, SFI, Addison-Wesley, Reading, MA.

HERZ, A. V. and HOPFIELD, J. J. (1995), *Earthquake Cycles and Neural Reverberations: Collective Oscillations in Systems with Pulse-coupled Threshold Elements*, Phys. Rev. Lett. *75*, 1222–1225.

HILL, D., EATON, J. P., and JONES, L. M. (1990), *Seismicity, 1980–86. The San Andreas Fault System, California*, USGS Prof. Paper 1515, U.S. GPO, Washington, D.C., 115–152.

HILL, D. P., JOHNSTON, M. J. S., LANGBEIN, J. O., and BILHAM, R. (1995), *Response of Long Valley caldera to the $M_W = 7.3$ Landers, California, Earthquake*, J. Geophys. Res. *100*, 12,985–13,005.

HILL, D. P., REASENBERG, P. A., MICHAEL, A., ARABAZ, W. J., BEROZA, G., BRUMBAUGH, D., BRUNE, J. N., CASTRO, R., DAVIS, S., DEPOLO, D., ELLSWORTH, W. L., GOMBERG, J., HARMSEN, S., HOUSE, L., JACKSON, S. M., JOHNSTON, M. J. S., JONES, L., KELLER, R., MALONE, S., MUNGUIA, L., NAVA, S., PECHMANN, J. C., SANFORD, A., SIMPSON, R. W., SMITH, R. B., STARK, M., STICKNEY, M., VIDAL, A., WALTER, S., WONG, V., and ZOLLWEG, J. (1993), *Seismicity Remotely Triggered by the Magnitude 7.3 Landers, California, Earthquake*, Science *260*, 1617–1623.

HOLMES, P., LUMLEY, J. L., and BERKOOZ, G., *Turbulence, Coherent Structures, Dynamical Systems and Symmetry* (Cambridge University Press, Cambridge, U.K., 1996).

HOTELLING, H. (1933), *Analysis of a Complex of Statistical Variables into Principal Components*, J. Educ. Psych. *24*, 417–520.

HOUSE, L. S., SYKES, L. R., DAVIES, J. N., and JACOB, K. H. (1981), *Identification of a possible seismic gap near Unalaska Island, eastern Aleutians, Alaska*. In *Earthquake Prediction: An International Review*, AGU Monograph, AGU, Washington, D.C., 81–92.

HUANG, Y., SALEUR, H., SAMMIS, C., and SORNETTE, D. (1998), *Precursors, Aftershocks, Criticality and Self-organized Criticality*, Europhys. Lett. *41*, 43–49.

JAUMÉ, S. C. and SYKES, L. R. (1999), *Evolving Towards a Critical Point: A Review of Accelerating Seismic Moment/Energy Release prior to Large and Great Earthquakes*, Pure Appl. Geophys. *155*, 279–306.

JONES, L. M. and HAUKSSON, E. (1997), *The Seismic Cycle in Southern California: Precursor or Response?* Geophys. Res. Lett. *24*(4), 469–472.

KAGAN, Y. Y. and JACKSON, D. D. (1992), *Seismic Gap Hypothesis, Ten Years After*, J. Geophys. Res. *96*, 21,419–21,431.

KANAMORI, H. (1981), *The nature of seismicity patterns before large earthquakes*. In *Earthquake Prediction: An International Review*, AGU Monograph, AGU, Washington, D.C., 1–19.

KATO, N., OHTAKE, M., and HIRASAWA, T. (1997), *Possible Mechanism of Precursory Seismic Quiescence: Regional Stress Relaxation due to Preseismic Sliding*. Pure Appl. Geophys. *150*, 249–267.

KERR, R. A. (1998), *A Slow Start for Earthquakes*, Science *279*, 985.

KING, G. C. P., STEIN, R. S., and LIN, J. (1994), *Static Stress Changes and the Triggering of Earthquakes*, BSSA *84*, 935–953.

KLEIN, W., RUNDLE, J. B., and FERGUSON, C. (1997), *Scaling and Nucleation in Models of Earthquake Faults*, Phys. Rev. Lett. *78*, 3793–3796.

LINDE, A. T., GLADWIN, M. T., JOHNSTON, M. J. S., GWYTHER, R. L., and BILHAM, R. G. (1996), *A Slow Earthquake Sequence on the San Andreas Fault*, Nature *383*, 65–68.

LYZENGA, G. A., RAEFSKY, A., and MULLIGAN, S. G. (1991), *Models of Recurrent Strike-Slip Earthquake Cycles and the State of Crustal Stress*, J. Geophys. Res. *96*, 21,623–21,640.

MAIN, I. G. (1999a), *Applicability of Time-to-failure Analysis to Accelerated Strain Release before Earthquakes and Volcanic Eruptions*, Geophys. J. Int. *139*, F1–F6.

MAIN, I. G. (1999b), *Is the Reliable Prediction of Individual Earthquakes a Realistic Scientific Goal?* Nature website debate, http://helix.nature.com/debates/earthquake/equake_frameset.html.

MARONE, C., VIDALE, J. E., and ELLSWORTH, W. L. (1995), *Fault Healing Inferred from Time-dependent Variations in Source Properties of Repeating Earthquakes*, Geophys. Res. Lett. *22*, 3095–3098.

MCGUIRE, J. J., IMHLE, P. F., and JORDAN, T. H. (1996), *Time-domain Observations of a Slow Precursor to the 1994 Romanche Transform Earthquake*, Science *274*, 82–85.

MOGHADDAM, B., WAHID, W., and PENTLAND, A. (1998), *Beyond Eigenfaces: Probabilistic Matching for face recognition*. In *Third IEEE Intl. Conf. on Automatic Face and Gesture Recognition*, 1–6.

MOGI, K. (1969), *Some Features of Recent Seismic Activity in and near Japan, (2) Activity before and after Great Earthquakes*, Bull. Earthquake Res. Inst., Tokyo Univ. *47*, 395–417.

MOGI, K. (1977), *Seismic activity and earthquake predictions*, Proc. Symp. on Earthquake Prediction, Seis. Soc. Japan, 203–214.

MOGI, K. (1979), *Two Kinds of Seismic Gaps*, Pure Appl. Geophys. *117*, 1172–1186.

MORI, H. and KURAMOTO, Y., *Dissipative Structures and Chaos* (Springer-Verlag, Berlin, 1998).

NANJO, K., NAGAHAMA, H., and SATOMURA, M. (1998), *Rates of Aftershock Decay and the Fractal Structure of Active Fault Systems*, Tectonophysics *287*, 173–186.

NIJHOUT, H. F. (1997), *Pattern formation in biological systems*. In *Pattern Formation in the Physical and Biological Sciences*, Lecture Notes V, SFI, Addison Wesley, Reading, MA, 269–298.

NORTH, G. R. (1984), *Empirical Orthogonal Functions and Normal Modes*, J. Atm. Sci. *41*(5), 879–887.

OUCHI, T. (1993), *Population Dynamics of Earthquakes and Mathematical Modeling*, Pure Appl. Geophys. *140*, 15–28.

PACHECO, J. F., SCHOLZ, C. H., and SYKES, L. R. (1992), *Changes in Frequency-size Relationship from Small to Large Earthquakes*, Nature *355*, 71–73.

PENLAND, C. (1989), *Random Forcing and Forecasting Using Principal Oscillation Pattern Analysis*, Monthly Weather Rev. *117*, 2165–2185.

PENLAND, C. and MAGORIAN, T. (1993), *Prediction of Niño 3 Sea Surface Temperatures Using Linear Inverse Modeling*, J. Climate *6*, 1067–1076.

PENLAND, C. and SARDESHMUKH, P. D. (1995), *The Optimal Growth of Tropical Sea Surface Temperature Anomalies*, J. Climate *8*, 1999–2024.

POLLITZ, F. F. and SACKS, I. S. (1997), *The 1995 Kobe, Japan, Earthquake: A Long-delayed Aftershock of the Offshore 1944 Tonankai and 1946 Nankaido Earthquakes*, BSSA *87*, 1–10.

PREISENDORFER, R. W., *Principle Component Analysis in Meteorology and Oceanography* (Elsevier, Amsterdam. 1988).

PRESS, F. and ALLEN, C. (1995), *Patterns of Seismic Release in the Southern California Region*, J. Geophys. Res. *100*, 6421–6430.

PRESS, W. H., FLANNERY, B. P., TEUKOLSKY, S. A., and VETTERING, W. T., *Numerical Recipes in C*, 2nd edit. (Cambridge University Press, Cambridge, U.K., 1992).

RICHTER, C. F., *Elementary Seismology*, (Freeman, San Francisco, 1958).

ROMANOWICZ, B. and RUNDLE, J. B. (1993), *On Scaling Relations for Large Earthquakes*, BSSA *83*, 1294–1297.

RUNDLE, J. B. (1988), *A Physical Model for Earthquakes: 2. Applications to Southern California*, J. Geophys. Res. *93*, 6255–6274.

RUNDLE, J. B. (1989), *Derivation of the Complete Gutenberg-Richter Magnitude-frequency Relation Using the Principle of Scale Invariance*, J. Geophys. Res. *94*, 12,337–12,342.

RUNDLE, J. B. (1993), *Magnitude-frequency Relation for Earthquakes Using a Statistical Mechanical Approach*, J. Geophys. Res. *98*, 21,943–21,950.

RUNDLE, J. B., GROSS, S., KLEIN, W., FERGUSON, C., and TURCOTTE, D. L. (1997), *The Statistical Mechanics of Earthquakes*, Tectonophysics *277*, 147–164.

RUNDLE, J. B. and KLEIN, W. (1995), *New Ideas about the Physics of Earthquakes*, Rev. Geophys. Space Phys. Suppl. (July) *283*, 283–286.

RUNDLE, J. B., KLEIN, W., and GROSS, S. (1999), *Physical Basis for Statistical Patterns in Complex Earthquake Populations: Models, Predictions and Tests*, Pure Appl. Geophys. *155*, 575–607.

RUNDLE, J. B., KLEIN, W., TIAMPO, K., and GROSS, S. (2000a), *Linear Pattern Dynamics in Nonlinear Threshold Systems*, Phys. Rev. E. *61*, 2418–2432.

RUNDLE, J. B., KLEIN, W., GROSS, S., and TIAMPO, K. F. (2000b), *Dynamics of seismicity patterns in systems of earthquake faults*. In *Geocomplexity and the Physics of Earthquakes*, AGU Monograph, Washington, D.C., 127–146.

SALEUR, H., SAMMIS, C. G., and SORNETTE, D. (1995), *Discrete Scale Invariance, Complex Fractal Dimensions, and Log-periodic Fluctuations in Seismicity*, J. Geophys. Res. *100*, 17,661–17,677.

SAVAGE, J. C. (1988), *Principal Component Analysis of Geodetically Measured Deformation in Long Valley Caldera, Eastern California, 1983–1987*, J. Geophys. Res. *93*, 13,297–13,305.

SAVAGE, J. C. (1993), *The Parkfield Prediction Fallacy*, BSSA *83*, 1–6.

SCHOLZ, C. H., *The Mechanics of Earthquakes and Faulting* (Cambridge University Press, Cambridge, U.K., 1990).

SCHWARTZ, D. P. and COPPERSMITH, K. J. (1984), *Fault Behavior and Characteristic Earthquakes – Examples from the Wasatch and San Andreas Fault Zones*, J. Geophys. Res. *89*, 5681–5698.

SIEH, K., STUIVER, M., and BRILLINGER, D. (1989), *A More Precise Chronology of Earthquakes Produced by the San Andreas Fault in Southern California*, J. Geophys. Res. *94*, 603–623.

STARK, M. A. and DAVIS, S. D. (1996), *Remotely Triggered Microearthquakes at The Geysers Geothermal Field, California*, Geophys. Res. Lett. *23*, (9), 945–948.

STEIN, R. S. (1999), *The Role of Stress Transfer in Earthquake Occurrence*, Nature *402*, 605–609.

STEIN, R. S., KING, G. C. P., and LIN, J. (1992), *Change in Failure Stress on the Southern San Andreas Fault System Caused by the 1992 Magnitude = 7.4 Landers Earthquake*, Science *258*, 1328–1331.

SWAN, F. H., SCHWARTZ, D. P., and CLUFF, L. S. (1980), *Recurrence of Moderate to Large Magnitude Earthquakes Produced by Surface Faulting on the Wasatch Fault Zone, Utah*, BSSA *70*, 1431–1462.

TIAMPO, K. F., RUNDLE, J. B., KLEIN, W., and GROSS, S. J. (1999), *Systematic Evolution of Nonlocal Space-time Earthquake Patterns in Southern California*, Eos Trans. AGU *80*, 1013.

TIAMPO, K. F., RUNDLE, J. B., KLEIN, W., McGINNIS, S., and GROSS, S. J., *Observation of Systematic Variations in Non-local Seismicity Patterns from Southern California*. In *Geocomplexity and the Physics of Earthquakes*, AGU Monograph, Washington, D.C. (2000).

TURCOTTE, D.L., *Fractals and Chaos in Geology and Geophysics*, 2nd edition (Cambridge University Press, Cambridge, U.K. 1997).

VAUTARD, R. and GHIL, M. (1989), *Singular Spectrum Analysis in Nonlinear Dynamics, with Applications to Paleodynamic Time Series*, Physica D *35*, 395–424.

WYSS, M. and HABERMAN, R. E. (1988), *Precursory Quiescence before the August 1982 Stone Canyon, San Andreas Fault, Earthquakes*, Pure Appl. Geophys. *126*, 333–356.

WYSS, M., SCHORLEMMER, D., and WIEMER, S. (2000), *Mapping Asperities by Minima of Local Recurrence Time: San Jacinto-Elsinore Fault Zones*, J. Geophys. Res. *105*, 7829–7844.

WYSS, M., SHIMAZIKI, K., and URABE, T. (1996), *Quantitative Mapping of a Precursory Seismic Quiescence to the Izu-Oshima 1990 (M 6.5) Earthquake, Japan*, Geophys. J. Int. *127*, 735–743.

WYSS, M. and WIEMER, S. (1999), *How Can One Test the Seismic Gap Hypothesis? The Case of Repeated Ruptures in the Aleutians*, Pure Appl. Geophys. *155*, 259–278.

YAMASHITA, T. and KNOPOFF, L. (1989), *A Model of Foreshock Occurrence*, Geophys. J. *96*, 389–399.

(Received February 20, 2001, revised June 11, 2001, accepted June 15, 2001)

To access this journal online:
http://www.birkhauser.ch

Pure appl. geophys. 159 (2002) 2469–2490
0033–4553/02/102469–22 $ 1.50 + 0.20/0

© Birkhäuser Verlag, Basel, 2002

| Pure and Applied Geophysics

Long-range Automaton Models of Earthquakes: Power-law Accelerations, Correlation Evolution, and Mode-switching

DION WEATHERLEY,[1] PETER MORA,[2] and MENG FEN XIA[3]

Abstract—We introduce a conceptual model for the in-plane physics of an earthquake fault. The model employs cellular automaton techniques to simulate tectonic loading, earthquake rupture, and strain redistribution. The impact of a hypothetical crustal elastodynamic Green's function is approximated by a long-range strain redistribution law with a r^{-p} dependance. We investigate the influence of the effective elastodynamic interaction range upon the dynamical behaviour of the model by conducting experiments with different values of the exponent (p). The results indicate that this model has two distinct, stable modes of behaviour. The first mode produces a characteristic earthquake distribution with moderate to large events preceeded by an interval of time in which the rate of energy release accelerates. A correlation function analysis reveals that accelerating sequences are associated with a systematic, global evolution of strain energy correlations within the system. The second stable mode produces Gutenberg-Richter statistics, with near-linear energy release and no significant global correlation evolution. A model with effectively short-range interactions preferentially displays Gutenberg-Richter behaviour. However, models with long-range interactions appear to switch between the characteristic and GR modes. As the range of elastodynamic interactions is increased, characteristic behaviour begins to dominate GR behaviour. These models demonstrate that evolution of strain energy correlations may occur within systems with a fixed elastodynamic interaction range. Supposing that similar mode-switching dynamical behaviour occurs within earthquake faults then intermediate-term forecasting of large earthquakes may be feasible for some earthquakes but not for others, in alignment with certain empirical seismological observations. Further numerical investigation of dynamical models of this type may lead to advances in earthquake forecasting research and theoretical seismology.

Key words: Critical point hypothesis, cellular automata, correlation evolution.

[1] QUAKES, The Univ. of Queensland, Brisbane, 4072, Australia.
E-mail: weatherley@quakes.earth.uq.edu.au
[2] QUAKES, The Univ. of Queensland, Brisbane, 4072, Australia.
E-mail: mora@quakes.earth.uq.edu.au
[3] State Key Laboratory of Nonlinear Mechanics, Institute of Mechanics, Chinese Academy of Sciences, Beijing, 100080, China. E-mail: xiam@lnm.imech.ac.cn, Department of Physics, Peking University, Beijing, 100871, China.

Introduction

It is well known that the regional or global distribution of earthquake magnitudes empirically obeys the Gutenberg-Richter law (GUTENBERG and RICHTER, 1956), a power-law event size distribution of the form:

$$N(M) = K_M M^{-b}$$

where K_M is an empirical constant and b is the so-called regional or global b value. Empirical values for the b value typically fall in the range $0.8 < b < 1.2$. If seismic energy is employed in place of magnitude, the corresponding empirical exponent is $b \sim 2/3$ (WU, 2000).

Seismologists became interested in cellular automata as possible analogues of earthquake fault dynamics when BAK and TANG (1989) demonstrated that even highly simplified, nearest neighbour automata produce power-law event size distributions. The sandpile automaton of BAK *et al.* (1987) was employed to demonstrate the emergence of power-law statistics as the result of the combined action of a large number of simple elements which interact only with nearby cells via a prespecified interaction rule. BAK *et al.* (1987) termed such behaviour Self-Organised Critical (SOC) behaviour.

SOC generated considerable debate amongst earthquake forecasting researchers. The SOC state is considered to be a minimally-stable state, characterised by a high mean energy in the system with very small mean energy fluctuations (BAK *et al.*, 1987). Events occur at random and are distributed according to a power-law size-distribution. Assuming the earth's crust is in an SOC state, GELLER *et al.* (1997) argued that earthquakes cannot be predicted. In response, WYSS *et al.* (1997) cited a variety of evidence for precursory activity prior to catatrophic failure to argue that further research is required before earthquake forecasting is proven to be unfeasible.

One type of precursory activity which has driven the forecasting debate in recent years is that of accelerating seismic moment release. BUFE and VARNES (1993) demonstrated that the 1989 Loma Prieta, California earthquake was preceded by an interval of about 30–50 years in which the rate of seismic activity accelerated. Employing cumulative Benioff strain release (square-root of the seismic moment release) as a measure of seismic activity, the researchers utilised an empirical time-to-failure relationship to make a retrospective forecast of the time-of-occurrence of the $M_w = 6.9$ Loma Prieta earthquake with a precision of 2 months. The earthquake magnitude was predicted to be $M_{pred} = 6.83$ (BUFE and VARNES, 1993). Subsequently, accelerating moment release has been identified prior to a considerable number of moderate to large earthquakes from a variety of tectonic settings. JAUME and SYKES (1999) presented a review of reported cases of accelerating moment release.

Power-law accelerations in energy release are an important feature of a class of dynamical systems known as Critical Point (CP) systems. In statistical physics,

Renormalisation Group Theory was developed to describe CP systems. This theory identifies accelerating energy release as the signature of the progressive development of long-range correlations in a physical property of the system (SORNETTE and SAMMIS, 1995). This concept of correlation evolution became the basis for the Critical Point Hypothesis (CPH) of earthquakes (SALEUR et al., 1996). Under this hypothesis, an earthquake rupture may only occur when long-range stress correlations exist along the length of a fault. The progressive formation of such correlations is driven by tectonic loading and the redistribution of stress due to smaller earthquakes in the neighbourhood of the fault. Earthquake rupture destroys the long-range stress correlations, returning the fault to a state far-from-failure until a sufficiently correlated stress field is reformed via the action of subsequent, smaller earthquakes.

Renormalisation group theory and cellular automata have a number of features in common. The earthquake forecasting debate fueled considerable research employing both these techniques (see MAIN, 1996 for a review). The majority of automata studied were nearest-neighbour models; interactions only occur between cells and those adjacent to them. WEATHERLEY et al. (2000) found that nearest neighbour automata may display behaviour in which the size of highly correlated regions increases prior to large events. However in these models, the formation of such correlated regions is driven predominantly by external (tectonic) loading rather than strain redistribution due to smaller to moderate events, as predicted by the CPH. SAMMIS and SMITH (1998) did identify correlation evolution (consistent with the CPH) in an automaton consisting of a discrete fractal heirarchy of cells, including one macroscopically large cell which failed during the largest events.

Applying Damage Mechanics principles, XIA et al. (1999) developed a 1-D fibre-bundle model displaying behaviour qualitatively similar to CP behaviour, known as Evolution Induced Catastrophy (WEI et al., 2000). In this model, strain energy from failed fibres was transferred to fibres within a local region whose size may be predefined or may depend upon the local pattern of damage. Unlike nearest neighbour models, the longer-range transfer of strain in the fibre-bundle model resulted in evolution qualitatively similar to CP behaviour.

KLEIN et al. (2000) examined a long-range automaton which is based upon a cellular automaton equivalent of the Burridge-Knopoff block-slider model (RUNDLE and JACKSON, 1997; RUNDLE and BROWN, 1991). KLEIN et al. employed a discretised stress Green's function with a r^{-3} dependance to characterise the long-range interactions. Long-range springs join blocks to a large number of surrounding blocks and r is defined as the distance between two blocks which are attached to one-another. This model results in Gutenberg-Richter event size-distributions with a b value of $b = 1.5$, in agreement with a coarse-grained, mean-field description of the model derived by KLEIN et al. (2000). Scaling of events in these models is associated with nucleation near a mean-field spinodal rather than critical point fluctuations (KLEIN et al., 1997).

Our motivation for designing the model presented in this paper, was to determine whether dynamical behaviour consistent with the CPH may be obtained in 2-D automata employing long-range transfer of strain energy from failed cells. A conceptual approach has been employed to design a model for the evolution of strain energy supported by contacting asperities within the plane of a fault. A 2-D grid of cells represents the contacting asperities. Each cell is predefined with a constant fracture energy, and has a variable strain energy. Tectonic loading of the fault is simulated by increasing the strain energy of all cells by a small amount each loading timestep. When the strain energy of a cell exceeds its fracture energy, the strain of the cell is transferred to cells within a square transfer region surrounding the failed cell.

An elastodynamic interaction function (similar to a Green's function) with a r^{-p} dependance determines the amount of strain energy transferred to cells a distance r from the failed cell. The form of the interaction function was chosen because linear elasticity predicts that, for a 2-D dislocation in a 3-D homogeneous elastic medium, the magnitude of the stress Green's function (T) decays as $T \sim r^{-3}$. Based upon this, one might expect $p = 3$ is a good choice of interaction exponent (or perhaps $p = 2$ for the 2-D case). However as mentioned by KLEIN *et al.* (2000), microcracks, fault gouge, and pore-fluids may mask the r^{-3} interaction, resulting in effectively longer-range static stress interactions. In addition, crack propagation theory predicts $1/\sqrt{r}$ dynamic stress concentrations at the rupture front of propagating cracks. MORA and PLACE (2002) report that correlation evolution can occur in particle-based elasto-dynamic systems, suggesting that $p \leq 1$ may be more appropriate for the interaction exponent. Furthermore, for elasto-plastic cases a $\log r$ term may appear. Hence, it is unclear on physical grounds what is the appropriate exponent for the interaction function. For this reason, we implement the model for an arbitrary choice of interaction exponent and then perform numerical experiments employing various values of the interaction exponent in the range $0.2 < p < 3$.

It is important to emphasise that the elastodynamic interaction range as defined by the interaction function (or the stress Green's function) is *not* equivalent to the correlation range as defined in the CPH. The CPH proposes that correlations in some physical property (stress, strain, strain energy, or other) evolve in the lead-up to a large event. This evolution is characterised by an increase in the average size of regions with correlated properties. Hence, it is often stated that the correlation range increases as a catastrophic earthquake becomes imminent. However, the elasto-dynamic interaction range defined by the stress Green's function of the Crust, is not expected to evolve in the lead-up to a large event. In our model, the interaction exponent is held constant during any given simulation and so, undergoes no evolution.

The results which follow testify that, even with a fixed elastodynamic interaction range, correlation evolution may occur prior to large events. It is apparent that the interaction exponent (p) is a tuning parameter governing the mode of dynamical behaviour displayed by the model. Small interaction exponents ($p < 1$) corresponding to effectively long-range interactions, result in dynamics dominated by characteristic, large earthquakes. These earthquakes are preceded by accelerating strain energy release, associated with global evolution of strain energy correlations within the system. Systems with intermediate-range interactions ($1 < p < 1.5$) appear to switch between this characteristic mode and a dynamical mode characterised by linear energy release and no significant correlation evolution. Linear energy release eventually dominates the dynamical behaviour of the model, as $p \to 2$. Implications of these results for theoretical seismology and earthquake forecasting research are discussed in the sections following the results.

Model Description

Basic Features of the Model

The model consists of a regular, square lattice containing N^2 cells. Each cell $i = 1, \ldots, N^2$ is assigned a variable, scalar strain energy ε_i which is initially set to zero. In addition, each cell is also assigned a constant, scalar fracture energy, ε_{fi} in the range $0.1 < \varepsilon_{fi} < 1.0$.

The fracture energies of cells are constrained to a statistical fractal distribution with a dimension, $D = 2.3$. The statistical fractal is generated using a well-known Fourier filtering technique (TURCOTTE, 1997). The value of the fractal dimension was chosen because "fractal statistics with a fractal dimension near 2.2 are a good approximation for many geological processes" (HUANG and TURCOTTE, 1988). This is based upon detailed studies of the earth's topography.

The strain energy of all cells is periodically increased by an amount $\Delta\varepsilon$, known hereafter as the strain increment. The value of the strain increment is determined dynamically as the minimum strain energy required to fail at least one cell in the model. Failure of a cell occurs when the strain energy of the cell equals or exceeds its fracture energy, i.e., $\varepsilon_i \geq \varepsilon_{fi}$. Strain redistribution from failed cells is described in detail below.

An external loading mechanism of this form is often employed in automata to simulate slow tectonic loading in the models. Use of a variable strain increment ($\Delta\varepsilon$) motivates the introduction of a model timescale (t_l) which is proportional to the strain increment. The model time ($t_\ell(n)$) of the n-th loading increment is given by

$$t_\ell(n) = \sum_{j=1}^{n} K\Delta\varepsilon(j) \; .$$

Defined in this manner, the model time ensures a constant rate of external loading in accord with tectonic loading of earthquake faults. Since model units are arbitrary, we set $K = 1$. Temporal statistics such as energy release are analysed with respect to model time rather than the number of loading timesteps.

Strain Redistribution

The strain energy of a failed cell is redistributed to neighbouring cells in the grid. In classical cellular automata, the strain energy of failed cells is transferred to the 4 (or 8) nearest (or next nearest) neighbours in the grid. Redistribution of strain energy may trigger the failure of neighbouring cells. These cells in turn, redistribute their strain energy possibly triggering further cell failures. It is this mechanism of cascading ruptures which results in power-law event-size statistics similar to those of natural earthquakes.

The models examined in this paper employ a strain redistribution mechanism which is a generalisation of the classical nearest neighbour mechanism. Unlike classical cellular automata, the strain energy from failed cells is redistributed to all cells within a square region ($s \times s$) surrounding the failed cell. This region is called the transfer (or interaction) region. The fraction of the released strain energy transferred to a given cell k ($= 1, \ldots, s^2$) is a function of the Cartesian distance of cell k from the failed cell at (x_f, y_f). To achieve this, each cell (k) within the transfer region is assigned an interaction coefficient, I_k:

$$I_k = \frac{r_k^{-p}}{I_o} \quad , \tag{1}$$

where p is the interaction exponent. If the strain energy of the failed cell is ε_f, then the strain energy transferred to cell k is given by $\varepsilon_f I_k$. The normalisation constant, $I_o = \sum_{k=1}^{s^2-1} r_k^{-p}$ ensures that strain energy is conserved during strain redistribution i.e.,

$$\sum_{k=1}^{s^2-1} I_k = 1 \ .$$

The numerical experiment described in the following section consists of a parameter space investigation in which the interaction exponent is varied in the range $0.2 < p < 3$. We chose to vary the interaction exponent because, as discussed in the Introduction, it is unclear on physical grounds what is an appropriate value for the interaction exponent. As the exponent decreases, the rate of decay of the interaction function decreases. In terms of strain redistribution, this translates into a greater fraction of the released strain being transfered to larger distances in the model (as the exponent p decreases). In other words, the effective range of elastodynamic interactions increases, as p decreases.

In the simulations described below, only non-zero values of p are employed. The transfer region size (s) is chosen so that the minimum strain redistributed is small compared with the minimum cell fracture energy (0.1) in most simulations. Under such circumstances, the finite size (and shape) of the transfer region has a negligible impact upon strain redistribution; the interaction function determines the spatial redistribution of strain energy.

Cell Healing and Dissipation

In order to simulate seismic cycles, a method for healing failed cells must be specified. During the design phase of the model, a variety of different cell healing mechanisms were tested. These included mechanisms in which failed cells rapidly heal and may support strain energy transferred from the rupture front, and slow-healing mechanisms in which failed cells cannot support strain until after the rupture has completed. Rapid healing of cells resulted in two features which the authors considered undesirable on physical grounds. The first is that cells may fail multiple times during a given rupture. While earthquake rupture is a complex phenomenon, it is unlikely that a given portion of the rupture surface re-activates multiple times during earthquake rupture.

The second undesirable feature of rapid healing mechanisms is that very small mean energy fluctuations often result. Past research (WEATHERLEY et al., 2000; JAUMÉ et al., 2000) has shown that CP-like behaviour is most evident in models displaying large mean energy fluctuations. Linear energy release and no significant global correlation evolution corresponds to models displaying very small mean energy fluctuations. Thus, motivated by the physical grounds outlined above and a desire to simulate critical point behaviour, we employ the following slow-healing mechanism.

The strain energy of a failed cell is reduced to zero and all of the stored strain is redistributed within the cell's transfer region i.e., no bulk dissipation of strain energy is employed in this model. Once failed, a cell cannot support strain energy until completion of the rupture which failed that cell; cells do not fail multiple times during a given rupture. Strain energy transferred to failed cells is dissipated from the model. As a result, subsequent to any given rupture, the strain energy of cells within the rupture zone is zero. This is somewhat analogous to a frictional instability in which the stress of slipping fault patches decreases to a constant, residual level regardless of the frictional strengths of the patches.

One possible interpretation for the dissipative healing mechanism employed, is that the energy dissipated by failed cells represents the strain energy lost due to other processes within the fault zone such as frictional heating, and pore-fluid depressurisation. Perhaps one can argue that such a mechanism is qualitatively justified on physical grounds. The results which follow show that this dissipative healing mechanism results in long-term, stable dynamical behaviour including global correlation evolution consistent with the Critical Point Hypothesis.

As in many previous automata, open boundary conditions are employed. All strain energy transferred across external boundaries is dissipated from the model. The transfer and dissipation of strain energy across model boundaries may be considered analogous to the loss of strain energy from a fault zone due to transport of strain (during rupture) to another portion of the fault zone outside the model region. While some may argue for closed boundaries, in the authors' experience this often leads to highly conservative models which evolve towards a state of SOC quite unlike the dynamics of the models described here, unless a large portion of released strain ($> 50\%$) is lost via bulk dissipation.

Numerical Experiment and Empirical Results

The model specified above is arbitrary in the choice of interaction exponent (p) and transfer region size (s). In a previous study of long-range automata the transfer region size was varied in order to characterise the change in system behaviour as the number of cells which receive strain energy from failed cells is increased (KLEIN et al., 2000). These simulations employed a constant interaction exponent (usually $p = 3$) in a model with a fast healing, bulk dissipation mechanism. Under these conditions, increasing the transfer region size is considered equivalent to increasing the interaction range within the system (KLEIN et al., 2000).

As discussed earlier, we wish to explore a hypothetical 'parameter space' of models with differing interaction exponents (p). In this we hope to determine whether the evolution of correlated regions of strain energy is affected by the fixed, effective elastodynamic interaction range within the system. Thus, for each simulation we employ a fixed transfer region size $s = 59$ and vary the value of p.

The transfer region size is chosen so that the finite size of the region does not result in significant truncation of the interaction function (Equation 1) for simulations employing an interaction exponent in the range, $0.5 < p < 3$. Simulations employing interaction exponents in the range $0.2 < p < 0.5$ are also considered, however these simulations suffer some truncation by the square transfer region as the interaction exponent, $p \rightarrow 0.2$. This truncation gives rise to obvious undesirable features in the dynamical behaviour for simulations employing very small exponent values.

The same fracture energy distribution is employed for all simulations. Preliminary studies with various fracture energy distributions and various fractal dimensions have shown that the dynamical behaviour illustrated here, is relatively robust with respect to the choice of fracture energy distribution and range of fracture energies.

A total of 50 simulations were performed, each with a unique interaction exponent chosen from the set $p \in \{0.2 + 0.05i \| i = 0, 1, \ldots, 56\}$. In each simulation, 10^6 loading timesteps were computed, resulting in a synthetic earthquake catalogue consisting of 10^6 events. The simulations were performed on a SGI Origin 3800 supercomputer with 16×400 MHz R12000 CPUs. To compute 10^6 loading timesteps and record all relevent model data typically consumes one hour of computation time on a single CPU of this supercomputer. The results presented in the following subsections often involve considerable post-processing of the recorded model data. In particular, to compute the correlation function of the strain energy field for only 500 timesteps required approximately five hours of CPU time. This placed obvious restrictions upon the depth of the investigation into strain energy correlation evolution.

In each of the following subsections, various aspects of the models and their dynamical behaviour are examined in detail. These serve to illustrate both the dynamical behaviour of the model for a given interaction exponent, and the change in dynamical behaviour as the interaction exponent is varied. A summary of the results, some discussion, and possible implications for seismology are discussed subsequent to the numerical results.

Rupture Properties and Event Statistics

Strain redistribution from failed cells may trigger the failure of nearby cells which in turn, trigger failure of other cells. In this fashion, a cascading chain reaction can result in the failure of a macroscopic portion of the model. The pattern of failed cells for a typical large event from a simulation with $p = 1.2$ is shown in Figure 1a along with snapshots of the strain energy in the model immediately before (Fig. 1b) and subsequent to (Fig. 1c) this event. The regions where the most cells failed in the event correspond to the mostly highly strained regions prior to the event. It is also evident that large events are capable of failing regions with a relatively low strain prior to the events; provided these regions reside near a highly strained cluster of cells which fail during the large event.

The size of an event in an automaton simulation is often defined as the number of cells which fail during a single rupture cascade. This is sufficient if all cells have the same fracture energy (i.e., in a uniform automaton). However, for automata with heterogeneous fracture energies such as those considered here, a more appropriate definition of the event size (E) is the sum of the strain energy redistributed from all failed cells during the rupture cascade. The event size includes all strain energy transferred to other cells in the model and the strain energy dissipated from the model during the event.

Seismologists became interested in cellular automata when BAK and TANG (1989) showed that even highly simplified automata give rise to complex dynamics resulting in a power-law event-size distribution not unlike the empirical Gutenberg-Richter distribution. Cumulative event-size distributions for simulations with various interaction exponents may be compared in Figure 2a and interval event-size distributions for simulations with $p = 1, 2, 3$ are given in Figures 2b–d, respectively.

It is apparent that simulations with a relatively large effective interaction range (small p) produce an overabundance of large events. This overabundance diminishes as the interaction range decreases. From the interval distributions, it is evident that a power-law (or Gutenberg-Richter) event-size distribution is obtained for $p \sim 2$. For this value of p, the slope of the interval distribution is approximately 1.5, the same value as the theoretical prediction by KLEIN et al. (2000) for their similar Burridge-Knopoff, long-range automata. For interaction exponents $p > 2$, the interval event-size distribution begins to roll-over, indicating that these simulations have an underabundance of moderate to large events. Little study of models with $p > 2$ has

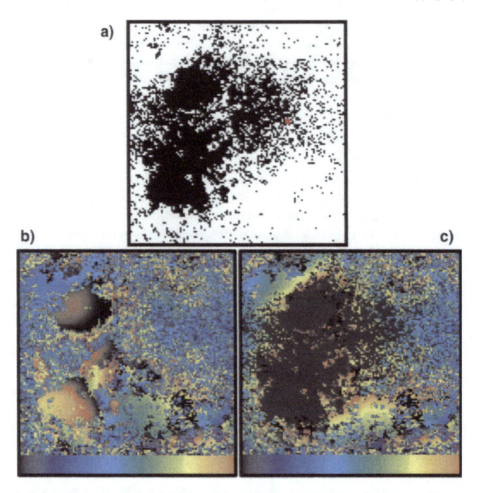

Figure 1
(a) Damage pattern for a large event in a simulation with $p = 1.2$. The event (whose epicenter is denoted by
a red square) failed a total of 5896 cells, releasing 3360.7 units of strain energy. Snapshots of the strain
energy in the model (b) before and (c) after this event. The snapshots illustrate that prior to large events,
there is considerable strain energy concentrated in the regions where the most cells fail. In the snapshots,
hotter (red) colours represent the highest strain energies (~ 1.0) and cooler (blue) represents the lowest
strain energies (~ 0.1).

been performed, although cursory investigation reveals a significant qualitative
change in rupture dynamics. Such behaviour may be of interest to researchers in
fields other than seismology.

Energy Conservation and Temporal Fluctuations

Each loading timestep, three processes alter the strain energy stored by cells in the
model. These are the loading mechanism, strain redistribution during failure, and the

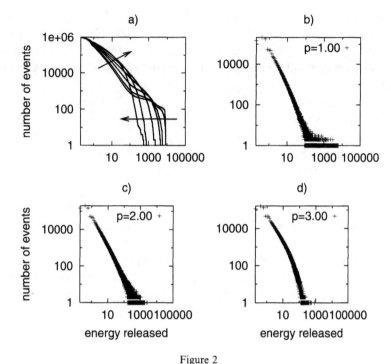

Figure 2

(a) Cumulative event-size statistics for simulations with a range of interaction exponents, $p \in [0.2, 3.0]$. Arrows indicate the direction of increasing p. (b–d) Interval distributions for simulations with $p = 1.0, 2.0, 3.0$, respectively. Notice that simulations with a longer effective interaction range ($p = 1.0$) tend to display an overabundance of large events. GR statistics with a slope of approximately 1.5 are obtained for $p = 2.0$.

resultant energy dissipation. The total stored strain energy (ε_T^n) at the end of the n-th timestep, may be expressed in terms of the total stored strain energy at the beginning of the n-th timestep (i.e., ε_T^{n-1}), the total increase in strain energy due to external loading (ε_L^n), and the total energy dissipated (ε_D^n) during the rupture. This relationship expresses the condition for energy conservation during a given loading timestep:

$$\varepsilon_T^n = \varepsilon_T^{n-1} + \varepsilon_L^n - \varepsilon_D^n .$$

Since the strain energy of each cell is incremented by $\Delta\varepsilon(n)$ in the nth timestep, $\varepsilon_L^n = N^2 \Delta\varepsilon(n)$. Also, by the definition of the mean stored strain energy $\varepsilon_T^n = N^2 \bar{\varepsilon}^n$. Using these substitutions, the energy conservation relation may be written in terms of the mean strain energy, the strain increment and the energy dissipated, i.e.,

$$\bar{\varepsilon}^n = \bar{\varepsilon}^{n-1} + \Delta\varepsilon(n) - \frac{\varepsilon_D^n}{N^2} . \tag{2}$$

Evidently, the mean strain energy of the system increases monotonically whilst the energy dissipated is less than the energy increase due to external loading (i.e.,

$\Delta\varepsilon(n) - \frac{\varepsilon_D^n}{N^2} > 0$). However, the mean energy will drop by an amount $\frac{\varepsilon_D^n}{N^2} - \Delta\varepsilon(n)$, if the energy dissipated in an event exceeds the loading energy. It may be concluded that the mean energy of the system should display a sawtooth behaviour, increasing gradually while only small events occur and dropping suddenly at the occurence of a large event.

Indicative time-series of the mean strain energy of the system are shown in Figures 3a–c for $p = 0.4, 1.2, 2.0$, respectively. These time-series consist of approximately 300,000 events and do not include transient behaviour due to the initial conditions; the time-series illustrate the long-term, stable dynamical behaviour of the model for various interaction ranges. The sawtooth mean energy evolution is particularly evident for the curves with the largest amplitude of fluctuations (Fig. 3a).

The temporal average of the mean strain energy at each timestep also displays a dependance upon the interaction exponent (Fig. 4a). Three regimes are apparent as the interaction exponent increases. The first regime ($p < 0.5$) contains models suffering truncation of the interaction function by the finite transfer region. The lower $\bar{\varepsilon}$ in this regime may be a reflection of the finite truncation however this is yet to be verified. During the second regime ($1/2 < p \le 2$), the average mean energy remains relatively constant around $\bar{\varepsilon} \sim 0.268$. In the final regime ($p \to 3$), the average mean energy rapidly decreases once more.

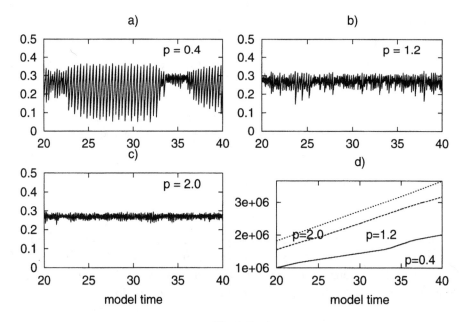

Figure 3
(a–c) Time-series of mean energy at the end of each timestep for approximately 300,000 timesteps of simulations with $p = 0.4, 1.2, 2.0$, respectively. (d) Corresponding plots of cumulative energy release from the same simulations. For models with a longer effective interaction range, mode-switching behaviour is clearly evident, diminishing as $p \to 2.0$.

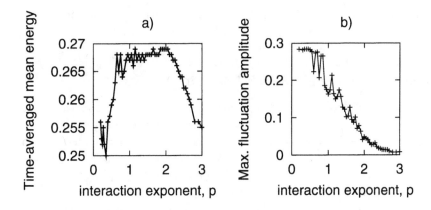

Figure 4
(a) Time-averaged mean energy for simulations with differing interaction exponents, p. (b) Estimates of the maximum amplitude of mean energy fluctuations for the same simulations.

In addition, a relationship between the amplitude of mean energy fluctuations and the interaction exponent, is clearly evident in Figures 3a–c. In an attempt to quantify this relationship, we have computed the maximum fluctuation amplitude for each simulation. This is achieved by recording the total energy dissipated (ε_D^n) and the strain increment ($\Delta\varepsilon(n)$) each timestep. For all the simulations, the range of strain increment sizes is approximately $0 < \Delta\varepsilon < 10^{-3}$, after the initial 100,000 timesteps. Recognising that the majority of the smaller events occurring in the simulations do not dissipate an appreciable amount of energy, the range in energy dissipated per timestep is $0 < \varepsilon_D \leq \varepsilon_{Dmax}(p)$.

The maximum energy dissipated depends upon the maximum event size which is dependant upon the choice of interaction exponent (see Fig. 2). We take $(\varepsilon_{Dmax(p)}/N^2)$ as a measure of the maximum fluctuation amplitude for a simulation with interaction exponent p (c.f. Equation 2). The maximum fluctuation amplitude displays a decreasing trend when plotted versus the interaction exponent (Fig. 4b).

Mode-switching Behaviour

Time-series of cumulative energy release are given in Figure 3d. These correspond to the mean energy time-series in Figures 3a–c. For the simulation with $p = 0.4$, two phases with a differing mean rate of energy release may be identified. Long, cyclic phases with large mean energy fluctuations and a relatively lesser mean rate of energy release are separated by relatively short intervals with a higher mean rate of energy release. The higher rate of energy release coincides with intervals in which the model displays relatively small mean-energy fluctuations (c.f. Fig. 3a). Such mode-switching behaviour is evident in all simulations with sufficiently large effective interaction range (with interaction exponents smaller than $p \sim 1.5$). However, as the interaction exponent increases, mode-switching becomes less obvious and linear energy release begins to dominate.

Accelerating Rate of Energy Release Prior to Large Events

Visual examination of cumulative energy release time-series revealed qualitative evidence for power-law accelerating sequences. This prompted an investigation to determine whether on average, the rate of energy release accelerates in the interval preceding large events in the model. We performed a simplified time-to-failure analysis of the cumulative energy release to determine the probability that a macroscopically large event is preceded by an interval in which the rate of energy release accelerates. We were particularly interested in whether the system-wide (global) energy release accelerates in the lead-up to a large event.

A power-law time-to-failure relationship is fit to the cumulative energy release in the interval preceding the 500 largest events in each of a selection of simulations. Following the methodology of BUFE and VARNES (1993), a power-law time-to-failure relationship of the following form is fit to energy release prior to large events:

$$\sum_{t=(t_f - \Delta t)}^{t < t_f} E(t) = A + B(t_f - t)^{-m}, \quad (m > 0) \tag{3}$$

where t_f is taken as the model time at which the large event occurs and A, B are empirical constants. The interval (Δt) and the power-law exponent (m) are varied in the ranges $0.1 < \Delta t < 0.4$ and $0.1 < m < 1.0$, and a goodness-of-fit parameter $(\mathscr{C})$ is computed for each choice of interval length and exponent value. A minimum interval length of $\Delta t = 0.1$ was imposed to ensure that all sequences contained at least 100 events.

The goodness-of-fit $(\mathscr{C})$ is defined as the ratio of the RMS error of the power-law fit to the RMS error of a linear fit for the same time interval. This parameter is chosen to maintain consistancy with past researchers examining natural accelerating seismic energy sequences (BUFE and VARNES, 1993). The fit with the smallest $\mathscr{C}$ is selected as the best power-law fit to the cumulative energy release for the given event, and the interval length, power-law exponent, and goodness-of-fit is recorded.

In the analysis, frequency histograms of the number of fits with given fit exponents (m) and interval lengths (Δt) were examined in addition to histograms of goodness-of-fit such as those of Figure 5. These histograms illustrate the general trend observed in the analysis: power-law fits to energy release are obtained more frequently in simulations with smaller values of p (i.e., with effectively longer-range elastodynamic interactions). Linear energy release is referred for simulations with a shorter interaction range.

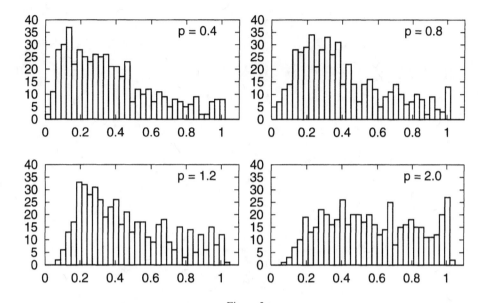

Figure 5

Frequency histograms of the number of power-law fits with a given goodness-of-fit, $\mathscr{C}$ for simulations with the given interaction exponent values. For small values of p, a large fraction of fits have $\mathscr{C} < 1$ indicating good power-law fits to energy release sequences. This population is progressively replaced by predominantly linear energy release sequences, as $p \rightarrow 2$.

Examples of power-law fits to energy release are given in Figure 6. Corresponding linear fits to these sequences are provided for comparison. Quite reasonable power-law accelerating sequences are picked by the regression algorithm for simulations with $p = 0.4$ and $p = 1.2$. Second-order fluctuations in the data are reminiscent of log-periodic corrections proposed to exist in systems displaying discrete scale invariance (SALEUR et al., 1996). The statistical fractal employed in these models has the property of discrete scale invariance.

Figure 6c illustrates a typical power-law fit to long energy release sequences in a simulation with $p = 2$. On average, energy release is near-linear. However one might argue there is an appearance of higher-order log-periodic fluctuations in energy release prior to large events even for $p = 2$. A possibility worth exploring is that large events in models with $p = 2$, might be predicted if a higher-order log-periodic correction were applied to Equation (3). However this is simply conjecture at present; confirmation would require further research outside the scope of this investigation.

Evolution of Strain Energy Correlations

A selection of simulations were rerun and a sequence of strain energy snapshots were recorded during the last 3000 timesteps prior to the 50 largest events in each simulation. This sequence spans a time interval of approx. 0.2 model time units prior

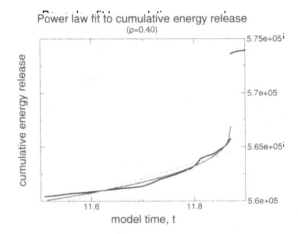

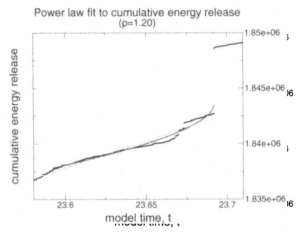

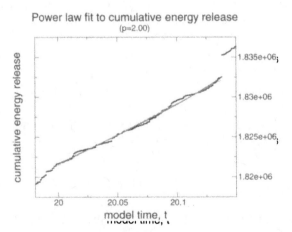

to each large event. A selection of such snapshot sequences are given in Figure 7. These correspond to the fitted event sequences in Figure 6.

During the accelerating sequences, small events occuring predominantly in the weak region (top, right) of the model redistribute strain into a broad, strong region (left) of the model. When the mean strain energy within this broad region reaches a sufficient level, a large event is triggered. Often the smaller events may trigger intermediate, moderate-sized events which fail only a portion of the highly strained region, prior to the large event. The evolution of the strain energy during such accelerating sequences is suggestive of the evolution expected if spatial strain correlations in the model evolve in a fashion consistent with the CPH.

This is confirmed by performing a correlation function analysis of the strain energy field using the strain energy snapshots. The correlation function $C(r,t)$ of the strain energy field is defined as the following:

$$C(r,t) = \frac{\langle (\varepsilon_i(t) - \bar{\varepsilon}(t))(\varepsilon_j(t) - \bar{\varepsilon}(t)) \rangle}{\langle (\varepsilon_i(t) - \bar{\varepsilon}(t))^2 \rangle} \tag{4}$$

where $r = \sqrt{(x_j - x_i)^2 + (y_j - y_i)^2}$ and $\langle \cdots \rangle$ represent the average over all pairs of cells a distance r apart. For the 128×128 model employed here, the range of distance scales is $0 \leq r < 182$. The discrete contributions to the correlation function due to each cell-pair are linearly interpolated onto this range to produce a correlation function $C(r)$ for each strain energy snapshot. We stress that the correlation function so computed is the *global* strain energy correlation function of the system. Correlation evolution analysis for subsets of the model region were not performed in this investigation.

The denominator of the correlation function normalises $C(r)$ to remove any dependance upon the mean strain energy $(\bar{\varepsilon}(t))$. This is required because, as we have seen, the mean energy of the system increases on average prior to large events. Without normalisation, an apparent evolution of the correlation function may simply reflect the mean energy evolution, without any increase in the degree of correlations at longer range; the specific mechanism proposed by the CPH.

During the analysis, individual energy release sequences were compared with corresponding snapshots and correlation functions. It was noticed that motion of the zero-crossings of $C(r,t)$ was related to the progressive growth of a broad, highly correlated region, prior to macroscopic events in the models. Space limitations precluded an in-depth analysis of this relationship.

◀

Figure 6

Three power-law time-to-failure fits to cumulative energy release for three sequences indicative of those from simulations with $p = 0.4, 1.2, 2.0$. The solid line represents the power-law fit and the dashed line is the best-fit linear approximation to the cumulative energy release data.

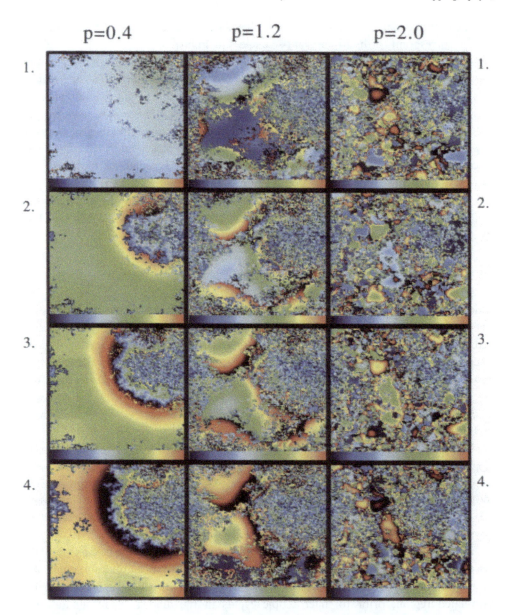

Figure 7

A sequence of four strain energy snapshots corresponding to the power-law fits in Figure 6. The first column is from a simulation with $p = 0.4$, the second column is obtained for $p = 1.2$, and the third is from a simulation with $p = 2.0$. The colour scale is the same as Figure 1.

Figure 8 attempts to illustrate the relative amount of correlation evolution which occurs in simulations with differing elastodynamic interaction ranges (differing p). The curve for each p value is obtained by first computing the mean correlation function immediately prior to the 50 largest events, and also 0.2 model time units

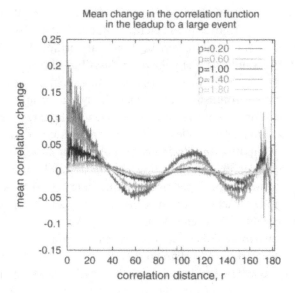

Figure 8

The mean change in correlations at each scale r in the lead-up to a large event, for various values of the interaction exponent. Global evolution of strain energy correlations occur prior to large events in simulations with longer-range elastodynamic interactions ($p \sim 1$). However little or no, correlation evolution precedes large events as $p \rightarrow 2$.

prior to these events. The mean correlation function from the earlier time is subtracted from the mean correlation function immediately prior to failure to give a mean change in correlations in the interval leading up to the largest 50 events.

It is quite evident in Figure 8 that there is evolution in the range $0.2 < p < 2$ and the amplitude of the mean correlation change increases as the effective interaction range increases (i.e., as $p \rightarrow 0$). On the other hand, a small effective interaction range ($p \sim 2$) results in little or no global correlation evolution prior to the largest 50 events. It is apparent from these results that correlation evolution consistent with the CPH accompanies accelerating energy release sequences prior to large events in models with effectively long-range elastodynamic interactions ($p \sim 1$). However, linear energy release sequences are accompanied by no significant global correlation evolution ($2 \leq p < 3$).

Discussion

The conceptual design of this model makes it difficult to claim that these models are a good representation of the physics occurring within the plane of a natural fault. However, the behaviour of these models may be somewhat analogous to natural earthquake faults, if the constitutive friction law governing fault rupture results in small events which predominantly move strain energy around within the fault region and large events which remove a considerable fraction of the stored strain energy from the region. With these characteristics, natural earthquake faults may display behaviour not too dissimilar to the behaviour described above.

The implications for prediction of large events in these models should be obvious: large events may only be forecast reliably when the system resides in the accelerating mode. In this case, a forecast of the time of occurrence of a large event may be made by fitting the cumulative energy release to a power-law time-to-failure relationship, or by periodically computing the strain energy correlation function, or both. On the other hand, large events occurring during the linear mode appear to display no systematic, global precursory behaviour upon which to base a forecast. The phenomenon of mode-switching indicates that, within any given simulation, both predictable and unpredictable events occur.

Similar mode-switching behaviour has been identified in other earthquake analogue models including a 3-D regional lithospheric model consisting of an upper brittle zone with a damage rheology, over a Maxwell viscoelastic substrate (BEN-ZION *et al.*, 1999). In agreement with the results here, one mode was characterised by Gutenberg-Richter scaling, while the other mode was charaterised by a characteristic earthquake distribution. According to BEN-ZION *et al.*, "the mode-switching phenomenon may also exist in responses of other systems with many degrees of freedom and nonlinear dynamics."

The stimulus which prompts a system with long-range interactions to switch from one stable mode to the other is unknown. However, it is thought that heterogeneity in the strain energy field is involved. In this model, accelerating energy release and correlation evolution appears to occur when the strain energy field is relatively smooth at small scales. Conversely, large small-scale heterogeneity in the strain energy field promotes linear energy release. Further research is planned to measure the degree of small-scale strain heterogeneity as a simulation approaches a switch between one type of behaviour and the other. It is hoped that the evolution of strain heterogeneity in the system may herald the transition between the different dynamical modes.

Conclusions

The model described in this paper displays two distinctly different stable dynamical modes. The first mode is characterised by small mean energy fluctuations, linear energy release, and a lack of global correlation evolution. The second mode is characterised by large mean energy fluctuations, and large events are preceded by accelerating energy release associated with evolution of the strain energy correlation function. It is apparent from the results presented that the dynamical behaviour of the model undergoes a transition from one stable mode to another, as the interaction range is varied. A small interaction range ($p \geq 1.5$) results in dynamics dominated by the first, linear mode whilst models with larger interaction ranges ($p < 1.5$) appear to mode-switch between the accelerating mode and the linear energy release mode. It is apparent that global correlation evolution consistent with the CPH occurs when the model resides in the second mode, suggesting that forecasts of macroscopic events are feasible under these circumstances.

Evidence for accelerating seismic moment release prior to a number of natural earthquakes has been reported, as has a few events which were not preceded by any acceleration in the rate of seismic moment release (JAUMÉ, 2000). These observations suggest that dynamical behaviour similar to that of the model, may be occurring within the earth's crust. In this case, earthquake prediction may be feasible only for some large earthquakes but not for others. Further research with models such as these may help to identify observable consequences of being in one mode or another, so that the dynamics and stress state of the earth's crust may be inferred for a given region, and "predictable" earthquakes may be forecast reliably.

Acknowledgments

This research was supported by the Australian Research Council, the University of Queensland, the Special Funds for Major State Basic Research Project and the National Natural Science Foundation of China (Grant No. 19732060 and No. 19972004), and the ARC IREX ACES International Visitors Program. The authors wish to thank W. Klein and Y. Ben-Zion for insightful discussions and advice during this research project. Thanks also to the reviewers for constructive criticism of the original manuscript.

REFERENCES

BAK, P. and TANG, C. (1989), *Earthquakes as a Self-organised Critical Phenomenon*, J. Geophys. Res. *94*, B11, 15,635–15,637.

BAK, P., TANG, C., and WIESENFELD, K. (1987), *Self-organised Criticality: An Explanation of 1/f Noise*, Phys. Rev. Lett. *59*, (4), 381–384.

BEN-ZION, Y., DAHMEN, K., LYAKHOVSKY, V., ERTAS, D., and AGNON, A. (1999), *Self-Driven Mode Switching of Earthquake Activity on a Fault System*, Earth Planet. Sci. Lett. *172*, (1–2), 11–21.

BUFE, C. G. and VARNES, D. J. (1993), *Predictive Modelling of the Seismic Cycle of the Greater San Francisco Bay Region*, J. Geophys. Res. *98*, (B6), 9871–9883.

GELLER, R. J., JACKSON, D. D., KAGAN, Y. Y., and MULARGIA, F. (1997), *Enhanced: Earthquakes Cannot be Predicted*, Science *275*, (5306), 1616.

GUTENBERG, B. and RICHTER, C. F. (1956), *Magnitude and Energy of Earthquakes*, Ann. Geofis. *9*, 1.

HUANG, J. and TURCOTTE, D. L. (1988), *Fractal Distributions of Stress and Strength and Variations of b-value*, Earth and Planet. Sci. Lett. *91*, 223–230.

JAUMÉ, S. C. and SYKES, L. R. (1999), *Evolving Towards a Critical Point: A Review of Accelerating Seismic Moment/Energy Release Prior to Large and Great Earthquakes*, Pure Appl. Geophys. *155*, 279–305.

JAUMÉ, S. C., WEATHERLEY, D., and MORA, P. (2000), *Accelerating Seismic Energy Release and Evolution of Event Time and Size Statistics: Results from Two Heterogeneous Cellular Automaton Models*, Pure Appl. Geophys. *157*, 2209–2226.

KLEIN, W., RUNDLE, J. B., and FERGUSON, C. D. (1997), *Scaling and Nucleation in Models of Earthquake Faults*, Phys. Rev. Lett. *78*, (19), 3793–3796.

KLEIN, W., ANGHEL, M., FERGUSON, D., RUNDLE, J. B., and SÁ MARTINS, J. S. (2000), *Statistical analysis of a model for earthquake faults with long-range stress transfer*. IN *Geocomplexity and the Physics of Earthquakes*, Geophysical Monograph 120, American Geophysical Union, 43–71.

MAIN, I. (1996), *Statistical Physics, Seismogenesis, and Seismic Hazard*, Rev. Geophys. *34*, (4), 433–462.

MORA, P. and PLACE, D. (2002), *Stress Correlation Function Evolution in Lattice Solid Elastodynamic Models of Shear and Fracture Zones and Earthquake Prediction*, Pure Appl. Geophys., *159*, 2413–2427.

RUNDLE, J. B. and BROWN, S. R. (1991), *Origin of Rate-dependence in Frictional Sliding*, J. Statis. Phys. *65*, 403–412.

RUNDLE, J. B. and JACKSON, D. D. (1997), *Numerical Simulation of Earthquake Sequences*, Bull. Seismol. Soc. Am. *67*, 1363–1377.

SALEUR, H., SAMMIS, C. G., and SORNETTE, D. (1996), *Discrete Scale Invariance, Complex Fractal Dimensions, and Log-periodic Fluctuations in Seismicity*, J. Geophys. Res. *101*, (B8), 17,661–17,677.

SAMMIS, C. G. and SMITH, S. W. (1998), *Seismic Cycles and the Evolution of Stress Correlation in Cellular Automaton Models of Finite Fault Networks*, Pure Appl. Geophys. *155*, 307–334.

SORNETTE, D. and SAMMIS, C. G. (1995), *Complex Critical Exponents from Renormalisation Group Theory of Earthquakes: Implications for Earthquake Predictions*, J. Phys. I France *5*, 607–619.

TURCOTTE, D. L., *Fractals in Geology and Geophysics* (Cambridge Univ. Press, 2nd ed., 1997).

WEATHERLEY, D., JAUMÉ, S. C., and MORA, P. (2000), *Evolution of Stress Deficit and Changing Rates of Seismicity in Cellular Automaton Models of Earthquake Faults*, Pure Appl. Geophys. *157*, 2183–2207.

WEI, Y. J., XIA, M. F., KE, F. J., YIN, X. C., and BAI, Y. L. (2000), *Evolution Induced Catastrophe and its Predictability*, Pure Appl. Geophys. *157*, (11/12).

WU, Z. L. (2000), *Frequency-size Distribution of Global Seismicity Seen from Broadband Radiated Energy*, Geophys. J. Int. *142*, 59–66.

WYSS, M., ACEVES, R. L., and PARK, S. K. (1997), *Cannot Earthquakes be Predicted?*, Science *278*, 487–490.

XIA, M. F, BAI, J., KE, F. J., and BAI, Y. L. (1999), *Sample-specificity and predictability of material failure*. In 1st ACES Workshop Proceedings (Univ. of Queensland, 1999).

(Received February 20, 2001, revised June 11, 2001, accepted June 15, 2001)

To access this journal online:
http://www.birkhauser.ch

Pure appl. geophys. 159 (2002) 2491–2509
0033–4553/02/102491–19 $ 1.50 + 0.20/0

❘ Pure and Applied Geophysics

Critical Sensitivity and Trans-scale Fluctuations in Catastrophic Rupture

MENG FEN XIA,[1,2] YU JIE WEI,[1] FU JIU KE,[1,3]
and YI LONG BAI[1]

Abstract — Rupture in the heterogeneous crust appears to be a catastrophe transition. Catastrophic rupture sensitively depends on the details of heterogeneity and stress transfer on multiple scales. These are difficult to identify and deal with. As a result, the threshold of earthquake-like rupture presents uncertainty. This may be the root of the difficulty of earthquake prediction. Based on a coupled pattern mapping model, we represent critical sensitivity and trans-scale fluctuations associated with catastrophic rupture. Critical sensitivity means that a system may become significantly sensitive near catastrophe transition. Trans-scale fluctuations mean that the level of stress fluctuations increases strongly and the spatial scale of stress and damage fluctuations evolves from the mesoscopic heterogeneity scale to the macroscopic scale as the catastrophe regime is approached. The underlying mechanism behind critical sensitivity and trans-scale fluctuations is the coupling effect between heterogeneity and dynamical nonlinearity. Such features may provide clues for prediction of catastrophic rupture, like material failure and great earthquakes. Critical sensitivity may be the physical mechanism underlying a promising earthquake forecasting method, the load-unload response ratio (LURR).

Key words: Critical sensitivity, trans-scale fluctuations, catastrophe transition, sample-specificity, heterogeneous media.

1. Introduction

As the humanity enters the New Millennium, it inherits the great achievements in the sciences and technology along with a traditional bandage, one of the greatest societal concerns, the problem of earthquake prediction. This subject has attracted considerable interest, and knowledge about earthquakes has significantly advanced in the past century. Nevertheless the self-similarity of earthquakes (GUTENBERG *et al.*, 1944) and self-organized criticality (SOC) (BAK *et al.*, 1987, 1988; BAK, 1994) aroused heated discussion about the predictability of earthquakes. Are we really on ground where the stresses are near failure everywhere and at all time? Most scientists

[1] State Key Laboratory of Nonlinear Mechanics (LNM), Institute of Mechanics, Chinese Academy of Sciences, Beijing 100080, China. E-mail: weiyj@lnm.imech.ac.cn, baiyl@lnm.imech.ac.cn
[2] Department of Physics, Peking University, Beijing 100871, China. E-mail: xiam@lnm.imech.ac.cn
[3] Department of Applied Physics, Beijing University of Aeronautics and Astronautics, Beijing 100083, China. E-mail: kefj@lnm.imech.ac.cn

agree on another viewpoint, that earthquake prediction remains a contemporary difficulty given the current knowledge (KNOPOFF, 2000; WYSS *et al.*, 1997). Clearly, further study of the physics of preparation for catastrophic rupture is required.

A series of recent works suggest that earthquake might depend sensitively on the details of heterogeneous structure and stress transfer in the earth's crust. (DIODATI *et al.*, 1991; LOCKNER *et al.*, 1991, 1992; GARCIMARTIN *et al.*, 1997; WANG *et al.*, 1998; LU *et al.*, 1998; HEIMPEL, 1997; STEIN, 1999; CURRAN *et al.*, 1997). This is quite similar to rupture in heterogeneous brittle media. Rupture appears to be a catastrophe transition (BAI *et al.*, 1994; WEI *et al.*, 2000) and the threshold of catastrophe shows uncertainty (XIA *et al.*, 1997; XIA *et al.*, 2000). It is insufficient to represent the rupture of disordered heterogeneous media by only macroscopically averaged properties (SAHIMI *et al.*, 1993; MEAKIN, 1991; IBNABDELJALIL *et al.*, 1997; CURTIN, 1997).

A large earthquake may be considered as a local catastrophic rupture in the earth's crust. The main underlying mechanism behind the complex behaviors of earthquakes and failure of disordered brittle materials might be attributed to the coupling between disordered heterogeneity on multiple scales (BEN-ZION *et al.*, 2000) and dynamical nonlinearity during nonequilibrium evolution (BAI *et al.*, 1994; CURRAN *et al.*, 1997; WEI *et al.*, 2000). In order to identify clues for prediction of earthquakes and material failure, a possible strategy is to explore general features of catastrophic rupture in heterogeneous brittle media.

Coupling effects between disordered heterogeneity on the mesoscopic scale and dynamical nonlinearity are so complex that direct experimental observations or theoretical conclusions are quite difficult. Even for numerical simulations, some reasonable simplification of physical concepts is a necessity. Recently, we have examined a coupled pattern mapping model (XIA *et al.*, 2000; WEI *et al.*, 2000) similar to the well-known fiber-bundle model (COLEMAN, 1958; DANIELS, 1945). However our model takes the coupling effects between mesoscopic disordered heterogeneity and dynamical nonlinearity due to stress redistribution into account. We found that such a model can reproduce distinctive features of rupture in complex heterogeneous media.

Notably, the model displays the catastrophe transition (BAI *et al.*, 1994; WEI *et al.*, 2000) and sample-specificity (XIA *et al.*, 1997, 2000), namely macroscopic uncertainty of observed rupture behavior. This is one of the roots for the difficulty in rupture prediction. We report here that the catastrophe transition presents some general features: critical sensitivity and trans-scale fluctuations. The critical sensitivity implies that the system may become significantly sensitive near the catastrophe transition. Trans-scale fluctuations refer to fluctuations which, at the catastrophe threshold, may be enhanced strongly and accordingly, the spatial scales of stress and damage fluctuations increase rapidly from the mesoscopic heterogeneity scale to the macroscopic scale. These general features provide insight into the essence of the catastrophe transition, and may provide clues for prediction of catastrophic ruptures, such as great earthquakes.

Section 2 briefly reviews the coupled pattern mapping model. Section 3 presents the evolution of the model behaviour. Critical sensitivity and trans-scale fluctuations are considered in sections 4 and 5, respectively. Section 6 contains a summary of the results and discussion.

2. Brief Review of Model

The model (BAI *et al.*, 1994; XIA *et al.*, 1997, 2000; WEI *et al.*, 2000) is a periodic lattice consisting of N mesoscopic units. Mesoscopic heterogeneity is modelled by assigning randomly an initial strength σ_{ci} to each unit i with $\{\sigma_{ci}\}$ given by distribution function $h(\sigma_c)$.

Mesoscopically, the system is specified by the damage pattern $X = \{x_i, i = 1, 2, \ldots, N\}$, the stress pattern $\Sigma = \{\sigma_i, i = 1, 2, \ldots, N\}$, and the initial strength pattern $\Sigma_c = \{\sigma_{ci}, i = 1, 2, \ldots, N\}$, where an intact or broken unit i is denoted by $x_i = 0$ or 1, respectively. σ_i is the stress on unit i and σ_{ci} is the initial strength of unit i. Macroscopic parameters are the damage fraction,

$$p = \frac{1}{N}\sum_{i=0}^{N} x_i \ , \tag{2.1}$$

and the nominal stress,

$$\sigma_0 = \frac{1}{N}\sum_{i=0}^{N} \sigma_i \ . \tag{2.2}$$

Samples with identical $h(\sigma_c)$ are considered to be identical macroscopically, although they are different from sample to sample mesoscopically due to disordered meso-heterogeneity. In the following calculations, we choose $h(\sigma_c)$ as a Weibull distribution function with a mean of 1 and a modulus, $m_c = 2$.

The pattern dynamics is defined by iterations of mappings between the coupled patterns X, Σ and Σ_c. The evolution of damage pattern X is determined by the stress pattern Σ and the strength pattern Σ_c, according to a mesoscopic failure condition. It is simply assumed that all units with $\sigma_i \geq \sigma_{ci}$ break simultaneously. The strength pattern Σ_c varies with damage pattern X as $\sigma_{ci}(1 - x_i)$. This means that a broken unit loses strength, and no longer supports any stress. The stress pattern Σ is determined from the damage pattern X according to a stress redistribution (SRD) rule, and it is assumed to be independent of history. It is convenient to represent the SRD rule with respect to a state with uniform stress ($\sigma_i = \sigma_0$).

In order to examine the effects of stress fluctuations, we considered various SRD rules as follows:
(1) Global mean field (GMF) model: The nominal stress σ_0 of broken units is shared by all intact units uniformly.

(2) Local mean stress concentration (LMSC) model: The nominal stress of a broken cluster is uniformly transferred to its two neighboring intact regions of size δ.

(3) Cluster mean field (CMF) model: The nominal stress of a broken cluster is equally redistributed to its two neighboring intact clusters and the stress within an intact cluster is uniform.

The LMSC and CMF models display stress fluctuations, whereas the GMF model is without stress fluctuations. The SRD rule represents the main dynamical nonlinearity in the model.

The evolution of the system is controlled by external loading, i.e., the nominal stress σ_0 increases from $\sigma_0 = 0$ to a failure threshold ($\sigma_0 = \sigma_{0f}$). We will consider quasi-static loading as a standard process: the nominal stress increment $\Delta\sigma_0$ is computed each loading step as the minimum increment necessary to break at least one unit.

3. Catastrophe Transition

We represent the evolution of a system by considering the time series of energy release ΔE. The energy release ΔE is calculated as a summation of the initial stored elastic energy of broken units in a mapping, or loading step $\Delta\sigma_0$. For simplicity, the dimensionless elastic modulus is assumed to be 2 and is identical for all units, and the stored elastic energy of a unit can be written as σ_i^2. Because the damage fraction p increases monotonously, the time series of ΔE can be shown as a plot of ΔE versus p.

Figure 1 shows time series of energy release ΔE for various SRD models. A distinct common feature is that the evolution presents a catastrophe transition at p_c, which corresponds to a threshold of nominal stress σ_{0f}. For a specified nominal stress below the threshold ($\sigma_0 < \sigma_{0f}$), the evolution remains in a globally stable (GS) mode with mesoscopic damage accumulation. At the threshold ($\sigma_0 = \sigma_{0f}$) however, the system falls into a condition of self-sustained catastrophic failure (CF), the main rupture appears and the system evolves to entire failure ($p = 1$) eventually. Such an evolution-induced mode transition demonstrates a general behavior of failure in heterogeneous brittle media. This behavior is called evolution induced catastrophe (EIC).

Generally speaking, rupture prediction is concerned particularly with the catastrophe transition and main rupture. Unfortunately, it is found that the threshold of catastrophe transition (σ_{0f}) shows uncertainty. Thus, it is impossible to predict catastrophic rupture in terms of a few macroscopic parameters such as the damage fraction p, the nominal stress σ_0 and parameters defining the properties of the strength distribution (e.g., modulus m_c, when the Weibull distribution function is applied). Figure 2 shows an ensemble distribution of catastrophe transition threshold (σ_{0f}) for samples which are identical macroscopically. For the GMF model without stress fluctuations, all samples exhibit identical threshold σ_{0f} determined by strength

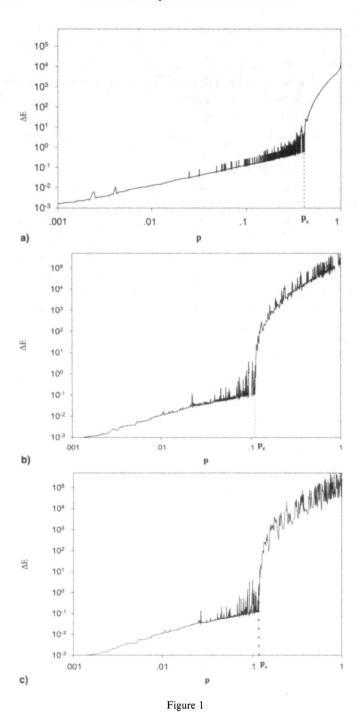

Figure 1
The energy release ΔE versus damage fraction p for a mapping with $N = 10,000$. (a) GMF model, (b)
LMSC model with $\delta = 5$, (c) CMF model.

distribution function $h(\sigma_c)$. In this case rupture prediction is deterministic, based only upon macroscopic parameters. However, for models with stress fluctuations, LMSC and CMF, the threshold σ_{0f} shows diversity, i.e., the threshold is different from sample to sample for macroscopical identical samples. Such a behavior is called sample-specificity (XIA *et al.*, 1997, 2000). Such diversity of macroscopic failure strength has been reported by SAHIMI *et al.* (1993) and BAI *et al.* (1994). Sample-specificity leads to macroscopic uncertainty, and rupture prediction becomes impossible based upon macroscopic parameters alone.

In order to circumvent this obstacle to rupture prediction, a possible strategy is to explore some universal features of the catastrophe transition in search of clues of use in rupture prediction.

A well-known general feature of threshold systems is that the size-distribution of events follows a power law. This has been observed in both material failure (DIODATI *et al.*, 1991; GARCIMARTIN *et al.*, 1997; LU *et al.*, 1998) and earthquakes (as evidenced by the empirical Gutenberg-Richter relation) (GUTENBERG and RICHTER, 1944). Our model also displays power-law event size-distribution. Figure 3 gives the statistics $N(\Delta E)$ of energy release ΔE calculated for each loading step. The log-log plot follows a rather respectable straight line ranging about $1.5 \sim 2$ decades. In this model, the power-law size-distribution is mainly attributed to events prior to the catastrophe transition. The exponent (*b* value) is not universal. A power law is suggestive of a dynamical system near criticality, corresponding with the catastrophe

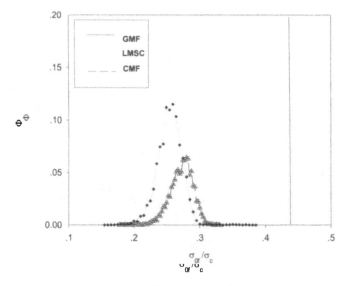

Figure 2
Ensemble distribution of the catastrophe transition threshold σ_{0f} for samples which are identical macroscopically. 10,000 samples with $N = 8000$ were examined. The solid line represents the GMF model. The dotted line represents the LMSC model and the dashed line represents the CMF model.

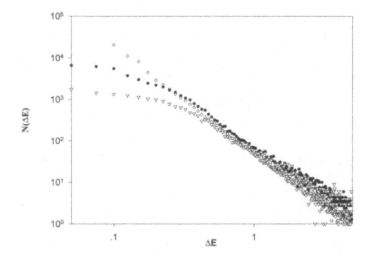

Figure 3
Statistics of energy release $N(\Delta E)$. ◇ GMF model, $N = 10,000$ for 5000 samples, the slope is nearly -2.10, • LMSC model, $N = 40,000$, $\delta = 5$, for 2000 samples, the slope is -2.34 approximately, ▽ CMF model, $N = 100,000$, for 3000 samples, the slope is about -2.59.

transition in the model. The region surrounding the transition point may be considered the critical region.

From the viewpoint of rupture prediction, however, a power-law size-distribution might provide a negative conclusion (GELLER et al., 1997).

However, there is observational evidence, while indirect, of natural seismicity suggesting the existence of general features prior to the catastrophe transition or main rupture. To monitor these general features may provide clues for rupture prediction. There have been a number of reports of accelerating seismic moment release (AMR) (SAMMIS et al., 1999; JAUMÉ et al., 1999) and a change in the rate of occurrence of moderate-sized earthquakes prior to large earthquakes in a variety of tectonic settings. Yin and co-workers (YIN et al., 1994) have reported numerous cases where a measure of the load-unload response ratio (LURR) has increased markedly prior to a number of moderate to large earthquakes. Recently, it was found that the optimal region size for AMR and for LURR is comparable, which may suggest that AMR and LURR have a common physical mechanism (YIN et al., 2002). These phenomena are considered to be evidence supporting Critical Point Hypothesis (CPH) (BOWMAN et al., 1998; TIAMPO et al., 2000) of earthquakes, which predicts the existence of a critical region with the progressive formation of long-range stress field correlation, a condition required for large earthquakes. From our model, we also found general features related to catastrophe transition or main rupture. They are called critical sensitivity and trans-scale fluctuations, which will be discussed in section 4 and section 5, separately. We will point out that there might be an essential relationship between CPH and catastrophe transition. In fact, the time series of energy release shown in Figure 1 displays an acceleration prior to catastrophe transition or main rupture. This might be similar to the AMR in earthquakes.

4. Critical Sensitivity

We report here a novel discovery of a general feature of systems in the vicinity of the catastrophe transition. This is called critical sensitivity, which means that the sensitivity of a system may strongly be enhanced in various aspects as the system approaches the catastrophe transition. In this section, we discuss two kinds of critical sensitivity, i.e., sensitivity to external loading and sensitivity to stochastic micro-damage.

There is evidence to suggest that the earth's crust displays critical sensitivity prior to large earthquakes. Seismologists agree that foreshocks are a symptom of some preparatory process prior to the main rupture in some cases (WYSS *et al.*, 1997; YIN *et al.*, 1994). Yin *et al.* proposed a promising method for earthquake forecasting called the load-unload response ratio (LURR) (YIN *et al.*, 1994, 1999; WANG *et al.*, 1998). LURR is defined as the ratio of activity of small-to-intermediate earthquakes during a loading phase to the activity during an unloading phase, where the loading and unloading result from earth tides. It is found that in most cases (more than 80%), LURR increases significantly before large earthquakes and fluctuates slightly about unity in stable regions.

The LURR method may be explained as follows. A large earthquake can be considered as a local catastrophic rupture in the earth's crust, and the increase in LURR prior to large earthquakes implies that a region of the crust has become significantly sensitive to external loading perturbation prior to catastrophe. Thus, we presume that the increase of LURR might be evidence for critical sensitivity prior to large earthquakes.

In order to explore the features of critical sensitivity, we performed simulations based on the coupled pattern mapping model.

The sensitivity of energy release to external loading can be measured by

$$S = \frac{\Delta E'}{\Delta \sigma_0'} \bigg/ \frac{\Delta E}{\Delta \sigma_0} \ , \tag{4.1}$$

where $\Delta E'$ (ΔE) is the energy release induced by increment $\Delta \sigma_0'$ ($\Delta \sigma_0$), and

$$\Delta \sigma_0' = \Delta \sigma_0 + \alpha \bar{\sigma}_c \ , \tag{4.2}$$

where $\bar{\sigma}_c$ is the average strength of units (the mean of the distribution function $h(\sigma_c)$) and α is a small parameter ($\alpha \sim 10^{-2} \rightarrow 10^{-3}$). According to the definition of the sensitivity, $S \gg 1$ indicates high sensitivity, and $S \sim 1$ implies an insensitive state.

The sensitivity of energy release to external loading is shown in Figure 4. Figure 4(a) shows the time series of S for a sample. At the initial stage, S maintains a low value (of order 1) but S increases significantly near the catastrophe transition point $p/p_c = 1$. Figures 4(b), (c) and (d) show the time series of S for 200 samples identical macroscopically based on SRD models of GMF, LMSC and CMF, respectively. We can see that, although normalized variable p/p_c is adopted, the time series of S are different from sample to sample, which is evidence for sample-specificity. However, in all cases there is a common trend that S increases significantly near the catastrophe transition. This is the hallmark of critical sensitivity.

Figure 5 shows ensemble statistics of maximum S and the statistics of p_M/p_c, where p_M is the damage fraction at which S takes the maximum value S_{max}. It is found that, for most samples, the sensitivity S arrives at its maximum when $0.7 \leq p/p_c \leq 1$, and S_{max} is usually one order of magnitude higher than the initial value $S \approx 1$. These results imply that critical sensitivity is a significant precursor of the catastrophe transition.

Now we examine the sensitivity of energy release to stochastic microdamage. We consider a model where deterministic dynamics given by the coupled patten mapping and the stochastic microdamage coexist (XIA et al., 1996). The stochastic micro-damage is modelled by the break of Δn units chosen randomly at the beginning of a loading step. The sensitivity of energy release to the stochastic microdamage can be measured by the ratio of energy release with a stochastic microdamage to that without the stochastic damage:

$$S^* = \frac{\Delta E^*/\Delta n}{\Delta E} ,$$
(4.3)

where ΔE and ΔE^* are energy releases under the identical external loading condition but without and with stochastic microdamage of size Δn, respectively. Figure 6 shows S^* for one sample and for 200 samples identical macroscopically. It is obvious that, like the behavior of S, S^* also displays the sample-specificity and critical sensitivity.

It is interesting to note that the critical sensitivity of energy release to stochastic microdamage is a sensitivity linking different scales: from mesoscopic events to macroscopic behavior. Such a sensitivity implies that a minor change on the mesoscopic level can be strongly amplified during nonlinear evolution and leads to significant macroscopic effect as the system approaches the catastrophe transition. In section 3, we represent sample-specificity, i.e., samples which are identical macro-scopically but different from sample to sample mesoscopically due to disordered meso-heterogeneity display diverse behavior of catastrophe transition. Sample-specificity is also a feature linking different scales: the macroscopic behavior is sensitively dependent on the details of mesoscopic structure. There is a common underlying physical mechanism behind sample-specificity and sensitivity of energy

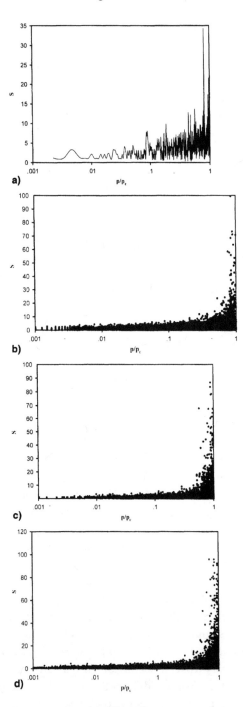

Figure 4

Time series of S, the critical sensitivity to external load pertubations ($N = 10,000$, $\alpha = 0.001$). (a) for a single sample, CMF model, (b) for 200 samples, GMF model, (c) for 200 samples, LMSC model with $\delta = 5$, (d) for 200 samples, CMF model.

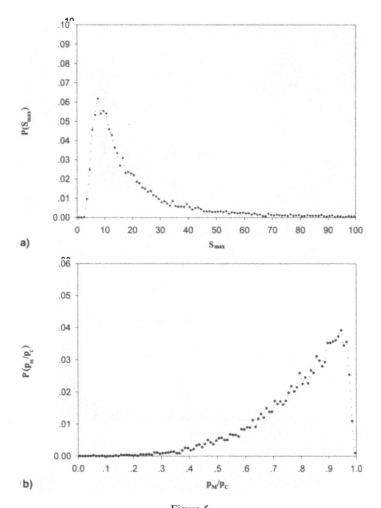

Figure 5
Ensemble statistics of (a) S_{max} and (b) p_M/p_c, for the CMF model with $N = 4000$ and $\alpha = 0.001$ (1000 samples).

release to stochastic microdamage. Such critical sensitivity is termed as trans-scale sensitivity.

The coupled pattern mapping model can be thought of driven nonlinear threshold systems, comprised of multitudes interacting, mesoscopic units subjected to a driving force. As the system approaches the catastrophe transition, increasingly more units are close to their threshold. It will be considerably easier to trigger larger cascade by perturbation at that time. This is the origin of critical sensitivity. Thus, critical sensitivity might be a general feature in widely used driven nonlinear threshold models. It is especially interesting to examine whether CPH also displays critical sensitivity.

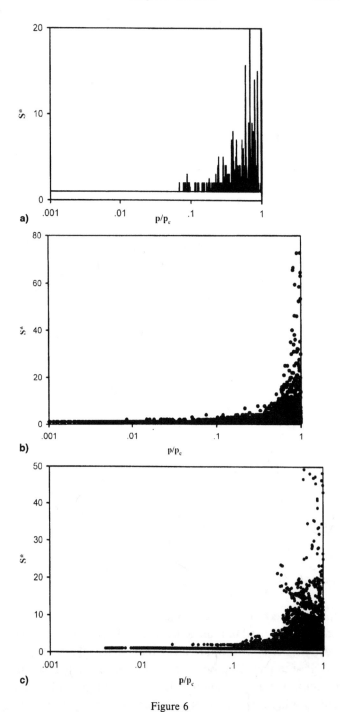

Figure 6
Time series of S^*, the critical sensitivity to stochastic microdamage ($N = 10,000, \Delta n = 1$), (a) for a single sample, GMF model, (b) for 200 samples, GMF model, (c) for 200 samples, CMF model.

5. Trans-scale Fluctuations

It is well known that fluctuations play a vital role in catastrophic rupture of heterogeneous brittle media. The rupture behavior of a system may not be completely representable by only its average macroscopic properties (XIA et al., 1997, 2000; IBNABDELJALIL et al., 1997; WEI et al., 2000; SAHIMI et al., 1993). However, the general features of fluctuations related to catastrophe is still an open question. We report here a class of fluctuations related to catastrophe. At catastrophe threshold ($\sigma_0 = \sigma_{0f}$) the stress fluctuations in system are enhanced significantly and the spatial scale of stress and damage fluctuations increases rapidly from the mesoscopic heterogeneity scale to the macroscopic scale; such fluctuations are called trans-scale fluctuations.

Stress fluctuations can be measured as the relative deviation of stress $\delta\sigma/\bar{\sigma}$, where

$$\delta\sigma = \left[\frac{1}{N(1-p)} \sum_{i=0}^{N} (\sigma_i - \bar{\sigma})^2 (1 - x_i) \right]^{1/2} \tag{5.1}$$

is standard deviation of stress supported by intact units, and

$$\bar{\sigma} = \frac{\sigma_0}{1-p} \tag{5.2}$$

is mean stress on intact units. Figure 7 shows $\delta\sigma/\bar{\sigma}$ versus p, corresponding to the time-series of the relative deviation of stress fluctuations. It is also found that time series of $\delta\sigma/\bar{\sigma}$ shows sample-specificity, i.e., the time-series are different from sample to sample for samples which are identical macroscopically. However, the time-series display a general trend that $\delta\sigma/\bar{\sigma}$ increases rapidly beyond the transition point p_c (keeping $\sigma_0 = \sigma_{0f}$) but prior to the main rupture. Ensemble statistics illustrate that, at the catastrophe threshold $\delta\sigma/\bar{\sigma}$ increases by about two orders of magnitude. This can be seen from the statistical distribution of maximum $\delta\sigma/\bar{\sigma}$ during GS regime and the catastrophe regime (shown in Fig. 8).

In order to reveal the characteristics of the stress pattern, we take a coarse-grained average of the stress pattern and examine its fluctuations. The approach is as follows: The system is divided into m cells with size C, then $C = N/m$. Denote $\langle \sigma \rangle_j$ to be the average stress over intact units in the j-th cell, and $\langle \sigma \rangle_j = 0$ if all units in the j-th cell are broken. Define the local damage fraction of the j-th cell as

$$P_j = \frac{1}{C} \sum_{i \in j} x_i \, , \tag{5.3}$$

where the summation is over all units in the j-th cell. The relative deviation of coarse-grained average stress can be calculated using

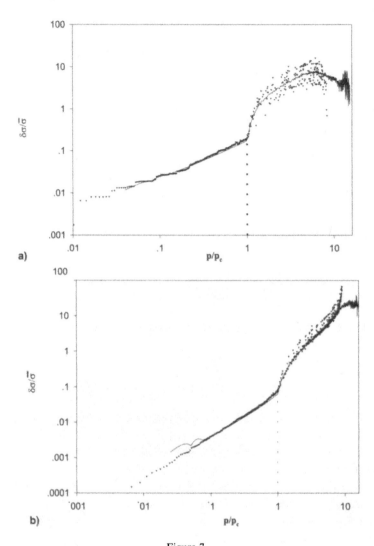

Figure 7
Relative deviation of stress fluctuations $\delta\sigma/\bar{\sigma}$ versus damage fraction p ($N = 4096$). The solid line is averaged over 200 samples while the triangle represents a single sample. (a) CMF model and (b) LMSC model with $\delta = 5$.

$$\frac{\delta\langle\sigma\rangle}{\bar{\sigma}} = \frac{1}{\bar{\sigma}}\left[\sum_j H(1 - P_j)(\langle\sigma\rangle_j - \bar{\sigma})^2 \bigg/ \sum_j H(1 - P_j)\right]^{1/2}, \qquad (5.4)$$

where $\bar{\sigma}$ is mean stress supported by intact units (given in Equation 5.2), and

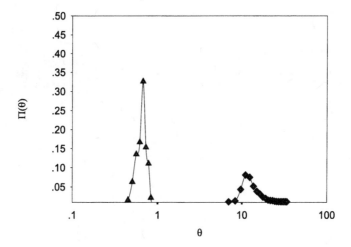

Figure 8

Ensemble distribution function (10,000 samples) of maximum values of $\delta\sigma/\bar{\sigma}$ for GS regime and catastrophic rupture regime ($N = 20,000$, CMF model), The two distribution functions are well-separated, indicating that stress fluctuations increase significantly in the catastrophe regime.

$$H(y) = \begin{cases} 0, & \text{for } y = 0 \\ 1, & \text{for } y > 0 \end{cases} . \tag{5.5}$$

The relative deviation of fluctuations in the coarse-grained average stress is shown in Figure 9(a). A distinct feature is that, for the GS regime, $\delta\langle\sigma\rangle/\bar{\sigma}$ decreases significantly with increasing coarse-grained scale C and $\delta\langle\sigma\rangle/\bar{\sigma} \approx 0$ for the macroscopic scale C. Qualitatively, stress pattern is macroscopically homogeneous in GS regime. At catastrophe threshold, macroscopic inhomogeneity of the stress pattern increases significantly.

The fluctuations of coarse-grained damage fraction $\{P_j, j = 1, 2, \ldots, m\}$ can be calculated by

$$\delta P = \left[\frac{1}{m} \sum_{j=1}^{m} (P_j - p)^2 \right]^{1/2}, \tag{5.6}$$

where P_j and p are defined in Equations (5.3) and (2.1), respectively. δP is shown in Figure 9(b).

In reality, between the catastrophe transition and main rupture, there is a time-interval, which may vary for different real systems. Trans-scale fluctuations occur during this interval and can be considered a significant indication of the catastrophe transition and an immediate precursor to main rupture. This may provide clues for prediction of the main rupture. The trans-scale fluctuations identified in the numerical simulations were compared to the statistics of damage events according to

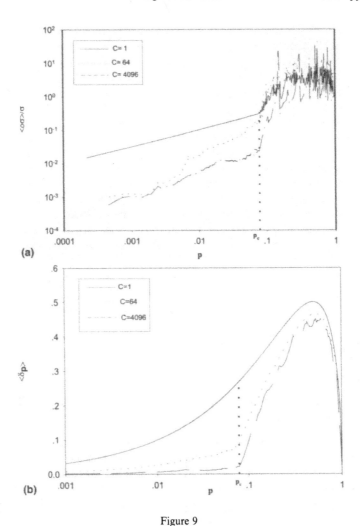

Figure 9
Trans-scale fluctuations: (a) the fluctuations of coarse-grained average stress $\langle\delta\sigma\rangle/\bar{\sigma}$ and (b) the fluctuations of coarse-grained average damage $\delta\langle p\rangle$ versus damage fraction p (CMF models, $N = 65,536$). The line type indicates the coarse-grained cell size C: the solid line represnts $C = 1$, the dotted line represents $C = 64$, and the dashed line represents $C = 4096$.

acoustic emission signals detected during the rupture process (LOCKNER *et al.*, 1991, 1992; GARCIMARTIN *et al.*, 1997).

Trans-scale fluctuations of stress imply that the correlation length of stress increases progressively from small scale to large scale at the catastrophe threshold. Long-range correlation of high stress is the condition required for main rupture to occur. This is very similar to the progressive formation of long-range stress field correlation predicted by Critical Point Hypothesis (CPH).

6. Summary and Discussions

Rupture in heterogeneous brittle media was examined based on a model called a coupled pattern mapping model. The rupture presents dynamical complexity, especially the catastrophe transition and sample-specificity. This complexity is responsible for the difficulty of rupture prediction. However, we have found clues for rupture prediciton from general features of catastrophic rupture: critical sensitivity and trans-scale fluctuations. These may be viewed as precursors of catastrophe transition and main rupture.

The underlying mechanism of the catastrophe transition, sample-specificity, critical sensitivity and trans-scale fluctuations, should be attributed to the coupling between heterogeneity and dynamical nonlinearity, which leads to a cascade of damage. The cascade is determined by coupling, competition and balance between strength heterogeneity and stress redistribution (SRD). In the globally stable (GS) regime, a cascade induced by SRD is limited to finite size by surrounding stronger units. Beyond catastrophe transition however, the cascade can no longer be inhibited and will continue without limit, i.e., the system evolves to an entirely failed state. Critical sensitivity implies that, near the catastrophe transition, the size of the cascade can be enhanced significantly by minor external stress perturbations. This is because, near catastrophe, more and more units support stresses close to their strength, and thus, a small disturbance may induce significant extra failure of events. In the catastrophe regime, the cascade evolves from the mesoscopic scale to the macroscopic scale, resulting in trans-scale fluctuations.

The coupling between heterogeneity and dynamical nonlinearity leads to trans-scale sensitivity resulting in sample-specificity. This makes the problem far more complex. Consequently, to explore the general features of the catastrophe transition is a problem of great importance. Although critical sensitivity and trans-scale fluctuations are features of a simple conceptual model, we are quite sure that they present general features of a class of real systems, including perhaps the earth's crust.

Our work suggests that there is an essential relationship between the catastrophe transition and the CPH. They display similar general features and might be attributed to a common underlying physical mechanism. A further work to compare the major features of the two theories would be very interesting.

Acknowledgments

This work is supported by the Special Funds for Major State Basic Research Project G2000077305 and National Natural Science Foundation of China (Grant No. 19732060, 19972004, 10047006).

REFERENCES

BAK, P., TANG, C., and WIESENFELD, K. (1987), *Self-Organized Criticality: An Explanation of 1/f Noise*, Phys. Rev. Lett. *59*, 381–384.

BAK, P., TANG, C., and WIESENFELD, K. (1988), *Self-Organized Criticality*, Phys. Rev. A *38*, 364–374.

BAK, P. *Self-Organized Criticality: A Holistic view of Nature*. In *Complexity: Metaphors, Models, and Reality* (eds. Cowan, G., Pines, D. and Meltzer, D.) (SFI Studies in Sciences of Complexity, Proc. Vol. XIX, Addison-Wesley 1994) pp. 477–493.

BAI, Y. L., LU, C.S., KE, F. J., XIA, M. F. (1994), *Evolution Induced Catastrophe*, Phys. Lett. A *185*, 196–199.

BEN-ZION, Y. and SAMMIS, C. G. (2001), *Characterization of Fault Zones*, Pure Appl. Geophys., (in press).

BOWMAN, D. D., OUILLON, G., SAMMIS, C. G., SORNETTE, A., and SORNETTE, D. (1998), *An Observational Test of the Critical Earthquake Concept*, J. Geophys. Res. *103*, 359.

CURRAN, D. R., SEAMAN, L., and SHOCKEY, D. A. (1997), *Dynamic Failure of Solids*, Phys. Rep. *147*, 253–388.

CURTIN, W. A. (1997), *Toughening in Disordered Brittle Materials*, Phys. Rev. B *55*, 11,270–11,276.

COLEMAN, B. D. (1958), *On The Strength of Classical Fibers and Fiber Bundles*, J. Mech. Phys. Solids *7*, 60–70.

DANIELS, H. E. (1945), *The Statistical Theory of The Strength of Bundles of Threads*, Proc. Roy. Soc. London A *183*, 405–435.

DIODATI, P., MARCHESONI, F., and PIAZZA, S. (1991), *Acoustic Emission from Volcanic Rocks: An Example of Self-organize Criticality*, Phys. Rev. Lett. *67*, 2239–2242.

GARCIMARTIN, A., GUARINO, L., BELLON and CILIBERTO, S. (1997), *Statistical Properties of Fracture Precursors*, Phys. Rev. Lett. *79*, 3202–3205.

GUTENBERG, B. and RICHTER, C. F. (1944), *Frequency of Earthquake in California*, Bull. Seismol. Soc. Am. *34*, 125–188.

GELLER, R. J., JACKSOM, D. D., KAGAN, Y. Y., and MULARGIA, F. (1997), *Earthquakes Cannot be Predicted*, Science *275*, 1616–1617.

HEIMPEL, M. (1997), *Critical Behaviour and the Evolution of Fault Strength during Earthquake Cycles*, Nature, *388*, 865–868.

IBNABDELJALIL, M. and CURTIN, W.A. (1997), *Strength and Reliability of Fiber-reinforced Composites: Local Load Sharing and Associated Size Effects*, Int. J. Solids and Structures *34*, 2649–2668.

JAUMÉ, S. C. and SYKES L. R. (1999), *Evolving Towards a Critical Point: A Review of Accelerating Seismic Moment/Energy Release Prior to Large and Great Earthquakes*, Pure Appl. Geophys. *155*, 279–305.

KNOPOFF, L. (2000), *The Magnitude Distribution of Declustered Earthquakes in Southern California*, Proc. Natl. Acad. Sci. USA *97*, 11,880–11,884.

LOCKNER, D. A., BYERLEE, J. D., PONOMAREV, A., and SIDORIN, A. (1991), *Quasi-Static Fault Growth and Shear Fracture Energy in Granite*, Nature *350*, 439–443.

LOCKNER, D. A. and BYERLEE, J. D., *Precursory AE Patterns Leading to Roch Fracture*. In *Proc. 5th Conf. on Acoustic Emmisions/Microseismic Active in Geologic Structure and Materials*, (ed. Hardy, H.R.) (Trans. Publ. 1992) pp. 1–4.

LU, C. S., TAKAYASU, H., TRETYAKOV, A. Y., TAKAYASU, M., and SYUMOYO, S. (1998), *Lattice Model of the Brittle Crust, Self-Organized Criticality in a Block*, Phys. Lett. A *242*, 349–354.

LANGER, J. S., CARLSON, J. M., MYERS, C. R., and SHAW, B. E. (1996), *Slip Complexity in Dynamic Models of Earthquake Faults*, Proc. Natl. Acad. Sci. USA *93*, 3825–3829.

MEAKIN, P. (1991), *Models for Material Failure and Deformation*, Science *252*, 226–234.

SAHIMI, M. and ARBABI, S. (1993), *Mechanics of Disordered Solid. III. Fracture Properties*, Phys. Rev. B *47*, 713–722.

SAMMIS, C. G. and SMITH S. W. (1999), *Seismic Cycles and the Evolution of Stress Correlation in Cellular Automaton Models of Finite Fault Networks*, Pure Appl. Geophys. *155*, 307–334.

STEIN, R. S. (1999), *The Role of Stress Transfer in Earthquake Occurrence*, Nature *402*, 605–609.

SWINBANKS, D. (1997), *Quake Panel Admits Prediction is 'Difficult'*, Nature *388*, 4.

TIAMPO, K. F., RUNDLE, J. B., GROSS, S. J., and KLEIN, K. (2000), *Karhunen-Loeve Expansion Analysis of Seismicity on the Southern California Fault System*, EOS Trans. AGU, *81*(48), Fall Meet. Suppl., Abstract NG61A-12, 2000.

TURCOTTE, D. L. (1999), *The Physics of Earthquakes: Is It a Statistical Problem?*, In Proc. 1-st ACES Workshop Proceedings. (ed. Mora, P.), 95–98 (The APEC Cooperation for Earthquake Simulation, Brisbane).

WANG, C. Y. and CAI, Y. E. (1997), *Sensitivity of Earthquake Cycles on the San Andreas Fault to Small Changes in Regional Compression*, Nature *388*, 158–161.

WANG, Y. C., YIN, X. C., and WANG, H. T. (1998), *The Experimental Simulation of Rocks on Load/Unload Response Ratio for Earthquake Prediction*, Earthquake Research in China (English Version) *12*, 367-372.

WEI, Y. J., XIA, M. F., KE, F. J., YIN, X. C., and BAI, Y. L. (2000), *Evolution-Induced Catastrophe and Its Predictability*, Pure Appl. Geophys. *157*, 1945–1957.

WYSS, M., ACEVES, R. L., PARK, S. K., GELLER, R. J., JACKSON, D.D., KAGAN, Y.Y., and MULARGIA F. (1997), *Cannot Earthquakes Be Predicted?*, Science *275*, 487–490.

XIA, M. F., BAI, Y. L., and KE, F. J. (1996), *A Stochastic Jump and Deterministic Dynamics Model of Impact Failure Evolution with Rate Effect*, Theor. Appl. Frac. Mech. *24*, 189–196.

XIA, M. F., KE, F. J., BAI, J., and BAI, Y. L. (1997), *Threshold Diversity and Trans-Scales Sensitivity in a Nonlinear Evolution Model*, Phys. Lett. A *236*, 60–64.

XIA, M. F., KE, F. J., WEI, Y. J., BAI, J., and BAI, Y. L. (2000), *Evolution Induced Catastrophe in a Nonlinear Dynamical Model of Materials Failures*, Nonlinear Dynamics *22*, 205–224.

YIN, X. C., CHEN, X. Z., SONG, Z. P., and YIN, C. (1994), *A New Approach to Earthquake Prediction: Load/Unload Response Ratio (LURR) Theory*, Pure Appl. Geophys. *145*, 701–705.

YIN, X. C., CHEN, X. Z., WANG, Y. C., WANG, H. T., PENG, K. Y., ZHANG, Y. X., and ZHUANG, J. C. (1999), *Development of a New Approach for Earthquake Prediction – The Load/Unload Response Ratio*. In Proc. 1-st ACES Workshop Proceedings. (ed. Mora, P.), pp. 325–330 (The APEC Cooperation for Earthquake Simulation, Brisbane).

YIN, X. C., MORA, P., PENG, K. Y., WANG, Y. C., and WEATHERLY, D. (2002), *Load-unload Response Ratio and Accelerating Moment/Energy Release, Critical Region Scaling and Earthquake Prediction*, Pure appl. geophys. *159*, 2511–2523.

(Received February 20, 2001, revised June 11, 2001, accepted June 25, 2001)

Pure appl. geophys. 159 (2002) 2511–2523
0033–4553/02/102511–13 $ 1.50 + 0.20/0

❙ Pure and Applied Geophysics

Load-Unload Response Ratio and Accelerating Moment/Energy Release Critical Region Scaling and Earthquake Prediction

Xiang-Chu Yin,[1,2] Peter Mora,[3] Keyin Peng,[1,2]
Yu Cang Wang,[1,3] and Dion Weatherley[3]

Abstract — The main idea of the Load-Unload Response Ratio (LURR) is that when a system is stable, its response to loading corresponds to its response to unloading, whereas when the system is approaching an unstable state, the response to loading and unloading becomes quite different. High LURR values and observations of Accelerating Moment/Energy Release (AMR/AER) prior to large earthquakes have led different research groups to suggest intermediate-term earthquake prediction is possible and imply that the LURR and AMR/AER observations may have a similar physical origin. To study this possibility, we conducted a retrospective examination of several Australian and Chinese earthquakes with magnitudes ranging from 5.0 to 7.9, including Australia's deadly Newcastle earthquake and the devastating Tangshan earthquake. Both LURR values and best-fit power-law time-to-failure functions were computed using data within a range of distances from the epicenter. Like the best-fit power-law fits in AMR/AER, the LURR value was optimal using data within a certain epicentral distance implying a critical region for LURR. Furthermore, LURR critical region size scales with mainshock magnitude and is similar to the AMR/AER critical region size. These results suggest a common physical origin for both the AMR/AER and LURR observations. Further research may provide clues that yield an understanding of this mechanism and help lead to a solid foundation for intermediate-term earthquake prediction.

Key words: LURR (Load-Unload Response Ratio), AMR (Accelerating Moment Release), AER (Accelerating Energy Release), CPH (Critical Point Hypothesis), earthquake prediction, critical region scaling.

1. Introduction

According to the critical point hypothesis (Keilis-Borok, 1990; Sornette and Sornette, 1990; Sornette and Sammis, 1995; Bowman *et al.*, 1998), the earth's crust is not perpetually in a critical state. The occurrence of a large or great earthquake in a region appears to dissipate a sufficient proportion of the

---

[1] State Key Laboratory of Nonlinear Mechanics, Chinese Academy of Sciences, Beijing, 100080, China. E-mails: xycin@public.bta.net.cn; yin@lnm.imech.ac.cn

[2] Center for Analysis and Prediction, China Seismological Bureau, Beijing, 100036, China.

[3] QUAKES, Department of Earth Sciences, The University of Queensland, Brisbane, Australia. E-mails: mora@quakes.uq.edu.au; wangyc@quakes.uq.edu.au; weatherley@quakes.uq.edu.au

accumulated energy to remove the crust from a critical state. Subsequently, tectonic loading drives the crust back towards the critical state. During the establishment of criticality, seismic moment release accelerates in the region surrounding the epicenter of the ensuing large or great earthquake. The Accelerating Moment/Energy Release (AMR/AER) sequences may be identified by fitting cumulative moment/energy release prior to a large or great earthquake to a power-law time-to-failure relation (BUFE and VARNES, 1993; BOWMAN *et al.*, 1998; JAUMÉ and SYKES, 1999). Such a fit provides an intermediate-term prediction of the time of occurrence of the large or great earthquake.

It has been suggested that the acceleration in seismic moment release is due to the establishment of long-range correlations in the regional stress field. Such long-range correlations prepare the region for a large earthquake (SYKES and JAUMÉ, 1990; RUNDLE *et al.*, 1999; SAMMIS and SMITH, 1999; MORA and PLACE, 2002). Once in the critical state, only a very small stress perturbation, such as that caused by the earth tides, may be sufficient to trigger earthquakes. Assuming the earth tides are sufficient to trigger earthquakes, especially moderate earthquakes, a parameter called the Load-Unload Response Ratio (LURR) may be used as a measure of the proximity to criticality (YIN and YIN, 1991; YIN, 1993; YIN *et al.*, 1994, 1995, 2000).

From the viewpoint of Damage Mechanics, the preparation process for an earthquake is the deformation and damage process of the focal media. LURR has been proposed as a measure of this process. LURR is typically defined as the ratio of Benioff Strain release during loading cycles compared to that during unloading cycles on optimally oriented (or specified) fault planes as induced by the earth tides. High LURR values (larger than unity) indicate that a region is prepared for a large or great earthquake.

Both high LURR values and observations of Accelerating Moment/Energy Release (AMR/AER) prior to large earthquakes have led different research groups to suggest intermediate-term earthquake prediction is possible. In recent years, a relationship between the magnitude of a large or great earthquake and the size of the region where a power-law time-to-failure function best fits cumulative moment release has been noted (BOWMAN *et al.*, 1998; JAUMÉ and SYKES, 1999). These results showed that AMR/AER exhibits a critical region size that scales with magnitude. Meanwhile, we have found that there is a correlation between LURR values and size of regions before large earthquakes. Thus, the question arises: Do AMR/AER and LURR have a common physical mechanism? In this paper, we compare the critical region – magnitude scaling relations for the two phenomena aiming to answer this question.

2. LURR as a Predictor of Large or Great Earthquakes

In previous years, a series of successful intermediate-term predictions have been reported for strong earthquakes in China and other countries using the LURR parameter (YIN and YIN, 1991; YIN, 1993; YIN *et al.*, 1994, 1995, 1996, 2000). While further research is required to analyze the null hypothesis and study the statistical likelihood of successful predictions, these results have provided encouragement that LURR anomalies may be a predictor of large or great earthquakes.

Several studies of LURR in Japan have been conducted for three regions – south Kanto region (34.5°–36°N; 139°–141°E), Tottori-Kobe region (circular region with center 35.3°N; 133.7°E and radius 300 km) and Tokai region (34°–35.5°N; 137.5°–139°E). The results of LURR for these three regions are shown in Figures 1–3, respectively. For the South Kanto region, high LURR values appeared during the second half of 1999 to the beginning of 2000 and then a series strong earthquakes with magnitude larger or equal to 6 occurred in this region. It is shown in Figure 2 that before the 1995 January Kobe earthquake (M 7.2) there is a significant LURR anomaly and it lasted for more than two years. The anomaly reappeared since the second half of 1999 afterwhich the Tottori earthquake occurred (October 6th, 2000, magnitude M 7.3). The LURR plot for the Tokai region of Figure 3 shows a high LURR value in mid-1995, followed by magnitude 5–6 earthquakes 1–2 years later. In mid-1998, the plot exhibited a very sharp spike in LURR with a maximum value reaching 23. Although the null hypothesis must be evaluated, the high LURR value is significant according to an analysis given in YIN *et al.* (2000). Based on this spike, we are expecting that some events with magnitude about M 6 or greater may soon occur in this region, probably in 2001 or 2002, although further research is required to estimate the likelihood. These results were presented at the 2nd ACES Workshop, Oct. 15–21, 2000, Hakone, and Japan.

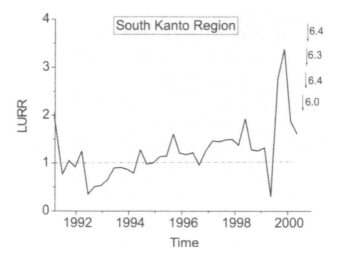

Figure 1
LURR value versus time in the south Kanto region.

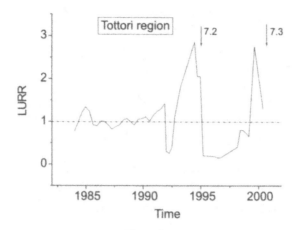

Figure 2
The LURR anomaly prior to the Kobe earthquake and the Tottori earthquake.

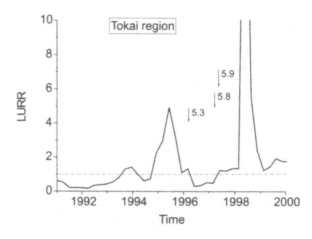

Figure 3
The variation of LURR in the Tokai region.

Recently, at the request of a seismologist from South Australia in December 2000 – David Love of the Department of Primary Industries – we analyzed the LURR and AMR within South Australia. Results at optimal radii are shown in Figures 4 and 5, respectively. These plots were transmitted to D. Love on February 8 and 9, 2000 along with additional analyses for different region sizes. According to the results, it appears that the crust of South Australia may be in a preparatory stage for a magnitude about 5.5–6 earthquake during the period 2001/3 to 2002/6 in the Burra-Peterborough region within about a 150 km radius of Peterborough. These results will be reported in detail elsewhere.

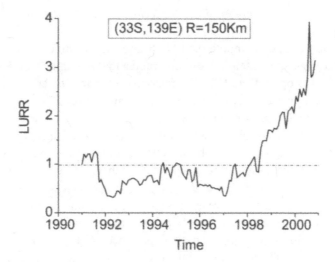

Figure 4
The variation of LURR in the Peterborough region, Australia (time window 1 year with sliding step 1 month. This plot is at the optimal radius that maximizes the peak LURR value.

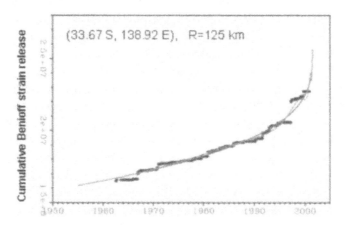

Figure 5
Power-law fits with and without log-periodic fluctuations using data from South Australia within a radius of 125 km around Burra (Burra is about 80 km south of Peterborough). This radius maximized the fit parameter $C = (RMS\ error\ power\ law\ fit)/(RMS\ error\ linear\ fit)$. According to the best-fit power law, the critical point time when the crust is primed for an earthquake will occur around mid-2001, and the predicted Benioff strain release at the critical point implies the expected magnitude is around 5.5. Using different radii around Burra, the predicted time fluctuates between mid-2001 and mid-2002 and magnitude fluctuated between 5.5 and 5.8.

3. A Test for a Common Physical Mechanism for LURR and AMR/AER

In the following, we compute the LURR anomaly for different region sizes before several large events to obtain the optimal radius that maximizes LURR. This is then compared to the optimal AMR/AER region size termed the critical region size.

First we conducted a retrospective AMR examination of several Australian and Chinese earthquakes with magnitude 5–7.9, including Australia's deadly Newcastle earthquake which occurred in 1989 and the devastating Tangshan earthquake which occurred in 1976. For each event, both LURR anomalies and best-fit power-law time-to-failure functions are computed using data within a range of different radii from the epicenter (i.e., using data within different region sizes). The best-fit power-law time-to-failure functions are defined here as those that optimize the goodness of fit parameter $C = (RMS\ error\ power\ law\ fit)/(RMS\ error\ linear\ fit)$. The optimal radius for AMR is defined as the radius which minimizes the fit parameter C. Figure 6 is a plot of the power law goodness of fit parameter C as a function of region size for the sequences prior to the 1997 M = 5.0 Burra earthquake, Australia, the 1989 M = 5.7 Newcastle earthquake, Australia, the 1995 M = 6.5 Wuding earthquake, China, the 1990 M = 7.0 Gonghe earthquake, China, and the 1976 M = 7.9 Tangshan earthquake, China. For small region sizes, the data shows considerable scatter. This is due to the paucity of seismic data for regions of these sizes. For the largest region sizes considered, cumulative moment release is not well represented by a power-law relation and the goodness of fit parameter becomes large. The optimal radii which specify the critical region size are 125, 125, 200, 250 and 650 km, respectively. Figure 7 shows typical power-law fits for these earthquakes for the AMR/AER sequences for their optimal radii.

Subsequently the LURR values for the same five Australian and Chinese earthquakes were computed using data within several different radii from the epicenter to compute the LURR critical region size (or optimal radius for LURR) which is defined as the radius that maximizes the peak LURR value just prior to the earthquake. The time windows for all cases are one year. Figure 8 shows the relation between peak LURR values and radii. The optimal radii for LURR are 75, 100, 200, 300, 600 km respectively for the magnitude 5.0, 5.7, 6.5, 7.0 and 7.9 earthquakes' analyses. These results show a clear correlation between LURR critical region size and magnitude.

The plots in Figure 9, delineate the LURR values for each event, using data within the LURR critical regions (optimal radii). High LURR values occur months to years prior to each event and some intermediate-term earthquake predictions have been made including the 1995 Wuding M = 6.5 earthquake (YIN *et al.*, 1995, 1996, 2000) and the 1990 Gonghe M = 7.0 earthquake (unpublished report, in Chinese).

Both AMR/AER and LURR exhibit a critical region size that scales with magnitude. Figure 10 shows the critical region size for AMR/AER versus the critical region size for LURR for the five earthquakes analyzed. A strong correlation is evident between the AMR/AER and LURR critical region sizes, suggesting these two observations have a common physical mechanism. Recent simulations demon-

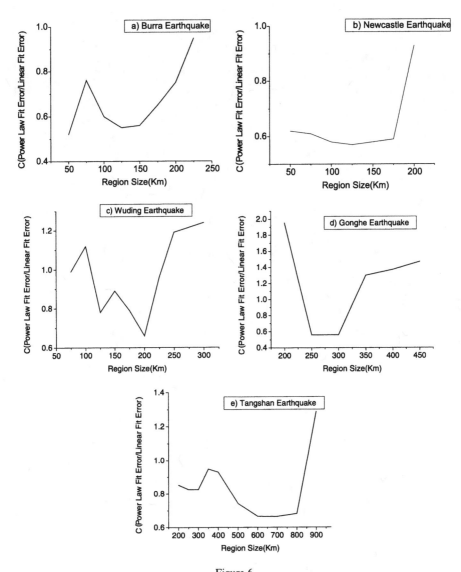

Figure 6

The goodness of power-law fit parameter C as a function of region size for the AMR/AER sequences prior to (a) the 1997 M = 5.0 Burra earthquake, Australia, (b) the 1989 M = 5.7 Newcastle earthquake, Australia, (c) the 1995 M = 6.5 Wuding earthquake, China, (d) the 1990 M = 7.0 Gonghe earthquake, China, and (e) the 1976 M = 7.9 Tangshan earthquake, China.

strate Accelerating Moment/Energy Release (MORA *et al.*, 2000) and an evolution in stress correlations prior to large events (MORA and PLACE, 2002; WEATHERLEY *et al.*, 2002) consistent with that predicted by the Critical Point Hypothesis. This suggests a mechanism that is CPH or CPH-like. If so, LURR may offer an approach to detect the critical sensitivity (WEI *et al.*, 2000; MORA *et al.*, 2002) of the crust as it

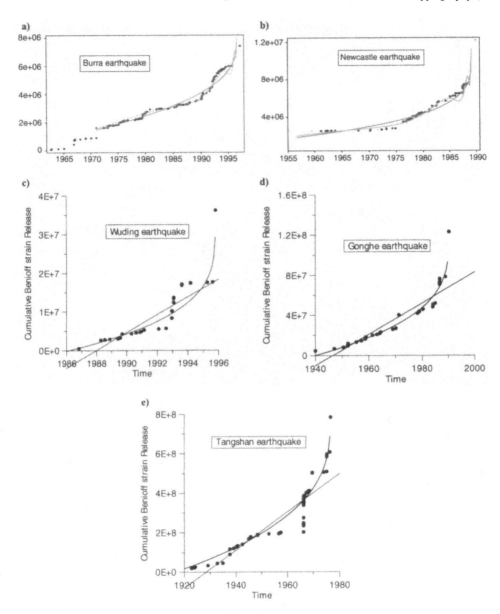

Figure 7
Typical power-law fits for these earthquakes for the radii, which minimize the fit parameter *C* in Figure 6.
In Figures 7a and 7b, the power-law fit and power-law fit with log-periodic fluctuations are shown. In
Figures 7c through 7e, the power law fit and linear fit are shown.

approaches a critical point in the lead-up to a large event. Furthermore, based on the
results presented in Figure 10, the critical region size – magnitude scaling relation for
AMR/AER and/or LURR provides a means to estimate the magnitude of an
oncoming earthquake.

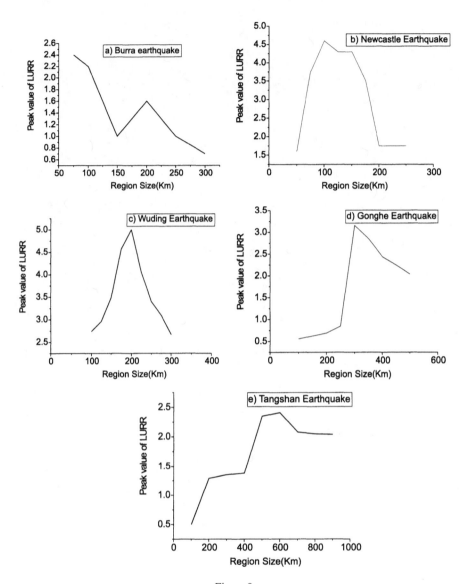

Figure 8
Peak LURR value versus region size for (a) the 1997 M = 5.0 Burra earthquake, Australia, (b) the 1989 M = 5.7 Newcastle earthquake, Australia, (c) 1995 M = 6.5 earthquake, China, (d) the 1990 M = 7.0 Gonghe earthquake, China, and (e) the 1976 M = 7.9 Tangshan earthquake, China.

4. Discussion and Conclusions

According to the critical point hypothesis (CPH), the occurrence of a large or great earthquake in a region removes the crust from a critical state. Tectonic loading

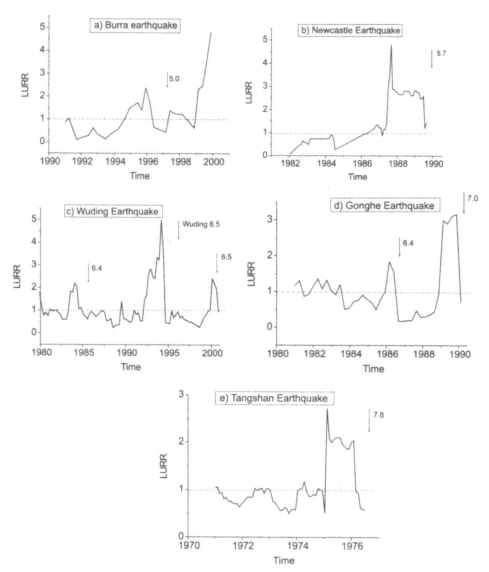

Figure 9

LURR value as a function of time for the five regions using data within the LURR critical region size (optimal radii that maximize peak LURR value) as determined by Figure 8. Namely, using region sizes of 75, 100, 200, 300 and 600 km for plots (a) through (e) respectively. The time windows for 5 events are 1 year.

drives the crust back towards the critical state. During the establishment of criticality, some phenomena appear including:

- Accelerating seismic activity of moderate-sized earthquakes (ELLSWORTH *et al.*, 1981; KEILIS-BOROK, 1990; SORNETTE and SORNETTE, 1990; SORNETTE and SAMMIS, 1995; KNOPOFF *et al.*, 1996 and BOWMAN *et al.*, 1998).

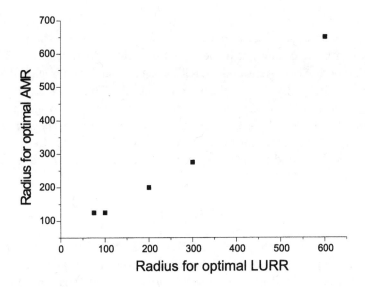

Figure 10
The critical region size for AMR/AER versus the critical region size for LURR for the five earthquakes analyzed.

- Accelerating seismic moment release (time-to-failure power law) AMR/AER (BUFE and VARNES, 1993; BOWMAN et al., 1998; JAUMÉ and SYKES, 1999).
- Establishment of long-range correlations in the regional stress field (SYKES and JAUMÉ, 1990; RUNDLE et al., 1999; SAMMIS and SMITH, 1999; MORA and PLACE, 2002).
- Critical sensitivity (WEI et al., 2000).
- Triggering earthquakes significantly by tidal stress (GRASSO and SORNETTE, 1998).
- Anomalous LURR (high value) (YIN and YIN, 1991; YIN, 1993; YIN et al., 1994, 1995, 2000).

On the other hand, from the view point of meso-mechanics (KRAJCINOVIC, 1996; KUKSHENKO et al., 1996), when a system – say, a specimen of heterogeneous media – is approaching the critical point (fracture), the micro-cracks grow in both number and size so that the interaction between them becomes increasingly intense, and consequently the system will become considerably more sensitive. That means a tiny external disturbance acting on it should induce a significant response. Now it is easy to understand the above-mentioned phenomena and they could be divided into two groups: The first group (first 3 phenomena) is due to the appearance of bigger cracks (moderate-sized events). The second group (last 3) is concerned with sensitivity and it is the consequence of the first phenomena (i.e., the increase in sensitivity occurs when crack interactions become stronger). In summation, all of the above-mentioned phenomena may have the same underlying physical mechanism under the framework of the critical point model.

The results presented here are preliminary since only five earthquakes spanning 2.9 magnitudes units have been analyzed. All five earthquakes exhibit both AMR/AER and high LURR values near the end of the AMR/AER sequence. The results establish that both AMR/AER and LURR have a similar critical region size that this critical region size scales with mainshock magnitude. This suggests that both the AMR/AER and LURR observations have a common physical origin and that the mechanism is CPH or CPH-like. Further work may provide clues that yield an understanding of the mechanism underlying AMR/AER and LURR, and will potentially lead to a solid foundation for intermediate-term earthquake prediction.

Acknowledgments

The authors wish to thank Dr. KOHJI HOSONO, Meteorological Research Institute of the Japan Meteorological Agency for supplying seismic data from Japan. We would also like to thank the Australian Geological Survey Organization and KEVIN MCCUE for provision of the Australian Earthquake Catalog, Russell Cuthbertson for assistance in data provision and formatting and DAVID LOVE of Primary Industries and Resources of South Australia for provision of the South Australian catalog. This research is supported by the Natural Sciences Foundation of China (Grant No. 19732060 and 40004002), the State Key Laboratory of Nonlinear Mechanics of the Chinese Academy of Sciences, the Center for Analysis and Prediction of the Chinese Seismological Bureau, the ARC IREX ACES International Visitors Program, the University of Queensland, and the Australian Research Council.

REFERENCES

BOWMAN, D. D., OUILLON, G., SAMMIS, C. G., SORNETTE, A., and SORNETTE, D. (1998), *An Observational Test of the Critical Earthquake Concept*, J. Geophys. Res. *103*, 24,359–24,372.

BUFE, C. G. and VARNES, D. J. (1993), *Predictive Modeling of the Seismic Cycle of the Greater San Francisco Bay Region*, J. Geophys. Res. *98*, No. B6, pp. 9871–9883.

ELLSWORTH, W. L., LINDH, A. G., PRESCOTT, W. H., and HEAD, D. G., *The 1906 San Francisco earthquake and the seismic cycle*. In *Earthquake Prediction: An International Review*, eds. Simpson, D. W., and Richards, P. G. (AGU, Washington, D. C. 1981) pp. 21–27.

GRASSO, J. and SORNETTE, D. (1998), *Testing Self-organized Criticality by Induced Seismicity*, J. Geophys. Res. *103*, B12, 29,965–29,987.

JAUMÉ, S. C. and SYKES, L. R. (1999), *Evolving Towards a Critical Point: A Review of Accelerating Seismic Moment/Energy Release Prior to Large and Great Earthquakes*, Pure Appl. Geophys. *155*, 279–306.

KEILIS-BOROK, V. (1990), *The Lithosphere of the earth as a large nonlinear system*. In *Quo Vadimus: Geophysics for the Next Generation*, Geophys. Monogr. ser. vol. 60 (ed. G. D. Garland and J. R. Apel), pp. 81–84, AGU, Washington, D. C.

KNOPOFF, L., LEVSHINA, T., KEILIS-BOROK, V. I., and MATTONI, C. (1996), *Increased Long-range Intermediate-magnitude Earthquake Activity prior to Strong Earthquakes in California*, J. Geophys. Res. *101*, 5779–5796.

KRAJCINIVIC, D. *Damage Mechanics*, (Elsevier, New York, 1996).

KUKSHENKO, V., TOMLIN, N., DAMASKINSKAYA, E., and LOCKNER, D. (1996), *A Two-stage Model of Fracture of Rocks*, Pure Appl. Geophys. *146*, 353–264.

MORA, P., PLACE, D., ABE, S., and JAUMÉ, S. (2000), *Lattice solid simulation of the physics of fault zones and earthquakes: The model, results and directions*. In *Geocomplexity and the Physics of Earthquakes* (eds. Rundle, J.B., Turcotte, D.L. and Klein, W.), pp. 105–125 (AGU, Washington).

MORA, P., WANG, Y. C., YIN, C., PLACE, D., and YIN, X. C. (2002), *Simulation of the Load-Unload Response Ratio and Critical Sensitivity in the Lattice Solid Model*, Pure Appl. Geophys. *159*, 2525–2536.

MORA, P. and PLACE, D. (2002), *Stress Correlation Function Evolution in Lattice Solid Elasto-dynamic Models of Shear and Fracture Zones and Earthquake Prediction*, Pure Appl. Geophys. *159*, 2413–2427.

RUNDLE, J. B., KLEIN, W., and GROSS, S. (1999), *A Physical Basis for Statistical Patterns in Complex Earthquake Populations: Models, Predictions and Tests*, Pure Appl. Geophys. *155*, 575–607.

SAMMIS, C. G. and SMITH, S. W. (1999), *Seismic Cycles and the Evolution of Stress Correlation in Cellular Automation Models of Finite Fault Networks*, Pure Appl. Geophys. *155*, 307–334.

SORNETTE, A. and SORNETTE, D. (1990), *Earthquake Rupture as a Critical Point: Consequences for Telluric Precursors*, Tectonophysics *179*, 327–334.

SORNETTE, D. and SAMMIS, C. G. (1995), *Complex Critical Exponents from Renormalization Group Theory of Earthquake Prediction*, J. Phys. I. France *5*, 607–619.

SYKES, L. R. and JAUMÉ, S. (1990), *Seismic Activity on Neighboring Faults as a Long-term Precursor to Large Earthquakes in the San Francisco Bay Area*, Nature *348*, 595–599.

WEATHERLEY, D., MORA, P., and XIA, M. (2002), *Long-range Automaton Models of Earthquakes: Power-law Accelerations, Correlation Evolution, and Mode Switching*, Pure Appl. Geophys. this issue.

WEI, Y. J., XIA, M. F., KE, F. J., YIN, X. C., and BAI, Y. L. (2000), *Evolution Induced Catastrophe and its Predictability*, Pure Appl. Geophys. *157*, 1945–1957.

YIN, X. C. and YIN, C. (1991), *The Precursor of Instability for Nonlinear Systems and its Application to Earthquake Prediction*, Science in China *34*, 977–986.

YIN, X. C. (1993), *New Approach to Earthquake Prediction*, PRERODA (Russia's "Nature"), 1, pp. 21–27 (in Russian).

YIN, X. C., YIN, C., and CHEN, X. Z. (1994), *The Precursor of Instability for Nonlinear Systems and its Application to Earthquake Prediction – the Load-Unload Response Ratio Theory*, Nonlinear Dynamics and Predictability of Geophysical Phenomena, AGU Geophysical Monograph 83 (eds. Newman, W. I., Gabrelov, A. M., and Turcotte, D.L.), pp. 55–60.

YIN, X. C., CHEN, X. Z., SONG, Z. P., and YIN, C. (1995), *A New Approach to Earthquake Prediction – The Load-Unload Response Ratio (LURR) Theory*, Pure Appl. Geophys. *145*, (3/4), 701–715.

YIN, X. C., CHEN, X. Z., SONG, Z. P., and WANG, Y.C. (1996), *The Temporal Variation in LURR in Kanto and other Regions in Japan and its Application to Earthquake Prediction*, Earthquake Research in China *10*, 381–385.

YIN, X. C., WANG, Y. C., PENG, K. Y., BAI, Y. L., WANG, H., and YIN, X. F. (2000), *Development of a New Approach to Earthquake Prediction: Load/Unload Response Ratio (LURR) Theory*, Pure Appl. Geophys. *157*, 1923–1941.

YIN, X. C., WANG, Y. C., PENG, K. Y., ZHANG, Y. X., and XIA, M. F. (2000), *New Developments of LURR Theory and its New Application*, International Workshop on Solid Earth Simulation and ACES WG Meeting, Abstract vol. (University of Tokyo, Jan 17–21, 2000).

(Received February 20, 2001, revised June 11, 2001, accepted June 25, 2001)

To access this journal online:
http://www.birkhauser.ch

Pure appl. geophys. 159 (2002) 2525–2536
0033–4553/02/102525–12 $ 1.50 + 0.20/0

❙ Pure and Applied Geophysics

Simulation of the Load-Unload Response Ratio and Critical Sensitivity in the Lattice Solid Model

PETER MORA,[1] YU CANG WANG,[1,2] CAN YIN,[1]
DAVID PLACE,[1] and XIANG-CHU YIN[2,3]

Abstract — The Load-Unload Response Ratio (LURR) method is an intermediate-term earthquake prediction approach that has shown considerable promise. It involves calculating the ratio of a specified energy release measure during loading and unloading where loading and unloading periods are determined from the earth tide induced perturbations in the Coulomb Failure Stress on optimally oriented faults. In the lead-up to large earthquakes, high LURR values are frequently observed a few months or years prior to the event. These signals may have a similar origin to the observed accelerating seismic moment release (AMR) prior to many large earthquakes or may be due to critical sensitivity of the crust when a large earthquake is imminent. As a first step towards studying the underlying physical mechanism for the LURR observations, numerical studies are conducted using the particle based lattice solid model (LSM) to determine whether LURR observations can be reproduced. The model is initialized as a heterogeneous 2-D block made up of random-sized particles bonded by elastic-brittle links. The system is subjected to uniaxial compression from rigid driving plates on the upper and lower edges of the model. Experiments are conducted using both strain and stress control to load the plates. A sinusoidal stress perturbation is added to the gradual compressional loading to simulate loading and unloading cycles and LURR is calculated. The results reproduce signals similar to those observed in earthquake prediction practice with a high LURR value followed by a sudden drop prior to macroscopic failure of the sample. The results suggest that LURR provides a good predictor for catastrophic failure in elastic-brittle systems and motivate further research to study the underlying physical mechanisms and statistical properties of high LURR values. The results provide encouragement for earthquake prediction research and the use of advanced simulation models to probe the physics of earthquakes.

Key words: Numerical simulation, Load-Unload Response Ratio method, earthquake prediction, lattice solid model, critical sensitivity.

---

[1] QUAKES, Department of Earth Sciences, The University of Queensland, Brisbane, 4072, Qld, Australia. E-mails: mora@quakes.uq.edu.au; wangyc@quakes.uq.edu.au; canyon@quakes.uq.edu.au; place@quakes.uq.edu.au

[2] LNM, Institute of Mechanics, Chinese Academy of Sciences, Beijing, 100080, China. E-mail: xcyin@public.bta.net.cn

[3] Center for Analysis and Prediction, China Seismological Bureau, Beijing, 100036, China. E-mail: yinxc@btamail.net.cn

Introduction

The Load-Unload Response Ratio (LURR) method is an intermediate-term earthquake prediction approach (YIN *et al.*, 1995, 2000) that has shown considerable promise. The method typically involves calculating the ratio of Benioff strain release during periods of loading and unloading as determined by calculating earth tide induced perturbations in the Coulomb Failure Stress on optimally oriented faults. In retrospective studies, high LURR values have been observed months to years prior to most events and some intermediate-term earthquake predictions have been made (YIN *et al.*, 2000).

The idea that motivated the LURR earthquake prediction approach is that when a system is stable, its response to loading is nearly the same as its response to unloading, whereas when the system is in an unstable state, the response to loading and unloading becomes quite different (YIN *et al.*, 1995, 2000). LURR is defined according to this difference. Suppose P and R are respectively the load and response of a system. If P undergoes a small change ΔP resulting in a small change to R of ΔR, then

$$X = \lim_{\Delta P \to 0} \frac{\Delta R}{\Delta P} ,$$ (1)

can be defined as the response rate, and LURR is defined as

$$\text{LURR} = \frac{X^+}{X^-} ,$$ (2)

where X^+ and X^- are response rates during loading and unloading. When a system is in a stable or linear state, $X^+ \approx X^-$ so LURR ≈ 1. When a system lies beyond the linear state, $X^+ > X^-$ and LURR > 1. Hence, LURR can be used as a criterion to judge the degree of stability of a system.

In earthquake prediction practice using LURR, loading and unloading periods are decided by calculating the earth tide induced perturbations in the Coulomb Failure Stress on optimally oriented faults (or a specified fault plane orientation), and LURR is often defined as ratio of cumulative Benioff strain release during loading compared to unloading. Specifically,

$$\text{LURR} = B^+/B^- .$$ (3)

where B^+ and B^-, respectively denote the cumulative Benioff strain release during loading and unloading. To avoid violent fluctuations due to poor statistics, the LURR values are computed from the cumulative Benioff strain release during loading and unloading summed over many load-unload cycles within a specified sliding time window (i.e., there are generally few events during a single load-unload cycle which lead to large statistical fluctuations in LURR between successive load-unload cycles). The length of the time windows must be chosen such that B^+

and B^- include enough earthquake events to average out the statistical fluctuations although not so long as to remove any time-varying signals with a physical origin. Typically, time windows of weeks to months are used in practice. In retrospective studies, high values of LURR have been observed a few months or years prior to most of the events and intermediate-term earthquake predictions have been made using this method (YIN *et al.*, 2000).

In recent years, accelerating seismic moment release (AMR) has been observed prior to many large earthquakes (BUFE and VARNES, 1993; BOWMAN *et al.*, 1998). Both AMR and high LURR may have a similar origin (YIN *et al.*, 2002) or LURR may be due to critical sensitivity before catastrophic events (WEI *et al.*, 2000). A physically based numerical simulation that is capable of reproducing LURR signals would provide a means to study the underlying physical mechanism for LURR signals.

The Lattice Solid Model (LSM) was developed to provide a basis to study the physics of rocks and the nonlinear dynamics of earthquakes (MORA and PLACE, 1994, 1998; PLACE and MORA, 1999, 2000, 2001; PLACE *et al.*, 2001). The LSM consists of a lattice of interacting particles. Intact material is modelled as particles linked by elastic-brittle bonds which can break if the separation exceeds a given threshold R_b relative to the equilibrium separation R_0, and frictional forces are applied to unbonded particles that come into contact. Using the LSM, fracture, shearing of rock, stick-slip behavior, dynamic rupture, and wave propagation are simulated with relative simplicity. Localization phenomena in fault gouge zones has been simulated (PLACE and MORA, 2000) and recent results have provided a comprehensive potential explanation for the Heat Flow Paradox (MORA and PLACE, 1998, 1999). Recent numerical experiments involving model systems subjected to compression have demonstrated that the LSM is capable of realistically simulating the fracturing behavior of rocks (PLACE and MORA, 2001; PLACE *et al.*, 2002). The lattice solid calculations compute the energies within the system (kinetic energy, energy lost to the artificial viscosity, fracture energy, energy lost to friction, external work done and potential energy). Numerical studies have verified that the sum of these energies is numerically conserved to a high precision (PLACE and MORA, 1999). Since the LSM is capable of realistically modeling fracture and slip events, it provides a means to study the underlying mechanism of LURR. In the following, we conduct simulations using the LSM with the aim of determining whether LURR signals can be reproduced as a first step towards this goal.

Numerical Experiments of LURR

In the present study, the model is initialized as a heterogeneous 2-D block made up of random-sized particles with diameters ranging from 0.2 to 1 model units. The system is subjected to uni-axial compression from rigid driving plates on the upper and lower edges of the model. Snapshots from a typical simulation are shown in

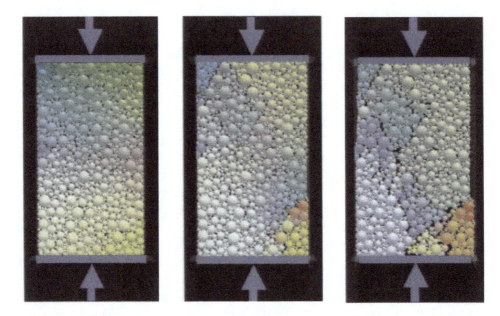

Figure 1
The setup for uniaxial compression numerical experiments illustrating formation of fractures in the random lattice solid model. Colors depict displacement (blue = left, red = right, green = down, yellow = up).

Figure 1. Two numerical experiments were conducted. In the first, loading is strain controlled and a constant driving rate is applied to the upper and lower edges of the model. In the second, stress control is used in which stress on the upper and lower edges is increased linearly and slowly until the sample fails. These two cases correspond to different effective rigidity outside the focal region and can yield different fracture behaviors. In both cases, a sinusoidal variation is added to the constant loading rate in order to simulate the stress perturbations induced by tidal forces. These variations resulted in stress of form

$$\Delta\sigma_{zz} = a \sin(2\pi t/T) \ . \tag{4}$$

so stress $\sigma_{zz} = \Delta\sigma_{zz} + kt$. The model initialization and parameters used in each experiment were identical. Tectonic loading is very slow. However, due to limited computer power, it is infeasible to use the observed tectonic driving rates in the numerical experiments. Therefore we use a higher loading rate $k \gg k_{\text{tectonic}}$. With the aim of ensuring the results remain meaningful with this higher loading rate, we selected parameters such that $|\Delta\sigma_{zz}| \ll \sigma_{zz}$, $\frac{d|\Delta\sigma_{zz}|}{dt} \gg k$ and $T_e \ll T \ll T_L$, where T_e is the synthetic earthquake rupture duration and T_L is the average time interval between large earthquakes. The values specified were loading rate $k = 30$ MPa/100,000 time steps, period of the sinusoidal perturbations $T = 4000$ time steps, amplitude of sinusoidal perturbation $a = 0.96$ MPa $=$ constant and breaking criterion

$R_b = 1.002\,R_0$. The elastic properties and model size were such that shear waves propagated vertically through the model in about 530 time steps, substantially shorter than period T. LURR values were calculated according to Equation (3) but using the cumulative energy release instead of cumulative Benioff strain release, i.e.,

$$\text{LURR} = E^+/E^- \; , \tag{5}$$

where E^+ and E^-, respectively denote the cumulative seismic energy release during loading and unloading within a given time window. These were obtained by summing total kinetic energy released during all loading or unloading cycles within the specified time window, where we define loading to be when $\frac{d\sigma_{zz}}{dt} \geq 0$ and unloading when $\frac{d\sigma_{zz}}{dt} < 0$. Specifically, the LURR value was calculated using

$$\text{LURR}(n) = \frac{E^+}{E^-} = \frac{\sum_{i=1}^{m} \Delta E^+(n-i+1)}{\sum_{i=1}^{m} \Delta E^-(n-i+1)} \; , \tag{6}$$

where m is the number of cycles in the time window (i.e., the LURR time window length is mT load-unload cycles), $n = m, m+1, m+2, \ldots$ is a time index (i.e., $t = nT$), and $\Delta E^+(j)$ and $\Delta E^-(j)$, respectively denote total kinetic energy released during the j-th load and unload cycle. The total kinetic energy release at any given instant t is the sum of the kinetic energy within the system and the energy lost to the artificial viscosity prior to time t, i.e., $E_{\text{TOTAL}}(t) = E_k(t) + E_v(t)$ where $E_k(t)$ denotes the kinetic energy at time t and $E_v(t)$ denotes the energy lost to the artificial viscosity prior to time t. The total energy release for the j-th load or unload cycle was therefore calculated using

$$\Delta E^+_{\text{TOTAL}}(j) = E_{\text{TOTAL}}(jT + T/4 + \Delta t_k) - E_{\text{TOTAL}}(jT - T/4 - \Delta t_k) \; , \tag{7}$$

and

$$\Delta E^-_{\text{TOTAL}}(j) = E_{\text{TOTAL}}(jT + 3T/4 - \Delta t_k) - E_{\text{TOTAL}}(jT + T/4 + \Delta t_k) \; , \tag{8}$$

where $\Delta t_k = T/2\pi \arcsin(kT/2a\pi)$ (Δt_k takes into account the time difference between the loading and unloading periods due to the linearly increasing stress).

Strain-controlled Compression Experiment

Figure 2 shows stress, kinetic energy, LURR value and the total energy released versus time step for the strain-controlled experiment. In this experiment, LURR is calculated using a sliding time window that is ten load-unload cycles long with a sliding increment of one cycle. Hence, LURR is calculated from Equation (6) using $m = 10$.

The sharp drops in stress and spikes of kinetic energy correspond to dynamic fracturing involving breaking of bonds and/or slip along fracture surfaces. These represent events in the simulation. Due to the artificial viscosity that is applied to

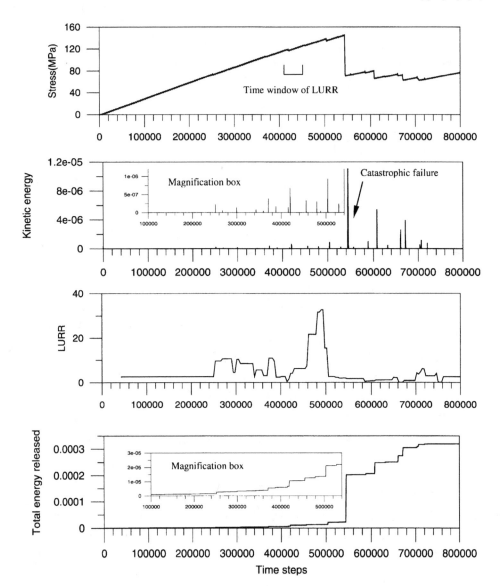

Figure 2
Results of the strain controlled experiment. From top to bottom, stress measured on the rigid driving plates, kinetic energy within the model, LURR value calculated from equations (6) through (8) and cumulative seismic energy release.

damp energy from the system, kinetic energy soon dies out after each event. It is this lost energy – termed viscous energy – summed with the instantaneous kinetic energy in the system that provides a measure of the total kinetic energy released by events ($E_{TOTAL} = E_v + E_k$) and allows LURR to be computed using equations (6) through (8).

The main rupture (catastrophic failure) of the sample occurred at 543,000 time steps and is seen as a sharp drop in stress, a large spike in kinetic energy, and a large step in total energy released. Just prior to failure, the stress reaches a peak of approximately 145 MPa which corresponds to the strength of the sample. The LURR value remains constant until 250,000 time steps. During this period, the sample remained intact and so the viscous energy was small and equaled the energy lost to the viscosity in loading the system at the specified rate. Due to the loading, there is a small amount of kinetic energy which causes viscous energy to accumulate throughout the simulation even with no dynamic fracturing events. The total energy release due to this effect is not linearly increasing but has a periodic variation due to the sinusoidal stress perturbations. For this reason, the measured energy release during loading and unloading was not identical prior to the first fracture at 250,000 time steps resulting in an LURR value of 2.6 rather than unity. Between 250,000 and 340,000 time steps, small infrequent events occurred and the LURR value typically fluctuated between 4.5 and 10.5. Larger and more frequent events started to occur at around 370,000 time steps. After this time, the LURR value dropped and typically remained at around 2.5 to 8.5 until 460,000 time steps, with the exception of a short spike and trough. From 460,000 to 490,000 time steps, the LURR value rose significantly up to a peak value of 32 and then dropped again. This rise and subsequent drop in LURR occurred in several steps, indicating that the peak value was the result of high seismic energy release during several successive load-unload cycles. After 500,000 time steps, the LURR value dropped to a relatively low value of 2 despite continuing moderate-sized events and a large event at around 505,000 time steps. At 543,000 time steps, the catastrophic failure event occurred at which time the LURR value was 1.7. The LURR value continued to drop, reaching a low of 0.4 at 585,000 time steps. The LURR remained low for the remainder of the simulation despite several large subsequent ruptures.

Based on these results, it appears that the response of the system to loading and unloading becomes very different during a certain period just prior to the catastrophic failure and that high LURR values have successfully detected this critically sensitive state. Interestingly, the LURR value dropped just before the catastrophic failure event, suggesting that as the system becomes sufficiently damaged prior to the large event, it becomes insensitive to stress perturbations. A similar behavior of a sharp rise in LURR and a subsequent drop immediately before a large earthquake has also been observed in earthquake prediction practice (YIN et al., 2000).

Stress-controlled Compression Experiment

Figure 3 shows stress, kinetic energy, LURR value and total energy released versus time step for the stress-controlled experiment, where LURR is calculated from Equation (6) using $m = 10$ as in the strain-controlled experiment.

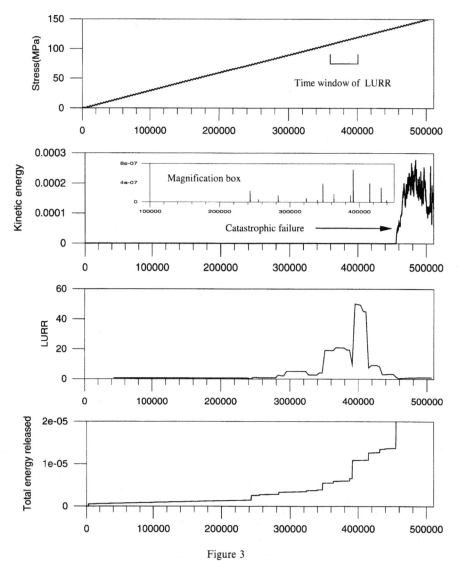

Figure 3
Results of the stress-controlled experiment. From top to bottom, stress measured on the rigid driving plates, kinetic energy within the model, LURR value calculated from equations (6) through (8) and cumulative seismic energy release.

The main rupture of the sample occurred at 455,000 time steps and is seen as a large rise in kinetic energy. Due to the use of stress control, the main rupture continues to grow and the sample fails catastrophically. Hence, kinetic energy does not drop once the main rupture occurs but remains at a high value. At the time of failure, the stress is around 137 MPa, somewhat lower than the peak stress in the strain-controlled experiment. The continued application of stress in this experiment probably explains the lower breaking strength of the sample than for the strain-controlled case. After a fracture event in the strain-controlled experiment, the internal configuration of the sample slightly changes as bonds break and surfaces slip, allowing the stress to drop and ruptures to arrest. As stress is built-up once again with continued strain loading, a new fracture will occur at the weakest point but due to the change in internal configuration, this may be in a different place than the previous fracture event. In contrast, the continued application of stress in the stress-controlled experiment tends to enhance the possibility for a rupture to runaway catastrophically.

In the stress-controlled experiment (Fig. 3), the LURR value remains low at around 1 to 2 before 280,000 time steps (the stable period), then increases rapidly up to 50 prior to the catastrophic failure event. Even if event statistics are small, it is remarkable that the LURR value rose and then dropped prior to catastrophic failure, thereby exhibiting very similar behavior as in the strain-controlled case. This strongly suggests that events preferentially occur during loading compared to unloading when the system approaches an unstable state followed by a change in response when the unstable state is reached. These results imply that LURR is capable of detecting this unstable or critical state prior to catastrophic failure.

Snapshots

In order to visualize the development of fractures and damage, snapshots from the two simulations are shown in Figure 4. The initial fracture event in each case (respectively at around 250,000 and 243,000, time steps for strain and stress control) corresponds to fracturing that occurs near the upper left corner of the sample. Subsequent to this event, small events occur and the displacement field evolves as the system deforms. In the final image in each sequence, one clearly observes large offsets on several fractures in the failed system. The stress-controlled case is already highly failed at 455,000 time steps, and rapidly evolves to rubble at subsequent time steps.

Discussion

Many other simulations with different arrangements of random particles were also made and all results were similar to those presented. Although a full parameter study is beyond the scope of this paper, simulations were conducted using values T, a and k a factor of two larger and smaller. These tests yielded similar results to those presented here, suggesting insensitivity to the specific choice of loading period, stress perturbation amplitude and loading rate at least within the range studied.

Based on the simulation results, LURR is capable of detecting the critically sensitive or unstable state just prior to catastrophic failure of elastic-brittle systems. In the simulations, one observes a sharp rise in LURR in the lead-up to the main

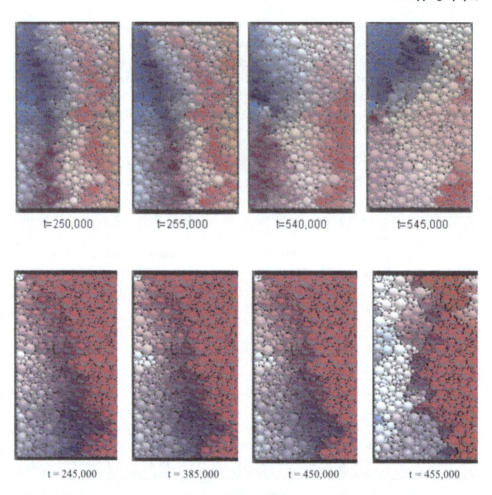

Figure 4
Top: Snapshots from the strain-controlled experiment. Bottom: Snapshots from the stress-controlled experiment. Colors depict horizontal and vertical displacements. Images are scaled individually to enhance visualization of major features of the deformation field and fractures with large offsets.

event and then a drop shortly prior to catastrophic failure followed by low LURR values thereafter. The initial drop in LURR is caused by large events that occur during the unloading cycle or very close to the peak stress and spanning across the boundary into the unloading cycle. In both cases, the catastrophic failure occurs during the unloading cycle near the peak stress although we believe the typical case would be for it to occur at random during loading or unloading near the peak stress. The results suggest that the system becomes insensitive to stress perturbations after sufficient damage has occurred or after catastrophic failure of the system. This behavior of high LURR values in the lead-up to events and a drop immediately prior to an event is frequently observed in LURR earthquake prediction practice (YIN et al., 2000).

The results provide encouragement for earthquake forecasting research and the use of advanced simulation models to probe the physical mechanisms involved. Recent simulations exhibit AMR (PLACE and MORA, 2000; MORA and PLACE, 2002) and an evolution in stress correlations prior to large events consistent with the Critical Point Hypothesis for earthquakes (MORA and PLACE, 2002). Observational studies (YIN et al., 2002) showing that the critical scaling regions for AMR and LURR are identical suggest LURR has a common physical origin as AMR. If so, AMR may be detecting the lead-up to a critical point whereas LURR may detect critical sensitivity once the system is very close to, or has reached, the unstable regime or critical point.

Conclusions

The lattice solid model has been used to simulate a 2-D elastic-brittle system being subjected to uniaxial compression in which stress perturbations are superimposed on the otherwise constant strain or stress loading rates. In each case, fractures develop and seismic energy is radiated within the model as the system is compressed until the sample fails catastrophically. The Load-Unload Response Ratio (LURR) that has been used for intermediate-term earthquake predictions is calculated in each case from the seismic energy release in the model during loading and unloading. The results show that LURR values become high and then drop prior to the main event, and remain low thereafter. These results reproduce LURR signals similar to those that have often been observed in earthquake prediction practice and suggest that LURR is correctly identifying the critically sensitive or unstable regime prior to the catastrophe in the model. This provides encouragement for the prospects of earthquake prediction using LURR, and motivates continued study of the LURR mechanism, earthquake forecasting research, and the use of advanced simulation models to probe the physics of earthquakes.

Acknowledgments

This research was funded by the National Natural Sciences Foundation of China (Grant No. 19732060 and 40004002), the Australian Research Council (ARC), The University of Queensland, and the ARC International Researcher Exchange Scheme 2000 (IREX). Computations were made on the Australian Solid Earth Simulator (ASES) thematic parallel supercomputer facility (phase I – 16 processor SGI Origin

3800) funded by the ARC, The University of Queensland, CSIRO, University of Western Australia and Silicon Graphics.

REFERENCES

BOWMAN, D. D., OUILLON, G., SAMMIS, C. G., SORNETTE, A., and SORNETTE, D. (1998), *An Observational Test of the Critical Earthquake Hypothesis*, J. Geophys. Res. *103*, 24,359–24,372.

BUFE, C. G. and VARNES, D. J. (1993), *Predictive Modeling of the Seismic Cycle in the Greater San Francisco Bay Region*, J. Geophys. Res. *98*, 9,871–9,833.

MORA, P. and PLACE, D. (1994), *Simulation of the Frictional Stick-slip Instability*, Pure Appl. Geophys. *143*, 61–87.

MORA, P. and PLACE, D. (1998), *Numerical Simulation of Earthquake Faults with Gouge: Towards a Comprehensive Explanation of the Heat Flow Paradox*, J. Geophys. Res. *103*, 21,067–21,089.

MORA, P. and PLACE, D. (1999), *The Weakness of Earthquake Faults*, Geophys. Res. Lett. *26*, 123–126.

MORA, P. and PLACE, D. (2002), *Stress Correlation Function Evolution in Lattice Solid Elasto-dynamic Models of Shear and Fracture Zones and Earthquake Prediction*, Pure Appl. Geophys. *159*, 2413–2426.

PLACE, D. and MORA, P. (1999), *A Lattice Solid Model to Simulate the Physics of Rocks and Earthquakes: Incorporation of Friction*, J. Comp. Phys. *150*, 1–41.

PLACE, D. and MORA, P. (2000), *Numerical Simulation of Localisation Phenomena in a Fault Zone*, Pure Appl. Geophys. *157*, 1821–1845.

PLACE, D. and MORA, P. (2001), *A random lattice solid model for simulation of fault zone dynamics and fracture processes*. In *Bifurcation and Localisation Theory for Soils and Rocks'99* (eds., Mühlhaus H-B., Dyskin, A. V. and Pasternak, E.), (AA Balkema, Rotterdam/Brookfield. 2001).

PLACE, D., LOMBARD, F., MORA, P., and ABE, S. (2002), *Simulation of the Micro-physics of Rocks Using LSMearth*, Pure Appl. Geophys. *159*, 1911–1932.

WEI, Y. J., XIA, M. F., KE, F. J., YIN, X. C., and BAI, Y. L. (2000), *Evolution Induced Catastrophe and its Predictability*, Pure Appl. Geophys. *157*, 1945–1957.

YIN, X. C., CHEN, X. Z., SONG, Z. P., and YIN, C. (1995), *A New Approach to Earthquake Prediction: The Load/Unload Response Ratio (LURR) theory*, Pure Appl. Geophys. *145*, 701–715.

YIN, X. C., WANG, Y. C., PENG, K. Y., and BAI, Y. L. (2000), *Development of a New Approach to Earthquake Prediction: Load/Unload Response Ratio (LURR) Theory*, Pure Appl. Geophys. *157*, 2365–2383.

YIN, X. C., MORA, P., PENG, K. Y., WANG, Y. C., and WEATHERLEY, D. (2002), *Load-Unload Response Ratio and Accelerating Moment (energy) Release Critical Region Scaling and Earthquake Prediction*, Pure Appl. Geophys. *159*, 2511–2523.

(Received February 20, 2001, revised June 11, 2001, accepted June 15, 2001)

To access this journal online:
http://www.birkhauser.ch

www.ingramcontent.com/pod-product-compliance
Lightning Source LLC
LaVergne TN
LVHW062302060326
832902LV00013B/2021